Study Guide
to accompany
Chemistry: The Practical Science

Paul Kelter, *University of Illinois at Urbana-Champaign*
Michael Mosher, *University of Nebraska at Kearney*
Andrew Scott, *Perth College, UHI Millennium Institute*
Contributing Writer Charles William McLaughlin, *University of Nebraska at Lincoln*

Gretchen M. Adams
University of Illinois at Urbana–Champaign

Frank J. Torre
Springfield College

Houghton Mifflin Company • Boston • New York

Vice President and Executive Publisher: George Hoffman
Vice President and Publisher: Charles Hartford
Marketing Manager: Laura McGinn
Marketing Assistant: Kris Bishop
Senior Development Editor: Rebecca Berardy Schwartz
Assistant Editor: Amy Galvin
Editorial Associate: Henry Cheek
Project Editor: Andrea Cava

Printed in the U.S.A.

ISBN-10: 0-618-73621-2
ISBN-13: 978-0-618-73621-8

123456789-POO- 10 09 08 07 06

TABLE OF CONTENTS

Preface

Chapter 1   The World of Chemistry   1

Chapter 2   Atoms: A Quest for Understanding   14

Chapter 3   Introducing Quantitative Chemistry   27

Chapter 4   Solution Stoichiometry and Types of Reactions   53

Chapter 5   Energy   81

Chapter 6   Quantum Chemistry: The Strange World of Atoms   95

Chapter 7   Periodic Properties of the Elements   113

Chapter 8   Bonding Basics   125

Chapter 9   Advanced Models of Bonding   146

Chapter 10   The Behavior and Applications of Gases   161

Chapter 11   The Chemistry of Water and the Nature of Liquids   180

Chapter 12   Carbon   198

Chapter 13   Modern Materials   214

Chapter 14   Thermodynamics: A Look at Why Reactions Happen   226

Chapter 15   Chemical Kinetics   241

Chapter 16   Chemical Equilibrium   257

Chapter 17   Acids and Bases   275

Chapter 18   Applications of Aqueous Equilibria   293

Chapter 19   Electrochemistry   315

Chapter 20   Coordination Complexes   329

Chapter 21   Nuclear Chemistry   344

Chapter 22   The Chemistry of Life   356

Preface

You may be asking yourself, "What can I do to be successful in learning chemistry?" From our experience as former students and now teachers in the field of general chemistry (where we observe and work with students on a daily basis), we have found taking the following steps helps:

- Read the textbook—both before you go to class and afterward. It is important to read the material before class so that you will be familiar with the terms and concepts that are covered. Equally important, you should reread the material after class. Go over your notes and clear up anything that you do not understand. This will help you make the most of your learning experience in class and help you understand the material as a whole better.

- Keep up with your homework on a daily basis. Trying to do all of your reading and complete all of your homework problems the day before the exam is not a good way to learn chemistry. This subject is quite abstract at times and requires a lot of problem solving, so you need time to digest and work on the material. Be sure to get help with anything you do not understand right away. Topics in chemistry build on each other, so it is important to lay a good foundation of knowledge.

- Focus on the conceptual aspects of chemistry, not just the mathematical equations. Many students try to do chemistry by "plugging and chugging" into equations. This approach is not an effective way to learn the material, especially at this level. Your instructor will find ways to work around this strategy to see if you truly understand the underlying chemistry.

- Get to know your instructor and your classroom peers. They are excellent resources and people who can answer your questions. Do not be afraid to ask them for help.

This *Study Guide* is meant to help you in your endeavor in learning chemistry. It is written to accompany the first edition of Kelter, Mosher, and Scott's *Chemistry: The Practical Science*. Our section descriptions and table and figure references match their textbook. Our descriptions and examples focus on concepts that students often find particularly difficult. The exercises at the end of each chapter provide problems in addition to those you can find in the textbook. In the *Study Guide,* you will find an assortment of both mathematical and nonmathematical problems to help you think about chemistry in different ways. The answers to all end-of-chapter exercises are also provided. Before you look at them, first try to work these problems out on your own or with your peers, and only then check your answers. This strategy will enhance both your problem-solving skills and your overall understanding of chemistry.

Of course, this *Study Guide* is not meant to be a substitute for the text. The textbook has a richness of information that the manual cannot approach. However, the *Study Guide* will provide you with another perspective and deepen your understanding.

Good luck this semester!

G.M.A.

F.J.T.

Chapter 1

The World of Chemistry

The Bottom Line

When you finish this chapter, you will be able to:

- Understand the importance of chemicals and the nature of atoms and molecules.
- Distinguish between a chemical change and a physical change.
- Use the scientific method to find out about nature.
- Use dimensional analysis and significant figures to accurately perform chemical calculations.
- Understand the meaning of accuracy and precision.

1.1   What Do Chemists Do?

Chemists, like all scientists, solve problems by defining key questions to ask. They then devise experiments as part of a scientific approach to learning. Analysis of the data from the experiments leads to conclusions that chemists hope will solve the problem.

Modern chemists work to develop countless products, including environmentally friendly alternatives to gasoline in automobiles, faster computer chips, and low-fat potato chips.

1.2   The Chemist's Shorthand

This section defines key terms that are commonly used by chemists. You will need to learn them.

- *Matter* is the material from which things are made. It is anything that has mass and occupies space.

- *Atoms* are tiny indivisible particles. They are the smallest identifying unit of an element.

- *Elements* are substances that contain only one kind of atom.

- *Molecules* are particles composed of two or more atoms bonded together.

- *Compounds* are substances containing different elements chemically bonded together.

- *Chemical composition* is the relative proportion of elements in a compound.

- A *physical property* is a characteristic of a substance that can be determined without changing its chemical composition. (Examples are color, or being a solid, liquid, or gas at room temperature.)

- A *physical change* occurs when a substance changes its physical properties without changing its chemical composition. (Examples are changing from a solid to a liquid or from a liquid to a gas.)

- A *chemical property* is a characteristic of a substance that can be determined as it undergoes a change in its chemical composition.

- A *chemical change* involves the conversion of one substance to another substance through a change in its chemical composition or reorganization of its atoms.

- A *homogeneous mixture* (or solution) is a mixture of substances that is uniform throughout.

- A *heterogeneous mixture* is a mixture of substances that has a different composition in different places.

Example 1   **Chemical and Physical Changes**

Identify each of these as a chemical or physical change:
   a) a candle burning
   b) ice melting in a drink
   c) dissolving salt in water
   d) rusting of metal on a car

First Thoughts

What are chemical changes? What are physical changes? Review their definitions.

Solution
   a) The conversion of the wax fuel to heat and light is evidence of a chemical change.
   b) Melting of ice changes the physical state of water from solid to liquid. This is a physical change.
   c) Dissolving salt in water is a physical change. The salt has not been converted into a different substance; it is just mixed with the water.
   d) The metal changes from a strong, shiny substance to a dull, brittle, discolored substance. This is evidence of a chemical change.

Example 2   **Heterogeneous and Homogeneous Mixtures**

Classify each of these as a heterogeneous or homogenous mixture:
   a) beach sand
   b) gasoline
   c) baking powder (which contains sodium hydrogen carbonate and potassium hydrogen tartrate)

Solution
   a) heterogeneous—it contains a mixture of silicon dioxide, calcite, and other minerals
   b) homogeneous—it is a solution of hydrocarbon liquids
   c) heterogeneous—it is a mixture of solids

1.3   The Scientific Method

The scientific method involves the following steps:
   1. Formulating a question that you want answered
   2. Finding out what is already known about your question
   3. Making observations and gathering data
   4. Creating a hypothesis (a possible explanation)
   5. Designing and performing experiments to test the hypothesis to see if it has predictive value

A *scientific law* summarizes the patterns that are observed in the collected data. Laws are descriptions of the behavior of the natural world. A *theory* is an attempt, based on the results of experiments, to explain observations that have been described as fact.

1.4   Units and Measurement

The SI units are based on the metric system. The metric system defines seven fundamental base units (see **Table 1.2** in the text). It is important to get a feel for these units and how they relate to the

English units with which we are so familiar. For example, a meter is slightly larger than a yard (39 inches versus 36 inches). A kilogram is 2.2 pounds.

To deal with very large or very small numbers, we attach prefixes to the base units (see **Table 1.3** in the text). For example, a megameter is 10^6 meters (1 million meters); the prefix *kilo-* indicates 10^3. A nanogram, by contrast, is 10^{-9} grams (one-billionth of a gram). It is important to learn the names of the prefixes, their abbreviations, and the quantity that each represents.

Example 3 Prefixes

Order the following prefixes from smallest to largest:

giga   pico   deci   milli   nano   kilo   micro   femto

Solution

femto (10^{-15}), pico (10^{-12}), nano (10^{-9}), micro (10^{-6}), deci (10^{-1}), kilo (10^3), giga (10^9)

The SI unit of mass is the kilogram (kg), which is often mistakenly referred to as "weight." In most chemistry laboratories, masses are reported in milligrams or grams. The SI unit of length is the meter, but typically chemical particles are measured in micrometers, nanometers, or picometers.

The SI unit of temperature is the kelvin (K, no degree sign). In laboratory work, temperature measurements are most often made in degrees Celsius ($^\circ$C). In medicine, degrees Fahrenheit ($^\circ$F) is commonly used as the unit of temperature. The relationships between the three temperature measurements are shown here:

$$T_K = T_C + 273$$
$$T_C = (T_F - 32)/1.8$$
$$T_F = 1.8(T_C) + 32$$

Example 4 Temperature Conversions

The average body temperature is 98.6°F. What is this temperature in Celsius and Kelvin?

Solution

$$T_C = (T_K - 32)/1.8 \text{ so } T_C = (98.6 - 32)/1.8 = 37.0^\circ C$$
$$T_K = T_C + 273 \text{ so } T_K = 37 + 273 = 310.0 \text{ K}$$

Derived units are combinations of SI units. They measure volume, density, force, pressure, energy, and velocity, among other things. **Table 1.4** in the text lists selected derived units. For volume, chemists often use the **liter (L),** which is equivalent to a cubic decimeter (dm^3), or the **milliliter (mL),** which is equivalent to a cubic centimeter (cm^3 or cc).

$$1 \text{ L} = 1000 \text{ mL} = 1000 \text{ cc} = 1000 \text{ cm}^3 = 1 \text{ dm}^3$$

Density is the mass of a substance that is present in a given volume of the substance.

$$\text{density} = \frac{\text{mass}}{\text{volume}}$$

Example 5 Density

A piece of metal has a volume of 20.2 cm^3 and a mass of 159 g. What is the density of the metal?

First Thoughts

Recall the formula for density. The units are the key as to where to substitute the appropriate numbers.

Solution

$$\text{density} = \frac{\text{mass}}{\text{volume}} = \frac{159 \text{ g}}{20.2 \text{ cm}^3} = 7.87 \text{ g/cm}^3$$

Further Insight

You must also be able to solve this equation for either the mass or the volume of a substance.

Example 6   **Volume from Density**

Ethanol, also known as ethyl alcohol, has a density of 0.789 g/cm^3. What volume will 30.3 g of ethanol occupy?

Solution

$$\text{density} = \frac{\text{mass}}{\text{volume}} \quad \text{so} \quad \text{volume} = \frac{\text{mass}}{\text{density}} = \frac{30.3 \text{ g}}{0.789 \text{ g/cm}^3} = 38.4 \text{ cm}^3$$

1.5   Conversions and Dimensional Analysis

Dimensional analysis is one of the most useful techniques that you will use in general chemistry problems. This approach applies conversion factors to cancel units until you have the proper unit in the proper place. Let's illustrate this idea with a simple example. First we need a table of conversion factors like **Table 1.6** in the text. (Other tables are available in reference books or online.)

Example 7   **Dimensional Analysis I**

Airlines allow each passenger to check two bags weighing 44 lb each. How many kilograms is this?

First Thoughts

We must first find the conversion factor between kilograms and pounds: 1 kg = 2.205 lb. Next we must set up the equation so that the units will cancel until we have the unit we desire. Units cancel when each is present in both the numerator and the denominator. The conversion factor 1 kg = 2.205 lb can be written as

$$\frac{1 \text{ kg}}{2.205 \text{ lb}} \quad \text{or} \quad \frac{2.205 \text{ lb}}{1 \text{ kg}}$$

Use whichever form will cause the units to cancel appropriately.

Solution

$$(2 \text{ bags})\left(\frac{44 \text{ lb}}{\text{bag}}\right)\left(\frac{1 \text{ kg}}{2.205 \text{ lb}}\right) = 40 \text{ kg}$$

Notice that the units cancel: bags with bag, and lb with lb. This leaves the unit kg in our answer. This simple but powerful method can be extended even further, as shown in the next two examples.

Example 8   **Dimensional Analysis II**

How many milliliters are in a half gallon of orange juice?

First Thoughts

Look up the conversion factors: 1 gal = 4 qt, 1 qt = 0.946 L, 1 L = 1000 mL. Set up the conversion factors so that the units cancel.

Solution

$$\left(\frac{1}{2}\ gal\right)\left(\frac{4\ qt}{1\ gal}\right)\left(\frac{0.946\ L}{1\ qt}\right)\left(\frac{1000\ mL}{1\ L}\right) = 1892\ mL$$

Example 9   **Dimensional Analysis III**

How many grams does 1.00 gallon of water weigh? The density of water is 1.00 g/mL.

Solution

$$(1.00\ gal)\left(\frac{4\ qt}{1\ gal}\right)\left(\frac{0.946\ L}{1\ qt}\right)\left(\frac{1000\ mL}{1\ L}\right)\left(\frac{1.00\ g}{1\ mL}\right) = 3.78 \times 10^3\ g$$

1.6   Uncertainty, Precision, Accuracy, and Significant Figures

Chemists must frequently measure quantities. Whenever a measurement is made, however, it introduces a certain degree of uncertainty. The only time you can get an **exact number** is if you can count separate units, such as the number of pages in this book. Defined quantities are also exact. For example, a dozen pencils is always 12 pencils.

If you find the mass of a pencil or measure the length of a pencil, your measurement is not exact. The mass of the pencil could be 2 g or 2.0 g or 2.00 g or 2.000 g, depending on the quality of the balance that you use. **Accuracy** is an indication of how close your measurement is to the true value. The accuracy of your pencil measurement, for example, depends on the quality of the balance and your lab technique. If you make repeated measurements, your average value answer is more likely to be accurate. **Precision** is an indication of how close the measurements are to one another.

Example 10   **Accuracy and Precision**

Two students measured the mass of the same pencil and recorded the following results:

	Student A	Student B
	2.22 g	2.30 g
	2.40 g	2.33 g
	2.35 g	2.32 g
	2.36 g	2.31 g
Average	2.33 g	2.32 g

Knowing that the true mass of the pencil is 2.32 g, comment on the accuracy and precision of the students' results.

First Thoughts

When you examine the data, you must consider how close the average is to the accepted value to determine its accuracy. How much the individual measurements deviate from the average is a measure of the precision of the data.

Solution

Student A's work yielded an accurate result but the deviations from the average show a lack of precision. Student B's work is both accurate and precise.

Significant figures are specific numbers in a measurement whose value we can trust. When a measured value is reported, it is assumed that all figures are correct except for the last one, for which there is an uncertainty of ±1.

In your text, **Table 1.7** lists the **significant figure rules. Table 1.8** gives the rules that you should use when doing calculations. These are important rules to learn! The following examples make use of the rules in both tables, so study them carefully.

When performing numerical calculations, there is always the question of when to round off a number. **Do not round off at intermediate stages** of a calculation, because you may lose figures that affect the accuracy of the final result. Instead, round off the **final** result to indicate the correct number of significant figures, but keep track of the significant figures along the way.

Example 11   Counting Significant Figures

How many significant figures are in the following values?
   a) 3063 m
   b) 0.0062 g
   c) 0.0420 g
   d) 1.21×10^2 L

First Thoughts

Review the rules in **Table 1.7**. Remember, if you are not certain about the proper number of significant figures, convert the number to scientific notation.

Solution

   a) 3063 m = 4 SF
   b) 0.0062 g = 2 SF; the leading zeros are not significant
   c) 0.0420 g = 3 SF; the leading zero is not significant but the trailing zero is significant
   d) 1.21×10^2 L = 3 SF

Example 12   Significant Figures in Calculations

Perform the following calculations, giving the final answer to the correct number of significant figures.
   a) 19.98 – 2.4
   b) 33 × 0.0212
   c) $4.1 \times 10^3/(8.28 – 3.141)$

First Thoughts

Review the rules in **Table 1.8**. Remember to round off only the final answer.

Solution

   a) 19.98 – 2.4 = 17.58, which rounds to **17.6**. With addition or subtraction, the answer contains the same number of decimal places as the datum with the least. In this case, 2.4 has one decimal place.
   b) 33 × 0.0212 = 0.6996, which rounds to **0.70**. The answer has two significant figures, like the datum 33.

c) $4.1 \times 10^3/(8.28 - 3.141) = 797.82$, which rounds to **$8.0 \times 10^2$**. The answer has two significant figures, like the datum 4.1×10^3.

1.7   The Chemical Challenges of the Future

This section of the text discusses how chemistry can address the issues faced by society now and in the future. It explains how our understanding of chemistry can be used to further understand and move forward in fields such as medicine, agriculture, pollution, global warming, materials, energy supply, nanotechnology, and life itself.

Exercises

Section 1.1

1. What does chemistry study?

2. Why must chemists, like other scientists, be inquisitive?

3. From the time you got up this morning until you went to chemistry class, can you identify six items that you used that somehow involved a chemist in their manufacture?

4. Look at a container containing aspirin or other analgesic. What is the unit in which the amount of active ingredient is measured? How do you think a chemist was involved in ensuring that each tablet contains that particular amount of active ingredient?

Section 1.2

5. What is the difference between an element and a compound?

6. Explain how nitrogen, which is found in the Earth's atmosphere, can be an element and a molecule but not a compound.

7. Why is the pure silver found in mines an element and not a molecule or a compound? (*Hint:* Search the Internet for information about silver.)

8. Which of the following are elements and which are compounds?
   a) helium
   b) carbon dioxide
   c) ammonia
   d) copper
   e) calcium chloride

9. What is the difference between a homogeneous mixture and a heterogeneous mixture?

10. Which of the following are homogeneous mixtures and which are heterogeneous mixtures?
    a) milk
    b) seawater
    c) soil
    d) tea
    e) orange soda

11. Classify each of the following as a chemical change or a physical change.
    a) burning wood in a fireplace
    b) boiling water in a pot
    c) rusting of a shovel left out in the snow
    d) making water from hydrogen and oxygen
    e) raining

Section 1.3

12. List the steps in the scientific method.

13. Define a hypothesis.

14. What is the difference between a theory and a hypothesis?

15. How do scientists arrive at a scientific law?

Section 1.4

16. Put the following in order from smallest to largest: gram, kilogram, nanogram, microgram.

17. Put the following in order from smallest to largest: Mm, mm, nm, dm, m.

18. What are the SI units for mass and for volume?

19. What are the SI units for time and for temperature?

20. Perform the following conversions:
    a) $52°F$ to $°C$
    b) $100.°C$ to $°F$
    c) $35°C$ to K
    d) $312 K$ to $°F$

21. Perform the following conversions:
    a) 225 mL to L
    b) 613 mg to g
    c) 0.0125 g to mg
    d) 2.25 kg to g

22. How many 15-g packets would a 1.75-kg container of sugar make?

23. Using scientific notation, express 1.25 g in micrograms.

24. Using scientific notation, express 250 µL in liters.

25. A piece of metal has a mass of 6.885 g. It is added to a graduated cylinder that contains 25.00 mL of water. Upon addition of the metal, the water level rises to 27.55 mL. What is the density of the metal?

26. A sugar solution has a density of 1.18 g/mL. What volume would a 5.56-g sample occupy?

27. An organic liquid has a density of 0.789 g/mL. What would be the mass of a 25.0-mL sample?

Section 1.5

28. The distance from the Earth to the Sun is 93 million miles. How many meters is this?

29. If you are traveling on the highway at 65.0 mph, what is your speed in km/hr?

30. Light travels at 186,000 mi/s. How many kilometers will it cover in exactly 1 min?

31. A very dense element has a density of 19.4 kg/L. What is the density in g/cm^3?

32. Mercury's density is 13.6 g/mL. What is its density in lb/gal?

33. What is a runner's speed in mph, if she ran 5.0 km in 21 min 45 s?

34. A 5-lb bag of sugar cost $1.29. What is the cost per kilogram?

35. Gasoline costs 0.99 euro (1 euro = US$1.26) per liter in Spain. What is the cost in U.S. dollars per gallon?

36. A baseball pitcher threw a pitch clocked at 98.6 mph. How many m/s is this?

Section 1.6

37. Give an example of an exact number.

38. Define accuracy and precision.

In the following problems, state the number of significant figures and indicate which rule applies.

39. 503

40. 0.00123

41. 0.001300

42. 6.440×10^{-2}

In the following problems, perform the indicated calculation and report the answer to the correct number of significant figures.

43. $17.26 + 2.338 =$

44. $29.5/3.4 =$

45. 5.303×2.11

46. $(2.17 \times 10^{-3})/(6.6 \times 10^9) =$

47. $(6.5 - 2.79) \times 4.90 =$

48. Which has more significant figures: 3.65×10^{-5} or 0.00037?

Section 1.7

49. What is nanotechnology?

50. One air pollutant of concern is ozone. What problem does ozone cause for the environment?

Answers to Exercises

1. Chemistry studies the matter of our universe.

2. For chemists to solve problems, they must formulate questions to study.

3. There are many examples of chemistry in action: soap, deodorant, toothpaste, the fabric of towels (or your clothes), any of the cleaning products that you see, the flavors of foods, the manufacture of plates, silverware, glasses, and so on.

4. A typical aspirin tablet contains 325 mg of active ingredient. A chemist at the pharmaceutical company analyzed the content of a sample of the aspirin tablet to ensure that the amount was correct.

5. Elements contain only one type of atom. Compounds contain different elements. Carbon (C) and nitrogen gas (N_2) are examples of elements. Water (H_2O) and ammonia (NH_3) are examples of compounds.

6. Nitrogen, N, is an element containing the atom nitrogen. Nitrogen, N_2, is the molecular form of nitrogen, a gas in our atmosphere. It is not a compound because it contains only one type of atom.

7. Silver, Ag, is found in mines as pure metallic silver. It is an element because it has only one type of atom.

8. a) element (found on the periodic table)
    b) compound of carbon and oxygen
    c) compound of nitrogen and hydrogen
    d) element (found on the periodic table)
    e) compound of calcium and chlorine

9. A homogeneous mixture has the same chemical composition throughout the mixture; a heterogeneous mixture does not.

10. a) homogeneous (milk is "homogenized")
    b) heterogeneous
    c) heterogeneous
    d) homogeneous
    e) homogeneous

11. a) chemical change
    b) physical change
    c) chemical change
    d) chemical change
    e) physical change

12. a) Formulate a question to be answered.
    b) Find out what is already known about your question.
    c) Make observations and gather data.
    d) Create a hypothesis.
    e) Experimentally test the hypothesis to see if it has predictive value.

13. A hypothesis is a plausible explanation for what is observed.

14. A hypothesis is a possible explanation to a problem or question; a theory is an attempt based on experiments to explain factual observations.

15. Scientists propose a law to summarize the patterns that are observed in many experiments.

16. nanogram, microgram, gram, kilogram

17. nm, mm, dm, m, Mm

18. kilogram (kg); cubic meter (m^3)

19. second (s); kelvin (K)

20. a) $11^{\circ}C$
    b) $212^{\circ}F$
    c) 308 K
    d) $102^{\circ}F$

21. a) 0.225 L
    b) 0.613 g
    c) 12.5 mg
    d) 2250 g

22. 116 packets of 15 g each

23. 1.25×10^6 μg

24. 2.50×10^{-4} L

25. 2.70 g/mL

26. 4.71 mL

27. 19.7 g

28. 1.5×10^{11} m

29. 105 km/h

30. 1.80×10^7 km/min

31. 19.4 g/cm^3

32. 114 lb/gal

33. 8.6 mph

34. $0.57 per kg

35. $4.72 per gal

36. 44.1 m/s

37. 1 dozen = 12 (exact number in a dozen)

38. Accuracy is a measure of how close your value is to the true value. Precision is a measure of how close the values are to one another.

39. 3 SF; between zeros are significant

40. 3 SF; leading zeros are not significant

41. 4 SF; trailing zeros are significant

42. 4 SF; trailing zeros are significant

43. 19.60

44. 8.7

45. 11.2

46. 3.3×10^{-13}

47. 18

48. 3.65×10^{-5} (3 SF)

49. Nanotechnology is the creation of materials, devices, and systems through the manipulation of individual atoms and molecules up to 100 nanometers in size.

50. The depletion of ozone from our atmosphere removes molecules that block harmful ultraviolet rays from reaching the Earth. This may lead to a rise in skin cancer.

Chapter 2

Atoms: A Quest for Understanding

The Bottom Line

When you finish this chapter, you will understand:

- How all matter is composed of atoms and how atoms are composed of protons, neutrons, and electrons.
- The laws of definite composition, multiple proportions, and combining volumes.
- The concept of atomic mass.
- How to write chemical formulas.
- The structure of the periodic table and why elements are in groups and periods.
- How to name chemical compounds.

2.1   Early Attempts to Explain Matter

As early as 600 B.C., Greek philosophers proposed the concept of small indivisible pieces of matter called "atoms." In the eighteenth century, two important laws were proposed:

Law of Conservation of Matter: Matter is neither created nor destroyed in a chemical reaction.

Law of Definite Composition: The components of a compound are always present in the same ratio by mass.

Example 1   Law of Definite Composition

A sample of ammonia contains 14.0 g of nitrogen and 3.03 g of hydrogen. If a second sample of ammonia is found to contain 49.0 g of nitrogen, how many grams of hydrogen must be present?

First Thoughts

If the law of definite composition is true, the ratio of the masses of elements in ammonia must remain constant.

Solution

Calculate the factor by which the amount of nitrogen was increased: $49.0/14.0 = 3.50$. Therefore, grams of hydrogen = 3.03 g $\times 3.50 = 10.6$ g of hydrogen in the second sample.

2.2   Dalton's Atomic Theory and Beyond

Dalton identified the **law of multiple proportions:** If two elements combine to form more than one compound, the masses of one element that combine with a fixed mass of the other element are in ratios of small whole numbers.

Example 2   Law of Multiple Proportions

Consider water, H_2O, and hydrogen peroxide, H_2O_2. Show how the ratio of the masses of oxygen that combine with 1.0 g of hydrogen will be a small whole number according to the law of multiple proportions.

Solution

For H_2O, take the ratio of the atomic masses: $16.0/2.02 = 7.92$. That is, 7.92 g of oxygen combines with each 1.0 g of hydrogen. The ratio for H_2O_2 is $32.0/2.02 = 15.84$. That is, 15.84 g of oxygen combines with each 1.0 g of hydrogen. The ratio of oxygen in the two compounds is

$$\frac{15.84 \text{ g}}{7.92 \text{ g}} = 2.00$$

In 1803, Dalton also proposed his **atomic theory** (summarized in **Table 2.2** of the text). Although all of his ideas were not entirely correct, his theory became the foundation of modern chemistry at the time. When reading this section of the text, concentrate on the work of Gay-Lussac and his law of combining volumes of gases. This law led to the refinement of atomic theory by introducing the concept of the *diatomic* nature of gases like O_2, N_2, and H_2.

2.3   The Structure of the Atom

As explained in the text, from the many experiments performed by scientists in the 1900s, the following facts about the atom became apparent:

1.   Atoms can be broken down into protons, neutrons, and electrons.
2.   The nucleus of the atom contains protons and neutrons.
3.   Electrons are found in the space around the nucleus.
4.   Protons are positively charged, neutrons are neutral in charge, and electrons are negatively charged.
5.   The total mass of an atom is related to the number of protons and neutrons in the atom.
6.   Radiation is a result of the decay of atoms. Decay products include alpha particles, beta particles, and gamma rays.

Review **Table 2.3** in the text. Note that the mass of the electron is negligibly small compared to the mass of a proton or neutron.

2.4   Atoms and Isotopes

The periodic table of the elements lists the elements in order of their atomic number Z, which is the number of protons in the nucleus. The mass number of an element is equal to the sum of the number of protons plus neutrons. The symbol for the mass number is A. The number of neutrons, n, in the nucleus can be calculated from the equation $n = A - Z$. Any element can be symbolized by

$$^{A}_{Z}X$$

The number of protons determines which element you are dealing with. **Carbon** has **6 protons.** If the element has **7 protons,** it would be **nitrogen**. Carbon is symbolized by

$$^{12}_{6}C$$

It contains **6 protons** and **6 neutrons** $(12 - 6)$. Because the atom is electrically neutral, it also contains **6 electrons.** If the symbol were

$$^{14}_{6}C$$

carbon would have 6 protons, 6 electrons, and $14 - 6 = 8$ neutrons. This atom would be an **isotope** of carbon. Isotopes have the same atomic number but *different* mass numbers.

Example 3   **Atomic Bookkeeping**

Fill in the missing information in the following table:

Element	Mass Number	Protons	Neutrons	Electrons
Calcium			20	20
Bromine		35	45	
Iron	56			26

First Thoughts

Remember the relationship between atomic number, mass number, and the number of neutrons. Also, recall that the symbol of the element (found in the periodic table) is determined from the number of protons (atomic number). Finally, in a neutral atom, the number of electrons equals the number of protons.

Solution

Element	Mass Number	Protons	Neutrons	Electrons
Calcium	40	20	20	20
Bromine	80	35	45	35
Iron	56	26	30	26

2.5 Atomic Mass

A **mass spectrometer** is an instrument that measures the mass and abundance of isotopes of an element in a sample. You should review its operation in your textbook. **Atomic masses,** which can be found in a periodic table or a table of atomic masses, are **weighted averages** of the mass and abundance of each isotope found in the element. Consider the following example.

Example 4   **Calculating the Atomic Mass Value**

A sample of neon gas contains three isotopes: neon-20, neon-21, and neon-22. The abundances of these isotopes are 90.920%, 0.2570%, and 8.820%, respectively. The isotopic masses, in the same order, are 19.99244 amu, 20.99395 amu, and 21.99138 amu. Calculate the atomic mass of neon.

First Thoughts

Each isotope's contribution to the total atomic mass is calculated by finding the percentage of the isotopic mass accounted for by each isotope. The sum of the contributions of all isotopes is then the atomic mass of neon.

Solution

atomic mass = $(0.90920 \times 19.99244) + (0.002570 \times 20.99395) + (0.08820 \times 21.99138) = 20.171$ amu

2.6   The Periodic Table

Elements in the periodic table are arranged in periods (rows) and groups (columns) according to their chemical properties. Most elements are metals (located toward the left side of the periodic table); fewer are nonmetals (grouped on the right side of the table). The boundary between the metals and the nonmetals is depicted by a heavy stair-step line on most periodic tables. The elements on either side of the boundary line are referred to as metalloids—they posses both metallic and nonmetallic properties. You should learn the common names of the most important groups (they are listed in **Table 2.5** of the text).

Look at the periodic table and familiarize yourself with its arrangement. Learn the symbols of the common elements, especially the ones that are derived from Latin names like sodium, potassium, lead, and iron. Elements in "A" groups are referred to as **main-group** elements. Those in the "B" groups are known as **transition** elements.

2.7   Ionic Compounds

Ions are atoms that have a charge. This charge can be either positive or negative. How do atoms attain a charge? They either lose or gain electrons, which are negatively charged. Recall that atoms are neutral; that is, they have the same number of electrons (– charge) and protons (+ charge). If an atom loses an electron, it attains a +1 charge (for example, K^+) and is called a **cation.** If an atom gains an electron, it attains a –1 charge (for example, F^-); negative ions are called **anions.**

Example 5   **Ions**

Complete the following table.

Symbol	Protons	Electrons	Charge
$_{11}Na^+$			
$_{34}Se^{2-}$			
$_{38}Sr^{2+}$			

Solution

Symbol	Protons	Electrons	Charge
$_{11}Na^+$	11	10	+1
$_{34}Se^{2-}$	34	36	-2
$_{38}Sr^{2+}$	38	36	+2

Ionic compounds are formed from the combination of metals and nonmetals. The chemical formula for an ionic compound represents the ratio of each element in that compound. For example, in $CaCl_2$ there is one calcium atom for every two chlorine atoms. How is this ratio determined? It makes use of the *principle of electron neutrality*. In other words, the charge on the calcium ion is neutralized by the charge on the chlorine ions. The calcium ion has a +2 charge, while each chlorine ion has a –1 charge. Therefore, two chlorine ions are needed to neutralize the calcium ion. **Table 2.6**

gives the charges on typical main-group ions. Notice that metals form positive ions, whereas nonmetals form negative ions.

Example 6   **Determining Formulas**

Predict the formula of a compound formed from (a) potassium and oxygen and (b) magnesium and sulfur.

First Thoughts

Look at the group number for each of the elements that you are using to remind yourself of the charge that they will form as ions.

Solution

a) K_2O: K is in Group IA and forms a +1 ion. O is in Group VIA and forms a –2 ion. Two K^+ ions neutralize the O^{2-}.
b) MgS: Mg is in Group IIA and forms a +2 ion. S is in Group VIA and forms a –2 ion. One Mg^{2+} ion neutralizes the S^{2-}.

2.8   Molecules

Molecules contain atoms rather than ions. They are represented by a molecular formula, such as CO_2. The atoms in a molecule are held together by **covalent bonds** formed by sharing electrons. The formula of a molecular compound indicates the number of atoms in the molecule, rather than the ratio of ions in an ionic formula.

2.9   Naming Compounds

Chemical nomenclature follows specific rules that are given in **Tables 2.8, 2.9,** and **2.10** in your text. Study these rules carefully and learn how to apply them. We will summarize each table here and give examples of how to use the rules.

Rules for Naming Binary Covalent Compounds

1.   The first element uses the exact element name.
2.   The second element uses the stem name plus the suffix **-ide.**
3.   Use Greek prefixes (see **Table 2.8**) for each element name to denote the subscript in the formula.

Example 7   **Naming Binary Molecular Compounds**

Name the following binary molecular compounds:

$$PCl_3 \quad CO_2 \quad NO \quad N_2O_4 \quad SF_6 \quad IF_5$$

Solution

PCl_3 phosphorous trichloride
CO_2 carbon dioxide
NO   nitrogen monoxide
N_2O_4 dinitrogen tetroxide
SF_6 sulfur hexafluoride
IF_5 iodine pentafluoride

Rules for Naming Binary Ionic Compounds

1.   The first element uses the exact element name.
2.   The second element uses the stem name plus the suffix **-ide.**

3. For metals that can have more than one stable charge (transition metals—"B" groups), show the charge as a Roman numeral in parentheses immediately after the metal name.

Example 8   **Naming Binary Ionic Compounds**

Name the following binary ionic compounds:

$$KCl \quad SnF_4 \quad AlCl_3 \quad BaS \quad Fe_2O_3 \quad MgBr_2$$

Solution

KCl	potassium chloride
SnF_4	tin(IV) fluoride (*not* tin tetrafluoride)
$AlCl_3$	aluminum chloride (*not* aluminum trichloride)
BaS	barium sulfide
Fe_2O_3	iron(III) oxide
$MgBr_2$	magnesium bromide

Further Insight

Naming compounds requires a thorough understanding of the elements in the periodic table. You must be able to distinguish between a metal and a nonmetal. Also, you must recognize that representative elements form ion charges related to their groups. Transition metals have more than one possible charge.

Polyatomic ions, as the name suggests, are ions that contain more than one atom. There is no way to predict the name of a polyatomic ion. Carefully study **Table 2.10** and memorize the formula, name, and charge of these important ions.

Example 9   **Naming Compounds Containing Polyatomic Ions**

Name the following compounds:

$$KMnO_4 \qquad Ba(OH)_2 \qquad CuSO_4 \qquad Ca(NO_3)_2 \qquad Na_3PO_4$$

Solution

$KMnO_4$	potassium permanganate
$Ba(OH)_2$	barium hydroxide
$CuSO_4$	copper(II) sulfate
$Ca(NO_3)_2$	calcium nitrate
Na_3PO_4	sodium phosphate

Exercises

Section 2.1

1. A sample of NO_2 contains 14.0 g of nitrogen and 32.00 g of oxygen. If a second sample contains 35.00 g of nitrogen, how many grams of oxygen are in the second sample?

2. A sample of HNO_3 contains 1.01 g of hydrogen, 14.01 g of nitrogen and 48.00 g of oxygen. How many grams of nitrogen and oxygen are present in a second sample containing 5.15 g of hydrogen?

Section 2.2

3. Carbon and oxygen can react to form carbon monoxide and carbon dioxide. In a sample of carbon monoxide, there are 12.01 g of carbon and 16.00 g of oxygen. In a sample of carbon dioxide, 12.01 g of carbon combines with 32.00 g of oxygen. What is the ratio of the masses of oxygen that combines with 12.01 g of carbon in each case, and how does this illustrate the law of multiple proportions?

4. How does Dalton's atomic theory relate to each of the following chemical statements?
   a) $2H_2 + O_2 \rightarrow 2H_2O$
   b) There are 6.022×10^{23} atoms in 1.008 g of hydrogen.

Section 2.3

5. Which of the following statements are true?

   a) The nucleus of an atom contains protons, neutrons, and electrons.
   b) The mass number is the number of protons and neutrons.
   c) Radiation is the result of the decay of atoms.
   d) Decay particles are named alpha, beta, and gamma.

Section 2.4

6. Identify the following elements: $_{36}^{85}X$, $_{22}^{48}X$, $_{52}^{128}X$.

7. How many protons, neutrons, and electrons do each of the following contain: ^{64}Cu, ^{238}U, ^{45}Sc?

8. Which of the following atoms are isotopes?

Atom	Protons	Neutrons
A	15	16
B	16	16
C	15	17
D	17	18

9. Naturally occurring Li is a mixture of ^{6}Li and ^{7}Li. How many protons and neutrons does each isotope have?

10. An atom contains 12 protons and 12 neutrons. What is its atomic symbol and name?

11. An atom has 40 protons and 51 neutrons. What is its atomic symbol and name?

12. Element X has 28 electrons and 30 neutrons. What is its name?

13. Element Y has 57 electrons and 102 neutrons. What is its name?

14. Element Z has 49 electrons. What is its name and how many protons does it have?

Section 2.5

15. Boron has two isotopes with the following masses and abundances: 10.0129 amu, 19.91%; and 11.0093 amu; 80.09%. What is boron's average atomic mass?

16. Magnesium's three isotopes have the following masses and abundances: 23.9850 amu, 78.99%; 24.9858, 10.00%; and 25.9826 amu, 11.01%. Calculate the average atomic mass for magnesium.

17. An element has two isotopes with the following masses and abundances: 84.9118 amu, 72.15%; and 86.9092 amu, 27.85%. Calculate the element's average atomic mass. Can you identify the element?

Section 2.6

18. What name is given to the elements in Group IIA?

19. What name is given to the elements in Group VIIA?

20. What name is given to the elements in Group VIA?

21. Elements in the "B" groups are referred to as ___?

22. What is the name of the metalloid found in Group IVA, Period 3?

23. Which Group VIA element is a metalloid?

24. Which halogen is in the fifth period?

25. Which noble gas is in the fourth period?

26. Which alkali metal is in the third period?

For Exercises 27–31, give the period and group for each element, and identify each as a metal, metalloid, or nonmetal.

27. Ge

28. Br

29. Ba

30. Xe

31. Ag

Section 2.7

For Exercises 32–36, give the number of protons and electrons for each ion.

32. Br^-

33. Rb^+

34. P^{3-}

35. Sr^{2+}

36. O^{2-}

37. Which ionic compound would be formed from sodium and sulfur?

38. Predict the compound formed from magnesium and phosphorus.

39. Give the ionic formula for the compound formed from aluminum and chlorine.

Section 2.9

40. Name the following compounds:
    a) CaS
    b) KCl
    c) $MgBr_2$

41. Write the formulas for the following compounds:
    a) cesium sulfide
    b) sodium chloride
    c) magnesium oxide

42. Name the following compounds:
    a) CO
    b) N_2O_5
    c) PCl_5
    d) NO_2

43. Write the formulas for the following compounds:
    a) xenon tetrafluoride
    b) nitrogen triiodide
    c) dinitrogen tetrafluoride

44. Name the following compounds:
    a) $FeCl_3$
    b) $CuBr_2$
    c) Cr_2O_3

45. Write the formulas for the following compounds:
    a) copper(I) chloride
    b) iron(III) oxide
    c) gold(III) bromide
    d) lead(IV) oxide

46. Name the following compounds:
    a) Na_2SO_4
    b) Li_3PO_4
    c) $CaCO_3$

47. Write the formulas for the following compounds:
    a) magnesium phosphate
    b) lead(II) carbonate
    c) potassium nitrate

48. Which of the following is *not* a polyatomic ion: hydroxide, hydride, or carbonate?

49. What is the charge on each of the following polyatomic ions:
    a) phosphate
    b) nitrate
    c) sulfate

50. Write the formula of the compound formed by iron(III) and cyanide.

Answers to Exercises

1. 80.0 g

2. 71.4 g nitrogen and 245 g oxygen

3. The ratio is 2:1, which are both whole numbers in accordance with the law of multiple proportions.

4. a) Chemical reactions involve the rearrangement of atoms.
   b) Each element is made of tiny particles called atoms.

5. a) False: Electrons are outside the nucleus.
   b) True
   c) True
   d) False: Gamma radiation is a ray, not a particle.

6. Kr, Ti, Te

7. 29p, 35n, 29e; 92p, 146n, 92e; 21p, 24n, 21e

8. Atoms A and C; same number of protons but different number of neutrons.

9. ^{6}Li has 3 protons and 3 neutrons. ^{7}Li has 3 protons and 4 neutrons.

10. $^{24}_{12}$Mg , magnesium

11. $^{91}_{40}$Zr , zirconium

12. Nickel

13. Lanthanum

14. Indium, 49 protons. Elements are neutral in charge and have the same number of protons and electrons.

15. 10.811 amu

16. 24.31 amu

17. 85.47 amu, Rb

18. Alkaline earth metals

19. Halogens

20. Chalcogens

21. Transition metals

22. Silicon

23. Tellurium, Te

24. Iodine, I

25. Krypton (Superman's home planet)

26. Sodium, Na

27. Period 4, Group IVA, metalloid

28. Period 4, Group VIIA, nonmetal

29. Period 6, Group IIA, metal

30. Period 5, Group VIIIA, nonmetal (noble gas)

31. Period 5, Group IB, metal

32. 35 protons, 36 electrons

33. 37 protons, 36 electrons

34. 15 protons, 18 electrons

35. 38 protons, 36 electrons

36. 8 protons, 10 electrons

37. Na_2S

38. Mg_3P_2

39. $AlCl_3$

40. a) calcium sulfide
    b) potassium chloride
    c) magnesium bromide

41. a) Cs_2S
    b) NaCl
    c) MgO

42. a) carbon monoxide
    b) dinitrogen pentoxide
    c) phosphorous pentachloride
    d) nitrogen dioxide

43. a) XeF_4
    b) NI_3
    c) N_2F_4

44. a) iron(III) chloride
    b) copper(II) bromide
    c) chromium(III) oxide

45. a) CuCl
    b) Fe_2O_3
    c) $AuBr_3$
    d) PbO_2

46. a) sodium sulfate
    b) lithium phosphate
    c) calcium carbonate

47. a) $Mg_3(PO_4)_2$
    b) $PbCO_3$
    c) KNO_3

48. hydride

49. a) −3
    b) −1
    c) −2

50. $Fe(CN)_3$

Chapter 3

Introducing Quantitative Chemistry

The Bottom Line

When you finish this chapter, you will be able to:

- Determine the formula mass (molecular weight) of an element or compound in atomic mass units (amu) or in grams per mole (g/mol).
- Explain and assess the mole concept and describe how it relates to Avogadro's number and the mass in grams.
- Determine the mass percent of an element in a compound and utilize this concept to predict empirical and molecular formulas.
- Illustrate how to balance chemical equations, including on the atomic level.
- Model how to solve stoichiometry problems, such as those involving limiting reactants and percent yields.

3.1   Formula Masses

One way to determine the mass of molecules is to calculate the formula mass. The term *formula mass* is often used interchangeably with the terms *formula weight* and *molecular mass*. No matter which term is used, the way in which you calculate this mass is the same: add up the atomic mass of each element in the molecule or compound.

Example 1   **Calculating Formula Mass**

What is the formula mass for the sugar glucose, $C_6H_{12}O_6$?

First Thoughts

Before calculating the formula mass of glucose, you need to determine the average mass of one carbon atom, one hydrogen atom, and one oxygen atom. These masses are found on the periodic table.

> 1 carbon atom = 12.01 amu
> 1 hydrogen atom = 1.008 amu
> 1 oxygen atom = 16.00 amu

Solution

When determining the formula mass of an entire glucose molecule, you need to take into account the number of each element in glucose before adding up the entire mass.

6 atoms of carbon @ 12.01 amu	= 72.06
12 atoms of hydrogen @ 1.01 amu	= 12.12
6 atoms of oxygen @ 16.00 amu	= 96.00
Formula mass of glucose	= 180.18 amu

Further Insight

Try calculating the formula mass of calcium phosphate, $Ca_3(PO_4)_2$. Remember that the subscript "2" means two phosphate ions (each of which consists of one phosphorus atom and four oxygen atoms). You should calculate a formula mass of 310.18 amu for calcium phosphate.

3.2   Counting by Weighing

The concepts presented in this section can be some of the most difficult to grasp in chemistry. We have to take a very abstract idea (atoms, which we cannot physically see) and relate them to concrete quantities we measure in the laboratory (mass in grams or kilograms). We recommend reading and analyzing this section in your text several times until you feel confident in your understanding and can thoroughly explain this material to someone else. Study **Figure 3.4** in the text to help you see the conversion from atomic mass units to grams. This conversion is very important because scientists do not have a scale that measures mass in atomic mass units!

Here is the bottom line: *When you measure out a mass in grams of an element that has the same numerical value as the formula mass in atomic mass units, you always have 6.022×10^{23} atoms of that element.* The same principle applies to compounds. Your text verified both of these cases with aspirin and calcium. The number 6.022×10^{23} is known as **Avogadro's number** and is extremely important to chemists. It links elements and compounds on the atomic scale to the laboratory scale.

Example 2   **How Many Atoms?**

How many atoms are in a 1.25-g sample of sulfur?

First Thoughts

Recall that 1 amu = 1.6605×10^{-24} g. You can use this relationship to convert your mass in grams (laboratory scale) to the mass in atomic mass units (atomic scale). You can then use the average molecular mass (formula mass) to convert from atomic mass units to atoms of sulfur.

Solution

$$(1.25 \text{ g S})\left(\frac{1 \text{ amu S}}{1.6605 \times 10^{-24} \text{ g S}}\right)\left(\frac{1 \text{ atom S}}{32.07 \text{ amu S}}\right) = 2.35 \times 10^{22} \text{ atoms of sulfur}$$

Found on the
periodic table

Further Insight

You could have also solved the problem by applying the concept that when you measure out 32.07 g of sulfur (which is the formula mass of sulfur), you have 6.022×10^{23} atoms of sulfur.

$$(1.25 \text{ g S})\left(\frac{6.022 \times 10^{23} \text{ atoms S}}{32.07 \text{ g S}}\right) = 2.35 \times 10^{22} \text{ atoms of sulfur}$$

These same principles apply when you are trying to find the number of molecules in a given sample.

Example 3   **How Many Molecules?**

How many molecules of ammonia (NH_3) are in a 4.14-g sample?

Solution

$$\left(4.14 \text{ g NH}_3\right)\left(\frac{1 \text{ amu NH}_3}{1.6605 \times 10^{-24} \text{ g NH}_3}\right)\left(\frac{1 \text{ molecule NH}_3}{17.034 \text{ amu NH}_3}\right) = 1.46 \times 10^{23} \text{ molecules of NH}_3$$

OR

$$\left(4.14 \text{ g NH}_3\right)\left(\frac{6.022 \times 10^{23} \text{ molecules NH}_3}{17.034 \text{ g NH}_3}\right) = 1.46 \times 10^{23} \text{ molecules of NH}_3$$

Avogadro's number (N_A) is given a special name: the **mole** (abbreviated as "mol"). One mole is equivalent to 6.022×10^{23} "anything." That is, 1 mole corresponds to 6.022×10^{23}, just as 1 dozen = 12 or a trio = 3. For example, 1 mole of marbles = 6.022×10^{23} marbles and 1 mole of baseballs = 6.022×10^{23} baseballs. In chemistry, we tend to use the mole concept when dealing with atoms, molecules, or ions. For example, 1 mole of oxygen atoms = 6.022×10^{23} oxygen atoms and 1 mole of methane (CH_4) = 6.022×10^{23} CH_4 molecules. **Figure 3.6** in the text shows three compounds, all of which contain 6.022×10^{23} molecules or "units," which is equivalent to 1 mole of each compound. Notice that even though the number of units is the same in each case, the three compounds take up different amounts of space and have different masses.

Now let's relate this concept to mass in grams, which is the practical unit of measurement in the laboratory. When you measure out the mass of a given element that is equal to its formula mass, you have 6.022×10^{23} atoms of that element. We also know that 6.022×10^{23} = 1 mol. Therefore, one mole of an element is *equal* to its mass in grams that matches its formula mass. For example, 16.00 g of oxygen represents 6.022×10^{23} oxygen atoms, which also equals 1 mole of oxygen. Thus we can say that 16.00 g of oxygen = 1 mol of oxygen. This quantity is known as the **molar mass** of oxygen (16.00 g/mol). This same idea applies to compounds. NaCl, for example, has a formula mass of 58.44 amu (Na = 22.99 amu and Cl = 35.45 amu), so 58.44 g of NaCl = 6.022×10^{23} "units" of NaCl = 1 mol of NaCl. The molar mass of NaCl is therefore 58.44 g/mol. Notice that we use the term "units" when talking about compounds that are not molecules.

3.3   Working with Moles

The purpose of this section is to apply what you have learned in Sections 3.1 and 3.2 by converting between atoms or "units" of a substance, moles, and mass in grams.

Example 4   **Atoms and Moles**

How many moles of silver do 7.41×10^{24} atoms of silver represent?

First Thoughts

Remember that 6.022×10^{23} atoms of silver = 1 mole of silver.

Solution

$$\left(7.41 \times 10^{24} \text{ atoms Ag}\right)\left(\frac{1 \text{ mol Ag}}{6.022 \times 10^{23} \text{ atoms Ag}}\right) = 12.3 \text{ mol Ag}$$

Further Insight

The quantity 7.41×10^{24} molecules of a substance like ethanol, C_2H_5OH, would represent the same number of moles. One mole of C_2H_5OH contains 6.022×10^{23} molecules of C_2H_5OH.

Example 5   Grams to Moles

How many moles of sodium hydroxide, NaOH, are present in a 35.1-g sample?

First Thoughts

We can use the molar mass of NaOH to convert from mass in grams to moles. The molar mass in g/mol will be the same as the formula mass in amu.

Formula Mass: 1 Na @ 22.99   = 22.99 amu
                           1 O @ 16.00   = 16.00 amu
                           1 H @ 1.01   = 1.01 amu
                           _____
                                        40.00 amu = 40.00 g/mol

Now we use this molar mass to convert to moles.

Solution

$$(35.1 \text{ g NaOH})\left(\frac{1 \text{ mol NaOH}}{40.00 \text{ g NaOH}}\right) = 0.878 \text{ mol NaOH}$$

Further Insight

You can take this problem a step further and determine the number of formula units present in the 35.1-g sample, especially given that you have already converted this value to moles.

$$(0.878 \text{ mol NaOH})\left(\frac{6.022\times10^{23} \text{ units NaOH}}{1 \text{ mol NaOH}}\right) = 5.29\times10^{23} \text{ units of NaOH}$$

Example 6   Moles to Grams

What is the mass (in kilograms) of a 5.22-mol sample of cholesterol, $C_{27}H_{46}O$?

First Thoughts

You can create a flowchart to help you figure out how to start this problem.

$$\text{moles} \xrightarrow{\frac{g}{mol}} \text{grams} \xrightarrow{\frac{kg}{g}} \text{kilograms}$$

The "g/mol" part of the flowchart is the molar mass of cholesterol and the "kg/g" is just a conversion factor from grams to kilograms.

Solution

Molar mass: 27 C @ 12.01   = 324.27
                       46 H @ 1.01   = 46.46
                       1 O @ 16.00   = 16.00
                       _____
                                        386.73 g/mol

Therefore

$$(5.22 \text{ mol } C_{27}H_{46}O)\left(\frac{386.73 \text{ g } C_{27}H_{46}O}{1 \text{ mol } C_{27}H_{46}O}\right)\left(\frac{1 \text{ kg}}{1000 \text{ g}}\right) = 2.02 \text{ kg } C_{27}H_{46}O$$

Further Insight

Calculate how many molecules of cholesterol are in this sample. You should get an answer of 3.14×10^{24} molecules of $C_{27}H_{46}O$.

3.4   Percentages by Mass

The mass percent is calculated by dividing the total mass of the component in which you're interested by the total mass of the entire sample. You then multiply this value by 100% to get the mass percent.

$$\text{Mass percent} = \frac{\text{total mass component}}{\text{total mass whole substance}} \times 100\%$$

Example 7   **Calculating Mass Percent I**

Calculate the mass percent of phosphorus in sodium phosphate, Na_3PO_4.

Solution

Express the mass of phosphorus as a percentage of the total mass of one formula unit (or one mole) of sodium phosphate:

$$\left(1 \text{ mol P}\right)\left(\frac{30.97 \text{ g P}}{1 \text{ mol P}}\right) = 30.97 \text{ g P}$$

$$\left(3 \text{ mol Na}\right)\left(\frac{22.99 \text{ g Na}}{1 \text{ mol Na}}\right) = 68.97 \text{ g Na}$$

$$\left(4 \text{ mol O}\right)\left(\frac{16.00 \text{ g O}}{1 \text{ mol O}}\right) = 64.00 \text{ g O}$$

$$163.94 \text{ g} = \quad \text{total mass of 1 mol} \\ Na_3PO_4 \text{ (molar mass)}$$

$$\text{mass percent phosphorus} = \frac{\text{mass total P}}{\text{mass total } Na_3PO_4} \times 100\% = \frac{30.97}{163.94} \times 100\% = 18.89\% \text{ P}$$

Example 8   **Calculating Mass Percent II**

What is the percent by mass of iron in iron(III) oxide?

Solution

The formula for iron(III) oxide is Fe_2O_3.

$$\left(2 \text{ mol Fe}\right)\left(\frac{55.85 \text{ g Fe}}{1 \text{ mol Fe}}\right) = 111.7 \text{ g Fe}$$

$$\left(3 \text{ mol O}\right)\left(\frac{16.00 \text{ g O}}{1 \text{ mol O}}\right) = 48.00 \text{ g O}$$

$$159.7 \text{ g} = \quad \text{total mass of 1 mol} \\ Fe_2O_3 \text{ (molar mass)}$$

$$\text{mass percent iron} = \frac{\text{mass total Fe}}{\text{mass total Fe}_2O_3} \times 100\% = \frac{111.7}{159.7} \times 100\% = 69.94\% \text{ Fe}$$

3.5　Finding the Formula

The empirical formula of a compound is the formula that contains the *simplest* whole-number ratio of atoms. To find the empirical formula, you must first convert the mass ratio of the atoms into a mole ratio. You then divide all of the mole values by the smallest value and determine the simplest whole-number ratio of the atoms that make up the compound.

Example 9　　　　**Determine the Empirical Formula**

A molecule contains 56.79% C, 6.56% H, 8.28% N, and 28.37% O by mass. What is the empirical formula for this molecule?

First Thoughts

We must first convert the mass ratio of each element into a mole ratio. In this example, the mass percent of each element is given. From **Section 3.4** in the text, we know that these mass percents must add to 100%. Let's assume that we have 100 g of our sample. This sample therefore contains 56.79 g C, 6.56 g H, 8.28 g N, and 28.37 g O. Remember, we are looking for the ratios between these atoms to determine the empirical formula, so assuming we have a 100-g sample makes this comparison easier.

Solution

$$(56.79 \text{ g C})\left(\frac{1 \text{ mol C}}{12.01 \text{ g C}}\right) = 4.729 \text{ mol C}$$

$$(6.56 \text{ g H})\left(\frac{1 \text{ mol H}}{1.01 \text{ g H}}\right) = 6.50 \text{ mol H}$$

$$(8.28 \text{ g N})\left(\frac{1 \text{ mol N}}{14.01 \text{ g N}}\right) = 0.591 \text{ mol N}$$

$$(28.37 \text{ g O})\left(\frac{1 \text{ mol O}}{16.00 \text{ g O}}\right) = 1.773 \text{ mol O}$$

Next, we divide all of these mole values by the smallest value to get the simplest whole-number ratio of the atoms that make up our molecule. In this example, the smallest value is 0.591.

$$\frac{4.729}{0.591} = 8 \qquad \frac{6.50}{0.591} = 11 \qquad \frac{0.591}{0.591} = 1 \qquad \frac{1.773}{0.591} = 3$$

The empirical formula for the molecule is **$C_8H_{11}NO_3$**.

Further Insight

In some cases, when you divide all of the mole ratios by the smallest value, you will not obtain whole numbers. When this occurs, an extra step is required to find the empirical formula. See Example 10 for a demonstration.

Example 10   **One More Step**

A sample contains 3.87 g of phosphorus and 5.00 g of oxygen. What is the empirical formula?

First Thoughts

The problem gives the mass of each element in grams, so we can immediately calculate these amounts in moles.

Solution

$$(3.87 \text{ g P})\left(\frac{1 \text{ mol P}}{30.97 \text{ g P}}\right) = 0.125 \text{ mol P}$$

$$(5.00 \text{ g O})\left(\frac{1 \text{ mol O}}{16.00 \text{ g O}}\right) = 0.3125 \text{ mol O}$$

Next, we divide both mole values by the smallest value:

$$\frac{0.125}{0.125} = 1 \qquad \frac{0.3125}{0.125} = 2.5$$

If we stopped here, then the formula would be **PO$_{2.5}$**. This is not the simplest *whole-number* ratio, however, so it cannot be the empirical formula. (Also, when have you ever seen a compound with partial atoms?)

We can convert this formula to the smallest whole-number ratio by multiplying both numbers by 2. The empirical formula is therefore **P$_2$O$_5$**.

Further Insight

In this case, we had to go one step further than in Example 9 to obtain the empirical formula. The important point to remember is that the empirical formula must contain the *simplest whole-number ratio* of atoms. In Example 9, dividing both mole values by the smallest number gave us a whole-number ratio immediately. In Example 10, we had to multiply the ratio by 2 to get a whole-number ratio. In other cases, the multiplier needed could be another number.

The empirical formula tells us only the ratio in which the elements are present in a compound. It does not tell us the molecular formula, which is the *actual* number of each type of element in a compound. To determine the molecular formula, we must know the compound's molar mass. We can then calculate the molar mass of our empirical formula. From these data, we can determine how many times the empirical molar mass must be multiplied to obtain the molar mass of the molecular formula.

Example 11   **Determining the Molecular Formula**

The molar mass of the compound in Example 10 is 283.88 g/mol. What is the molecular formula of this compound?

Solution

We have already determined that the empirical formula of this compound is P$_2$O$_5$. Let's calculate the molar mass of P$_2$O$_5$ so that we can compare this value to 283.88 g/mol.

$$\left(2 \text{ mol P}\right)\left(\frac{30.97 \text{ g P}}{1 \text{ mol P}}\right) = 61.94 \text{ g P}$$

$$\left(5 \text{ mol O}\right)\left(\frac{16.00 \text{ g O}}{1 \text{ mol O}}\right) = \underline{80.00 \text{ g O}}$$

141.94 g P_2O_5 in 1 mol of this compound
(molar mass = 141.94 g/mol)

The molar mass of the molecular formula is twice the value of the molar mass of the empirical formula. Therefore, you must multiply the number of atoms in P_2O_5 by 2. The molecular formula is **P_4O_{10}**. To check your answer, calculate the molar mass of P_4O_{10}. Did you get 283.88 g/mol?

Example 12   **More Practice with Molecular Formulas**

Determine the molecular formulas based on the empirical formulas and molar masses given:

a) $C_4H_5N_2O$ (194.20 g/mol)

b)   SN (184.156 g/mol)

c) CH_2O (180.156 g/mol)

d) NH_3 (17.034 g/mol)

Solution

a) $C_8H_{10}N_4O_2$

b) S_4N_4

c) $C_6H_{12}O_6$

d) NH_3

3.6   Chemical Equations

Chemical reactions are crucial to understanding the process of change and are often depicted as chemical equations. For example, the reaction between hydrogen gas and oxygen gas to form water could be represented like this:

$$H_2(g) + O_2(g) \rightarrow H_2O(g)$$

The reactants in this chemical equation are hydrogen and oxygen gases. The product is water. Each compound is in a gaseous state.

A chemical equation also helps us determine how much of each reactant we need to produce a certain amount of product or how much product should result from the reaction of a given amount of reactants. Before we can perform these calculations, however, we must make sure the chemical equation is *balanced*. During a reaction, atoms cannot be created or destroyed, but rather are simply rearranged. This is known as the law of conservation of matter (or the law of conservation of mass).

Let's examine the equation above to see whether it follows this law as written. The reactant side contains two hydrogen atoms and two oxygen atoms. The product side contains two hydrogen atoms but only one oxygen atom. The reaction represented in this equation cannot occur unless we alter the *proportions* in which the hydrogen and oxygen gases react and water is formed. We alter these proportions by placing **coefficients** in front of the compounds until the atoms balance on both sides.

Let's balance the oxygen atoms by placing a "2" in front of water:

$$H_2(g) + O_2(g) \rightarrow 2H_2O(g)$$

Now there are two oxygen atoms on each side of the equation, but we have two hydrogen atoms on the reactant side and four hydrogen atoms on the product side. Let's place a "2" in front of the hydrogen gas to balance the equation:

$$2H_2(g) + O_2(g) \rightarrow 2H_2O(g)$$

Looking at this equation on the atomic scale allows us to better visualize how the atoms balance:

As this diagram indicates, two hydrogen molecules are needed for every one molecule of oxygen (which would then produce two water molecules). On a larger scale, two moles of hydrogen are needed for every one mole of oxygen (which would then produce two moles of water).

It is very important to use only coefficients to balance equations! *Never* change the subscripts in the formula to balance a chemical equation. Why would it be wrong to balance the equation like this?

$$H_2 + O_2 \rightarrow H_2O_2$$

It is wrong because you have changed the product from water to hydrogen peroxide—a very different compound!

Example 13   Balancing an Equation

Hydrogen gas and nitrogen gas react to form gaseous ammonia. What is the balanced equation for this process?

First Thoughts

You first need to write a chemical equation before you can balance it. The chemical equation is

$$H_2(g) + N_2(g) \rightarrow NH_3(g)$$

Solution

There are two nitrogen atoms on the reactant side and one nitrogen atom on the product side. Let's balance the nitrogens by placing a "2" in front of ammonia:

$$H_2(g) + N_2(g) \rightarrow 2NH_3(g)$$

(visualize on the atomic scale)

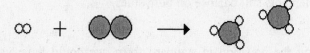

Now the nitrogens are balanced but the hydrogens are not (there are two H atoms on the reactant side and six H atoms on the product side). Place a "3" in front of the hydrogen gas to balance the hydrogens:

$$3H_2(g) + N_2(g) \rightarrow 2NH_3(g)$$

(visualize on the atomic scale)

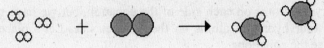

Further Insight

Sometimes balancing chemical equations requires you to use fractions as coefficients. In the standard form of chemical equations, fractions are not used, so you will have to get rid of them. See Example 14.

Example 14   Balancing Equations in Standard Form

Ethane (C_2H_6) reacts with oxygen in the air to produce carbon dioxide and water. What is the balanced equation for this reaction in standard form?

First Thoughts

Write the chemical equation for the reaction: $C_2H_6 + O_2 \rightarrow CO_2 + H_2O$

When an element appears in more than one compound on the side of an equation (in this case, oxygen on the product side), it is usually best to balance this element *last*. Start with a simpler atom, like carbon.

Solution

Place a "2" in front of CO_2 to get two carbons on each side:

$$C_2H_6 + O_2 \rightarrow 2CO_2 + H_2O$$

Place a "3" in front of H_2O to get six hydrogens on each side:

$$C_2H_6 + O_2 \rightarrow 2CO_2 + 3H_2O$$

The only element left to balance is oxygen. There are two oxygen atoms on the reactant side and seven oxygen atoms on the product side. We could place "7/2" or "3.5" in front of O_2 to get seven oxygen atoms on both sides. This approach would be perfectly okay in terms of balancing the equation because the important consideration is the *ratio* in which the reactants react with one another. However, when balancing an equation in standard form, we must use the smallest *whole-number* coefficients. Therefore, we multiply each coefficient by "2" to make "7/2" a whole number.

$$C_2H_6 + \frac{7}{2}O_2 \rightarrow 2CO_2 + 3H_2O$$

$$\underline{\times 2}$$

$$2C_2H_6 + 7O_2 \rightarrow 4CO_2 + 6H_2O$$

Notice how all of the elements still balance on both sides.

Further Insight

If we weren't balancing the equation in standard form, we could have used larger coefficients to balance the equation, such as

$$6C_2H_6 + 21O_2 \rightarrow 12CO_2 + 18H_2O$$

The atoms still balance *and* the ratios *between* the compounds are still the same (21/6 can be reduced to 7/2 when looking at the ratio between O_2 and C_2H_6).

3.7 Working with Equations

Besides balancing chemical equations, we can use them to do stoichiometry problems. The text gives excellent advice on the key parts of stoichiometry problem solving: the balanced equation plus where you start, where you want to end up, and how you get there.

Example 15 **Determining Products from Reactants**

What mass of iron(III) oxide can be produced from reacting 155 g of solid iron with excess oxygen gas?

First Thoughts

The first step to solving this problem is to write and balance the chemical equation:

$$Fe(s) + O_2(g) \rightarrow Fe_2O_3(s) \qquad \text{(unbalanced)}$$

$$4Fe(s) + 3O_2(g) \rightarrow 2Fe_2O_3(s) \qquad \text{(balanced)}$$

Remember, the balanced equation tells us the *ratio* in which the reactants must react with each other to form a certain amount of product. The coefficients can represent the number of molecules (or units) present *or* they can represent moles. These ratios are how we "move" from one part of the reaction to another. For example, the ratio between Fe and O_2 is 4 to 3. This relationship can be written in two different ways when doing calculations because this ratio *never* changes for this reaction:

$$\frac{4 \text{ mol Fe}}{3 \text{ mol } O_2} \quad \text{or} \quad \frac{3 \text{ mol } O_2}{4 \text{ mol Fe}}$$

Solution

Now that we have a balanced equation and understand the importance of ratios, let's look at where we start, where we want to end up, and how we get there.

Where We Start: Where We Want to End Up:

155 g of Fe   Mass of Fe_2O_3 produced

How We Get There:

To "move" from one part of the equation to another, we must convert the mass of iron to moles so that we can use the mole ratio between Fe and Fe_2O_3.

$$\left(155 \text{ g Fe}\right)\left(\frac{1 \text{ mol Fe}}{55.85 \text{ g Fe}}\right) = 2.78 \text{ mol Fe}$$

Found on the periodic table

Now we can use the mole ratio between Fe and Fe_2O_3 to determine the moles of Fe_2O_3 produced (we have plenty of oxygen gas available to react, so the amount of iron present will determine how much Fe_2O_3 is created).

$$\left(2.78 \text{ mol Fe}\right)\left(\frac{2 \text{ mol } Fe_2O_3}{4 \text{ mol Fe}}\right) = 1.39 \text{ mol } Fe_2O_3$$

We need the mass of Fe_2O_3 formed, so we convert the moles to grams using the molar mass of Fe_2O_3.

$$(1.39 \text{ mol Fe}_2O_3)\left(\frac{159.7 \text{ g Fe}_2O_3}{1 \text{ mol Fe}_2O_3}\right) = 222 \text{ g Fe}_2O_3$$

A flowchart summarizes our thought process to solve this problem:

$$\text{g Fe} \xrightarrow{\frac{\text{mol Fe}}{\text{g Fe}}} \text{mol Fe} \xrightarrow{\frac{\text{mol Fe}_2O_3}{\text{mol Fe}}} \text{mol Fe}_2O_3 \xrightarrow{\frac{\text{g Fe}_2O_3}{\text{mol Fe}_2O_3}} \text{g Fe}_2O_3$$

We could have solved this problem in one step by using our flowchart for guidance:

$$(155 \text{ g Fe})\left(\frac{1 \text{ mol Fe}}{55.85 \text{ g Fe}}\right)\left(\frac{2 \text{ mol Fe}_2O_3}{4 \text{ mol Fe}}\right)\left(\frac{159.7 \text{ g Fe}_2O_3}{1 \text{ mol Fe}_2O_3}\right) = 222 \text{ g Fe}_2O_3$$

Solving this problem in one step or a series of steps yields the same answer. Choose whichever method works best for you when solving such problems.

Further Insight

Besides determining how much Fe_2O_3 is produced, you can calculate how much oxygen gas is required to react with all 155 g of Fe by applying these same principles. The flowchart is

$$\text{g Fe} \xrightarrow{\frac{\text{mol Fe}}{\text{g Fe}}} \text{mol Fe} \xrightarrow{\frac{\text{mol O}_2}{\text{mol Fe}}} \text{mol O}_2 \xrightarrow{\frac{\text{g O}_2}{\text{mol O}_2}} \text{g O}_2$$

The solution is

$$(155 \text{ g Fe})\left(\frac{1 \text{ mol Fe}}{55.85 \text{ g Fe}}\right)\left(\frac{3 \text{ mol O}_2}{4 \text{ mol Fe}}\right)\left(\frac{32.00 \text{ g O}_2}{1 \text{ mol O}_2}\right) = 66.6 \text{ g O}_2 \text{ required}$$

Example 16   **Determining Reactants Required from a Known Product Quantity**

Consider the reaction of copper with silver nitrate:

$$Cu(s) + AgNO_3(aq) \rightarrow Ag(s) + Cu(NO_3)_2(aq)$$

a) What mass of copper must be used to produce 23.0 g of silver?
b) How many moles of silver nitrate must be used to produce 23.0 g of silver?

Solution

The balanced equation is

$$Cu(s) + 2AgNO_3(aq) \rightarrow 2Ag(s) + Cu(NO_3)_2(aq)$$

a) <u>Where We Start:</u>   <u>Where We Want to End Up:</u>

23.0 g Ag   Mass of Cu needed

<u>How We Get There:</u>

$$(23.0 \text{ g Ag})\left(\frac{1 \text{ mol Ag}}{107.9 \text{ g Ag}}\right)\left(\frac{1 \text{ mol Cu}}{2 \text{ mol Ag}}\right)\left(\frac{63.55 \text{ g Cu}}{1 \text{ mol Cu}}\right) = 6.77 \text{ g Cu USED}$$

b)

$$(23.0 \text{ g Ag})\left(\frac{1 \text{ mol Ag}}{107.9 \text{ g Ag}}\right)\left(\frac{2 \text{ mol AgNO}_3}{2 \text{ mol Ag}}\right) = 0.213 \text{ mol AgNO}_3$$

Note that there is more than one way to solve part b, such as calculating the moles of Cu used and using the mole ratio to determine the moles of $AgNO_3$ consumed. We just show one possible solution here.

Let's take what we have learned about chemical equations and stoichiometry one step further. When two or more compounds react, one of them will usually run out before the other(s). This compound, which limits how much product is formed, is called the **limiting reactant** (or limiting reagent). Your text gives a great practical example of this concept, especially in **Figures 3.13** and **3.14** (constructing a hamburger). Let's do some examples with some familiar equations that demonstrate the limiting reactant concept.

Example 17   Which Reactant Is Limiting?

Hydrogen and oxygen gases react with each other to form water. If 10.0 g of hydrogen and 10.0 g of oxygen are used, which reactant is limiting?

First Thoughts

We must first write and balance the equation:

$$2H_2(g) + O_2(g) \rightarrow 2H_2O(g)$$

Solution

There are many different ways to determine which reactant is limiting. Let's focus on two of the most common methods. You can choose whichever works best for you.

Method 1:   Determine how much water would be produced by each reactant if it was fully consumed and the other reactant was present in excess. The reactant that forms less water is limiting.

$$\left(10.0 \text{ g H}_2\right)\left(\frac{1 \text{ mol H}_2}{2.016 \text{ g H}_2}\right)\left(\frac{2 \text{ mol H}_2O}{2 \text{ mol H}_2}\right) = 4.96 \text{ mol H}_2O$$

$$\left(10.0 \text{ g O}_2\right)\left(\frac{1 \text{ mol O}_2}{32.00 \text{ g O}_2}\right)\left(\frac{2 \text{ mol H}_2O}{1 \text{ mol O}_2}\right) = 0.625 \text{ mol H}_2O$$

The maximum amount of water that can be formed is 0.625 mol—then the oxygen gas runs out. Therefore, oxygen is the limiting reactant.

You can also calculate the mass of H_2O produced because the maximum amount of moles of product is 0.625.

$$\left(0.625 \text{ mol H}_2O\right)\left(\frac{18.016 \text{ g H}_2O}{1 \text{ mol H}_2O}\right) = 11.3 \text{ g H}_2O \text{ formed}$$

Method 2:   Determine the amount of oxygen needed to react with the given amount of hydrogen, and then compare this amount to the amount of oxygen given in the problem.

$$\left(10.0 \text{ g H}_2\right)\left(\frac{1 \text{ mol H}_2}{2.016 \text{ g H}_2}\right)\left(\frac{1 \text{ mol O}_2}{2 \text{ mol H}_2}\right) = 2.48 \text{ mol O}_2 \text{ needed}$$

Compare the amount of oxygen needed to what you are given.

$$\left(10.0 \text{ g O}_2\right)\left(\frac{1 \text{ mol O}_2}{32.00 \text{ g O}_2}\right) = 0.313 \text{ mol O}_2 \text{ given}$$

Because you need 2.48 mol O_2 to react with all of the hydrogen and you are given only 0.313 mol O_2, oxygen must be limiting.

Further Insight

Not all reactions will have one limiting reactant. Sometimes the reactants are mixed in perfect stoichiometric amounts so that all of the reactants are used up in the reaction.

Example 18   How Much Is Left Over?

Hydrogen and nitrogen gases react to form ammonia. If 5.00 g of hydrogen gas and 10.0 g of nitrogen gas are mixed,
a) How many grams of ammonia are produced?
b) How many grams of the excess reactant are left over?

Solution

Balanced equation: $3H_2(g) + N_2(g) \rightarrow 2NH_3(g)$

a) Let's use Method 1 to solve this problem.

$$\left(5.00 \text{ g H}_2\right)\left(\frac{1 \text{ mol H}_2}{2.016 \text{ g H}_2}\right)\left(\frac{2 \text{ mol NH}_3}{3 \text{ mol H}_2}\right) = 1.65 \text{ mol NH}_3$$

$$\left(10.0 \text{ g N}_2\right)\left(\frac{1 \text{ mol N}_2}{28.02 \text{ g N}_2}\right)\left(\frac{2 \text{ mol NH}_3}{1 \text{ mol N}_2}\right) = 0.714 \text{ mol NH}_3$$

The maximum amount of ammonia that can be produced before the nitrogen runs out is 0.714 mol. Therefore, nitrogen is the limiting reactant. Next, we use the molar mass of ammonia to convert to grams:

$$\left(0.714 \text{ mol NH}_3\right)\left(\frac{17.034 \text{ g NH}_3}{1 \text{ mol NH}_3}\right) = 12.2 \text{ g NH}_3 \text{ produced}$$

Even though a larger amount of N_2 was present, it was the limiting reactant. The ratios in which the compounds react are crucial to chemical reactions.

b) To determine the amount of excess reactant left over, we must first calculate how much was used up in the reaction. Because nitrogen is the limiting reactant, hydrogen must be present in excess.

$$\left(10.0 \text{ g N}_2\right)\left(\frac{1 \text{ mol N}_2}{28.02 \text{ g N}_2}\right)\left(\frac{3 \text{ mol H}_2}{1 \text{ mol N}_2}\right) = 1.07 \text{ mol H}_2 \text{ used up}$$

How many moles of hydrogen did we start with before the reaction took place?

$$\left(5.00 \text{ g H}_2\right)\left(\frac{1 \text{ mol H}_2}{2.016 \text{ g H}_2}\right) = 2.48 \text{ mol H}_2 \text{ to start}$$

Therefore, to find the excess reactant:

$$2.48 \text{ mol H}_2 \text{ to start}$$

$$- \quad \underline{1.07} \text{ used up}$$

$$1.41 \text{ mol H}_2 \text{ left over}$$

$$\left(1.41 \text{ mol H}_2 \text{ left over}\right)\left(\frac{2.016 \text{ g H}_2}{1 \text{ mol H}_2}\right) = 2.84 \text{ g H}_2 \text{ left over}$$

In our study of chemical reactions, we will find that most reactions are *not* 100% efficient. In reality, the amount of product we calculate will form on paper is rarely what we produce in the laboratory due to "real world" factors such as side reactions or poor lab technique. Our "calculated" amount of product is the **theoretical yield.** The amount of product actually obtained in the lab is the **actual yield.**

Use the following equation to calculate percent yield:

$$\text{Percent yield} = \frac{\text{actual yield}}{\text{theoretical yield}} \times 100\%$$

Example 19   **Percent Yield**

Consider the following reaction: $2H_2(g) + CO(g) \rightarrow CH_3OH(l)$

Suppose 25.0 g of carbon monoxide is reacted with an excess of hydrogen and yields 16.7 g of methanol. What is the percent yield?

First Thoughts

Before we can calculate the percent yield, we must figure out the theoretical yield for CH_3OH (how much methanol *should be* created under ideal circumstances).

Because hydrogen is present in excess, carbon monoxide must be the limiting reactant.

$$\left(25.0 \text{ g CO}\right)\left(\frac{1 \text{ mol CO}}{28.01 \text{ g CO}}\right)\left(\frac{1 \text{ mol CH}_3\text{OH}}{1 \text{ mol CO}}\right)\left(\frac{32.042 \text{ g CH}_3\text{OH}}{1 \text{ mol CH}_3\text{OH}}\right) = 28.6 \text{ g CH}_3\text{OH}$$

Solution

Now that we know our theoretical and actual yields, we can calculate the percent yield:

$$\text{Percent yield} = \frac{16.7 \text{ g}}{28.6 \text{ g}} \times 100\% = 58.4\% \text{ yield}$$

Further Insight

You can also calculate the percent yield by comparing the actual and theoretical yields in terms of moles. The answer will still be the same, but it requires more steps in this case.

Actual yield: $\left(16.7 \text{ g CH}_3\text{OH}\right)\left(\frac{1 \text{ mol CH}_3\text{OH}}{32.042 \text{ g CH}_3\text{OH}}\right) = 0.521 \text{ mol CH}_3\text{OH}$

Theoretical yield: $\left(25.0 \text{ g CO}\right)\left(\frac{1 \text{ mol CO}}{28.01 \text{ g CO}}\right)\left(\frac{1 \text{ mol CH}_3\text{OH}}{1 \text{ mol CO}}\right) = 0.8925 \text{ mol CH}_3\text{OH}$

$$\text{Percent yield} = \frac{0.521 \text{ mol}}{0.8925 \text{ mol}} \times 100\% = 58.4\% \text{ yield}$$

There is a lot to grasp in this chapter, both conceptually and mathematically. We encourage you to do as many problems as possible to work out your difficulties. As you practice these types of problems, consult your text, this study guide, your teacher, and your peers for guidance. You can do it!

Exercises

Section 3.1

1. Determine the formula mass for each of the following compounds.
   a) KCl
   b) CH_4
   c) $MgBr_2$
   d) N_2O_5

2. Determine the formula mass of each of the following compounds.
   a) sodium phosphate
   b) carbon dioxide
   c) calcium hydroxide
   d) lithium oxide
   e) boron trifluoride

3. A compound has the formula SeF_x with a formula mass of 116.96 amu. What is the value of x?

Section 3.2

4. How many atoms present are in a 10.57-g sample of phosphorus?

5. How many SO_2 molecules are present in a 33.4-g sample of this compound?

6. How many oxygen atoms are present in 147.0 g of P_2O_5?

7. A sample of methane (CH_4) contains 5.72×10^{25} molecules. What is the mass (in kilograms) of this sample?

Section 3.3

8. a) How many moles of zinc are in 4.96×10^{24} atoms of zinc?
   b) What is the mass in grams of this zinc sample?

9. Which substance contains a larger number of molecules: 5 moles of water (H_2O) or 5 moles of glucose ($C_6H_{12}O_6$)? Choose the best answer.

   a) Glucose contains a larger number of molecules because it contains more atoms compared to water.
   b) Glucose contains a larger number of molecules because glucose is larger than water and thus takes up more space.
   c) Water contains a larger number of molecules because you can fit more water molecules in a 5.00-mol container versus glucose.
   d) Both substances contain the same number of molecules because both have the same number of atoms.
   e) Both substances contain the same number of molecules because the numbers of moles are identical and correspond to Avogadro's number.

10. a) How many moles of magnesium chloride, $MgCl_2$, are present in a 75.1-g sample?
    b) How many "units" of $MgCl_2$ exist in this sample?

11. What is the mass (in grams) of a 6.13-mol sample of aluminum hydroxide?

12. A 4.59-mol sample of silver nitrate contains _____ atoms of oxygen.

13. How many grams are present in a 0.555-mol sample of acetic acid, $HC_2H_3O_2$?

14. How many moles of NH_3 are represented by each of these samples?
    a) 250.0 g NH_3
    b) 250 molecules (exactly) of NH_3
    c) 250 atoms of H (exactly) in the NH_3

15. How many moles of humulone molecules, $C_{21}H_{30}O_5$, are present in 375 mg of humulone?

16. Determine the mass in grams of a single Cl_2 molecule.

17. How many atoms of carbon are present in 3.00 g of glycine, $C_2H_5O_2N$?

Section 3.4

18. Calculate the mass percent of potassium in potassium bromide, KBr.

19. Determine the percent composition by mass of the following compounds:
    a) CH_2O
    b) Ag_2SO_4
    c) $Fe_3(PO_4)_2$

20. What is the mass percent of nitrogen in each of the following compounds?
    a) dinitrogen tetroxide
    b) copper(II) nitrate
    c) nitric acid

21. Arrange the following compounds in order of *decreasing* mass percent of hydrogen.
    a) vitamin C, $C_6H_8O_6$
    b) caffeine, $C_8H_{10}N_4O_2$
    c) formaldehyde, CH_2O

22. Which of the following has the *greatest* percent by mass of oxygen?
    a) calcium phosphate
    b) sodium carbonate
    c) copper(II) oxide
    d) diboron trioxide
    e) cobalt(II) nitrate

23. An unknown compound is 69.94% Fe by mass. The substance contains two iron atoms. What is the molar mass of this substance?

Section 3.5

24. A compound contains 24.27% C, 4.07% H, and 71.65% Cl. What is the empirical formula for this compound?

25. What is the empirical formula of the compound that contains 22.1% Al, 25.4% P, and 52.5% O?

26. Calculate the empirical formula of the compound that contains 89.2 g Au and 10.9 g O.

27. Succinic acid is 40.7% carbon and 5.12% hydrogen, with the rest consisting of oxygen. The experimental value found for the molar mass is approximately 118 g/mol.
    a) Determine the empirical formula for succinic acid.
    b) What is the molecular formula?

28. A compound containing carbon and hydrogen was analyzed and found to consist of 83.65% carbon by mass. Its molecular weight is 86 amu. What is its molecular formula?

29. A compound is composed of element X and hydrogen. Analysis shows the compound to be 80% X by mass, with three times as many hydrogen atoms as X atoms per molecule. Which element is element X?

    a) C   b) He   c) F   d) S   e) none of these

Section 3.6

30. Balance the following equations and show that they are balanced on an atomic level (draw a picture for each).
    a) $CH_4 + O_2 \rightarrow CO_2 + H_2O$
    b) $SiO_2 + C \rightarrow Si + CO$
    c) $NH_3 + O_2 + CH_4 \rightarrow HCN + H_2O$

31. Which of the following statements are *false* regarding balancing a chemical equation?

    a) Subscripts must never be used to balance an equation because then you are changing the compound(s) in the reaction.
    b) Coefficients can never be fractions because you cannot have "part" or a fraction of a molecule.
    c) The number and type of atoms in the reaction must balance on both the reactant and product sides.
    d) Although the physical state of each compound in the reaction is important, it is not crucial to balancing the equation.
    e) Chemical reactions obey the law of conservation of matter.

32. Balance the following equations:
    a) $Fe(s) + O_2(g) \rightarrow Fe_2O_3(s)$
    b) $C_2H_2(g) + O_2(g) \rightarrow CO_2(g) + H_2O(g)$
    c) $Mg(s) + H_3PO_4(aq) \rightarrow Mg_3(PO_4)_2(s) + H_2(g)$

33. Balance each of the following equations in standard form:
    a) $NH_3(g) + HCl(g) \rightarrow NH_4Cl(s)$
    b) $NiCl_2(aq) + Na_3PO_4(aq) \rightarrow Ni_3(PO_4)_2(s) + NaCl(aq)$
    c) $NI_3(s) \rightarrow N_2(g) + I_2(g)$
    d) $NH_3(g) + O_2(g) \rightarrow NO(g) + H_2O(g)$

34. Balance the following equation in standard form and determine the sum of the coefficients:
    $LiAlH_4(s) + AlCl_3(s) \rightarrow AlH_3(s) + LiCl(s)$

Section 3.7

35. Glucose ferments to produce ethanol and carbon dioxide according to the *unbalanced* equation: $C_6H_{12}O_6 \rightarrow C_2H_6O + CO_2$.

    a) How many grams of ethanol are produced from 541 g of glucose?
    b) How many moles of glucose are required to form 133 g of carbon dioxide?

36. Tin(II) fluoride is added to some dental products to help prevent cavities. Of course, manufacturers must make the tin(II) fluoride first before they can add it to their products.

    $$Sn(s) + HF(aq) \rightarrow SnF_2(aq) + H_2(g)$$

    How many grams of tin(II) fluoride can be made from 44.0 g of hydrogen fluoride if there is plenty of tin available to react?

Use the following *unbalanced* equation for Exercises 37 and 38.

$$Al_4C_3 + H_2O \rightarrow Al(OH)_3 + CH_4$$

37. How many moles of water are needed to react with 250. g of Al_4C_3?

38. How many moles of $Al(OH)_3$ will be produced when 0.600 mol of CH_4 is formed?

39. Consider the following reaction, where X represents an unknown element:

    $$6X(s) + 2B_2O_3(s) \rightarrow B_4X_3(s) + 3XO_2(g)$$

    If 175 g of X reacts with diboron trioxide to produce 2.43 mol of B_4X_3, what is the identity of X?

    a) Ge   b) Mg   c) Si   d) N   e) C

40. Consider the reaction $3Fe(s) + 4H_2O(g) \rightarrow Fe_3O_4(s) + 4H_2(g)$. How many grams of steam must react to produce 375 g of Fe_3O_4?

41. A model car requires 1 body, 4 wheels, and 2 doors.
    a) If you have 7 bodies, 26 wheels, and 9 doors, how many model cars can you make?
    b) If you have 11 bodies, 31 wheels, and 25 doors, how many model cars can you make?

42. Carbon monoxide gas and oxygen gas combine to form carbon dioxide gas.
    a) If 50.0 g of carbon monoxide and 50.0 g of oxygen are reacted, which reactant is limiting?
    b) What mass of carbon dioxide is produced?

43. Which of the reaction mixtures in (a)–(e) would produce the *greatest* amount of product, assuming the following reaction goes to completion?

$$N_2(g) + 3H_2(g) \rightarrow 2NH_3(g)$$

    a) 3 moles of N_2 and 3 moles of H_2
    b) 1 mole of N_2 and 6 moles of H_2
    c) 5 moles of N_2 and 3 moles of H_2
    d) 1 mole of N_2 and 3 moles of H_2
    e) Each would produce the same amount of product.

44. Consider the equation $A + 3B \rightarrow 4C$. If 3.0 moles of A is reacted with 6.0 moles of B, which of the following is *true* after the reaction is complete?

    a) A is the leftover reactant because you need only 2 moles of A and you have 3 moles.
    b) A is the leftover reactant because for every 1 mole of A, 4 moles of C is produced.
    c) B is the leftover reactant because you have more moles of B than A.
    d) B is the leftover reactant because 3 moles of B reacts with every 1 mole of A.
    e) Neither reactant is left over.

45. Consider the following reaction:

$$16Ag(s) + S_8(s) \xrightarrow{\text{heat}} 8Ag_2S(s)$$

Identify the limiting reagent in each of the following reaction mixtures:
    a) 100 atoms of Ag and 100 molecules of S_8
    b) 0.68 mol Ag and 0.25 mol S_8
    c) 12.3 g Ag and 247 mg S_8

46. Solid magnesium (Mg) and solid iodine (I_2) react to form magnesium iodide (MgI_2):

$$Mg(s) + I_2(s) \rightarrow MgI_2(s)$$

    a) What mass of MgI_2 is produced from 50.0 g of Mg and 75.0 g of I_2?
    b) What mass of which reactant is left unreacted?

47. For the reaction $C_7H_{16}(g) + 11O_2(g) \rightarrow 7CO_2(g) + 8H_2O(g)$, determine how many moles of carbon dioxide and water are formed when 3.00 moles of each reactant is used. How many moles of the excess reactant are left over?

48. Carbon disulfide is produced by the reaction of carbon and sulfur dioxide:

$$5C(s) + 2SO_2(g) \rightarrow CS_2(g) + 4CO(g)$$

a) What is the percent yield for the reaction if 91.0 g carbon produces 106 g carbon monoxide?

b) What is the percent yield for the reaction if 25.0 g sulfur dioxide produces 11.9 g carbon disulfide?

49. Consider the following reaction:

$$3Ca(s) + N_2(g) \rightarrow Ca_3N_2(s)$$

Suppose 106.6 g calcium is reacted with 17.2 g N_2, and 83.3 g calcium nitride is produced. What is the percent yield of the reaction?

50. Ethylene oxide, C_2H_4O, is made by the oxidation of ethylene, C_2H_4:

$$2C_2H_4(g) + O_2(g) \rightarrow 2C_2H_4O(g)$$

Ethylene oxide is used to make ethylene glycol for antifreeze. If you wanted to obtain 20.0 g of ethylene oxide and the percent yield is typically 59.5%, how much ethylene should you start with?

Answers to Exercises

1. a) 74.55 amu
    b) 16.042 amu
    c) 184.11 amu
    d) 108.02 amu

2. a) Na_3PO_4: 163.94 amu
    b) CO_2: 44.01 amu
    c) $Ca(OH)_2$: 74.096 amu
    d) Li_2O: 29.882 amu
    e) BF_3: 67.81 amu

3. $x = 2$

4. 2.055×10^{23} atoms P

5. 3.14×10^{23} molecules SO_2

6. 3.118×10^{24} atoms O

7. 1.52 kg CH_4

8. a) 8.24 mol Zn
    b) 539 g Zn

9. e

10. a) 0.789 mol $MgCl_2$
    b) 4.75×10^{23} "units" $MgCl_2$

11. 478 g $Al(OH)_3$

12. 8.29×10^{24} atoms O

13. 33.3 g $HC_2H_3O_2$

14. a) 14.68 mol NH_3
    b) 4.151×10^{-22} mol NH_3
    c) 1.384×10^{-22} mol NH_3

15. 1.03×10^{-3} mol $C_{21}H_{30}O_5$

16. 1.177×10^{-22} g Cl_2

17. 4.81×10^{22} atoms of C

18. 32.86% K

19. a) 40.00% C, 6.71% H, 53.29% O
    b) 69.20% Ag, 10.28% S, 20.52% O
    c) 46.87% Fe, 17.33% P, 35.81% O

20. a) 30.45% N
    b) 14.94% N
    c) 22.23% N

21.   c   >   a   >   b
    6.714% H   5.723% H   4.152% H

22. d

23. 159.7 g/mol

24. CH_2Cl

25. $AlPO_4$

26. Au_2O_3

27. a) $C_2H_3O_2$
    b) $C_4H_6O_4$

28. C_6H_{14}

29. a

30. a) $CH_4 + 2O_2 \rightarrow CO_2 + 2H_2O$

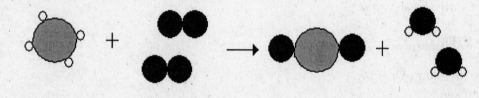

    b) $SiO_2 + 2C \rightarrow Si + 2CO$

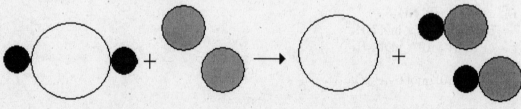

c) $2NH_3 + 3O_2 + 2CH_4 \rightarrow 2HCN + 6H_2O$

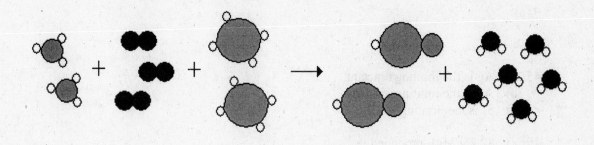

31. b

32. a) $4Fe(s) + 3O_2(g) \rightarrow 2Fe_2O_3(s)$

b) $C_2H_2(g) + \dfrac{5}{2}O_2(g) \rightarrow 2CO_2(g) + H_2O(g)$

OR

$2C_2H_2(g) + 5O_2(g) \rightarrow 4CO_2(g) + 2H_2O(g)$

c) $3Mg(s) + 2H_3PO_4(aq) \rightarrow Mg_3(PO_4)_2(s) + 3H_2(g)$

33. a) $NH_3(g) + HCl(g) \rightarrow NH_4Cl(s)$

b) $3NiCl_2(aq) + 2Na_3PO_4(aq) \rightarrow Ni_3(PO_4)_2(s) + 6NaCl(aq)$

c) $2NI_3(s) \rightarrow N_2(g) + 3I_2(g)$

d) $4NH_3(g) + 5O_2(g) \rightarrow 4NO(g) + 6H_2O(g)$

34. $3LiAlH_4(s) + AlCl_3(s) \rightarrow 4AlH_3(s) + 3LiCl(s)$

Sum of the coefficients = 3 + 1 + 4 + 3 = 11

35. a)   277 g C_2H_6O
    b)   1.51 mol $C_6H_{12}O_6$

36. 172 g SnF_2

37. 20.8 mol H_2O

38. 0.800 mol $Al(OH)_3$

39. e

40. 117 g steam

41. a)   4 model cars
    b)   7 model cars

42. a)   CO is the limiting reactant.
    b)   78.6 g CO_2

43. e

44. a

45. a)   Ag is the limiting reactant.
    b)   Ag is the limiting reactant.
    c) S_8 is the limiting reactant.

46. a)   82.2 g MgI_2 produced
    b)   42.8 g Mg unreacted

47. 1.91 mol CO_2 formed; 2.18 mol H_2O formed; 2.73 mol C_7H_{16} left over

48. a)   62.4%
    b)   80.1%

49. 91.5%

50. 21.4 g C_2H_4

Chapter 4

Solution Stoichiometry and Types of Reactions

The Bottom Line

When you finish this chapter, you will be able to:

- Describe what an electrolyte is and differentiate between strong electrolytes, weak electrolytes, and nonelectrolytes.
- Describe how to perform dilutions.
- Solve problems involving the molarity of a solution, and explain how molarity relates to chemical reactions such as precipitation and acid–base reactions.
- Write molecular, complete ionic, and net ionic equations and predict whether a precipitate will form in a reaction.
- Identify redox (oxidation–reduction) reactions, and assign oxidation states to atoms in an element, compound, or ion.

4.1   Water: A Most Versatile Solvent

This chapter deals mainly with solution stoichiometry—chemical reactions that reflect the concentrations of the reactants and products. Here are some key points concerning solutions:

- A solution is a homogeneous mixture that is created by dissolving a **solute** in a **solvent.**
- The solute consists of atoms, molecules, or ions that dissolve in the solvent.
- The solvent is the substance that dissolves the solute. The solvent is a compound that typically makes up the majority of the solution.
- When a solute dissolves in water, it forms an **aqueous solution.**
- Your text explains in detail why solutes dissolve in solvents. **Figure 4.4** illustrates the hydration sphere concept.
- When a solute dissolves in a solvent like water to release free ions, we call these ions **electrolytes.** A solution can be a strong electrolyte, a weak electrolyte, or a nonelectrolyte.
- **Strong electrolytes** conduct electricity very well because there are a lot of ions to "carry" an electrical charge. Strong acids, strong bases, and most ionic compounds are strong electrolytes. For example, the strong electrolyte potassium chloride completely dissociates into ions when it dissolves:

$$KCl(s) \rightarrow KCl(aq) \rightarrow K^+(aq) + Cl^-(aq)$$

- **Weak electrolytes** do not conduct electricity as well as strong electrolytes do, because ion formation occurs to a much smaller extent (although they still may dissolve completely in water). Ammonia dissolved in water is an example of a weak electrolyte. The ammonia does not proceed completely to products, so not as many ions are found in solution.

$$NH_3(aq) + H_2O(l) \rightleftharpoons NH_4^+(aq) + OH^-(aq)$$

- **Nonelectrolytes** do not conduct any electricity because they dissolve in water without forming ions. Table sugar, $C_{12}H_{22}O_{11}$, dissolves in water and forms a hydration cage, but does not produce any ions:

$$C_{12}H_{22}O_{11}(s) \rightarrow C_{12}H_{22}O_{11}(aq)$$

Table 4.1 in your text shows which types of compounds form strong electrolytes, weak electrolytes, and nonelectrolytes when dissolved in water (although keep in mind there are exceptions).

Example 1   **Identifying Electrolytes**

Classify the following compounds as strong electrolytes, weak electrolytes, or nonelectrolytes.

a) $MgCl_2$
b) $C_6H_{12}O_6$ (glucose)
c) HF
d) $H_2CO_3^-$
e) HCl

Solution

a) strong electrolyte (ionic compound—dissociates into ions)
b) nonelectrolyte (does not produce ions when dissolved)
c) weak electrolyte (weak acid—partially produces ions)
d) weak electrolyte (weak acid—partially produces ions)
e) strong electrolyte (strong acid—dissociates into ions)

4.2   The Concentration of Solutions

The concentration of a solution—that is, the amount of solute per volume of solution—is critical to relating solutions to stoichiometry. One of the most useful units of concentration is **molarity,** which is represented as moles of solute per liter of solution.

$$\text{Molarity } (M) = \frac{\text{moles of solute}}{\text{liter of solution}}$$

By using this unit of concentration, you can determine the number of moles of solute, which you can then convert to mass in grams or apply to a chemical equation. You can also determine the volume of solution needed to obtain a specific number of moles of solute from a given concentration.

Example 2   **Moles of Solute**

How many moles of potassium bromide are contained in 2.0 L of a 0.100 M solution of KBr?

Solution

The relationship between moles and concentration is

$$\text{Molarity } (M) = \frac{\text{moles of solute}}{\text{liter of solution}}$$

The molarity and volume of solution are given in the problem, so we can use this equation to solve for moles of KBr.

$$0.100\ M = \frac{n \text{ moles KBr}}{2.0\ L}$$

$$n = (0.100\ M)(2.0\ L) = \left(\frac{0.100\ mol}{L}\right)(2.0\ L)$$

$$n = 0.20 \text{ mol KBr}$$

We could have also solved this problem by using dimensional analysis and ensuring that the proper units cancel:

$$(2.0 \text{ L})\left(\frac{0.100 \text{ mol}}{1 \text{ L}}\right) = 0.20 \text{ mol KBr}$$

Example 3   Concentration of Acetic Acid

What is the molarity of a 300.0 mL acetic acid solution containing 55.0 g of $HC_2H_3O_2$?

First Thoughts

To find the concentration (in terms of molarity), we must convert the mass of $HC_2H_3O_2$ to moles using the molar mass of acetic acid and convert the volume of the solution from milliliters to liters.

Solution

Determine the moles of acetic acid:

$$(55.0 \text{ } HC_2H_3O_2)\left(\frac{1 \text{ mol } HC_2H_3O_2}{60.052 \text{ g } HC_2H_3O_2}\right) = 0.916 \text{ mol } HC_2H_3O_2$$

Determine the volume in liters:

$$(300.0 \text{ mL})\left(\frac{1 \text{ L}}{1000 \text{ mL}}\right) = 0.3000 \text{ L soln.}$$

Therefore, the molarity of the acetic acid solution is

$$\text{Molarity} = \frac{0.916 \text{ mol}}{0.3000 \text{ L}} = 3.05 \text{ } M \text{ } HC_2H_3O_2$$

Further Insight

You could have solved this problem by using dimensional analysis:

$$\frac{\text{g } HC_2H_3O_2}{\text{mL soln}} \xrightarrow{\frac{\text{mol}}{\text{g}}} \frac{\text{mol } HC_2H_3O_2}{\text{mL soln}} \xrightarrow{\frac{1000 \text{ mL}}{1 \text{ L}}} \frac{\text{mol } HC_2H_3O_2}{\text{L soln}}$$

$$\left(\frac{55.0 \text{ g } HC_2H_3O_2}{300.0 \text{ mL soln.}}\right)\left(\frac{1 \text{ mol } HC_2H_3O_2}{60.052 \text{ g } HC_2H_3O_2}\right)\left(\frac{1000 \text{ mL}}{1 \text{ L}}\right) = 3.05 \text{ } M \text{ } HC_2H_3O_2$$

Example 4   Volume Required

What volume of solution (in milliliters) is required to make a 0.50 M solution of sodium phosphate? The mass of the sodium phosphate in solution is 16.0 g.

First Thoughts

The formula of sodium phosphate is Na_3PO_4; it has a molar mass of 163.94 g/mol.

Solution

Let's use a flowchart and dimensional analysis to solve this problem.

$$g \ Na_3PO_4 \xrightarrow{\frac{mol \ Na_3PO_4}{g \ Na_3PO_4}} mol \ Na_3PO_4 \xrightarrow{\frac{1 \ L \ Na_3PO_4}{mol \ Na_3PO_4}} L \ Na_3PO_4 \xrightarrow{\frac{1000 \ mL}{1 \ L}} mL \ Na_3PO_4$$

$$\left(16.0 \ g \ Na_3PO_4\right)\left(\frac{1 \ mol \ Na_3PO_4}{163.94 \ g \ Na_3PO_4}\right)\left(\frac{1 \ L}{0.50 \ mol \ Na_3PO_4}\right)\left(\frac{1000 \ mL}{1 \ L}\right) = 195 \ mL \ of \ soln.$$

Molarity of Na_3PO_4

Further Insight

We can take our understanding of solution concentration another step further by applying this concept to the solute ions that completely dissociate in a solvent.

Example 5   **Moles of Ions**

How many moles of Mg^{2+} ions will be present in 125 mL of 0.700 M $MgCl_2$? How many moles of Cl^- ions will be present? What is the total number of moles of ions present in the solution?

First Thoughts

The best way to solve this problem is to first figure out how many moles of $MgCl_2$ are present and then analyze how $MgCl_2$ breaks down into its ions in solution.

Solution

First, determine the moles of $MgCl_2$:

$$mL \ soln. \xrightarrow{\frac{1 \ L}{1000 \ mL}} L \ soln. \xrightarrow{\frac{mol \ MgCl_2}{L \ soln.}} mol \ MgCl_2$$

$$\left(125 \ mL \ soln.\right)\left(\frac{1 \ L}{1000 \ mL}\right)\left(\frac{0.700 \ mol \ MgCl_2}{1 \ L}\right) = 0.0875 \ mol \ MgCl_2$$

Molarity of $MgCl_2$

Next, write the dissociation of $MgCl_2$ in solution as a balanced chemical equation:

$$MgCl_2(s) \xrightarrow{add \ H_2O} Mg^{2+}(aq) + 2Cl^-(aq)$$

Just as we did in Chapter 3, we use the stoichiometric ratio between $MgCl_2$ and Mg^{2+} to determine the moles of Mg^{2+} present in solution.

$$\left(0.0875 \ mol \ MgCl_2\right)\left(\frac{1 \ mol \ Mg^{2+}}{1 \ mol \ MgCl_2}\right) = 0.0875 \ mol \ Mg^{2+}$$

Similarly, we use the mole ratio between $MgCl_2$ and Cl^- to determine the moles of Cl^- present.

$$0.0\left(875 \ mol \ MgCl_2\right)\left(\frac{2 \ mol \ Cl^-}{1 \ mol \ MgCl_2}\right) = 0.175 \ mol \ Cl^-$$

Therefore, the total number of moles of ions in solution is

$$\begin{array}{r} 0.0875 \ mol \\ + \ \underline{0.175 \ mol} \\ 0.263 \ mol \quad = \quad total \ ions \ in \ soln. \end{array}$$

Further Insight

Keep in mind that the only time we can analyze the ions in this manner is when we're dealing with strong electrolytes like most ionic compounds and strong acids and bases.

The concentration in molarity is very useful when we are doing solution stoichiometry, such as with the $MgCl_2$ problem above. Concentration can also be measured in other units, however, such as parts per million (ppm) or parts per billion (ppb). These units of measurement are useful when we're analyzing solutions with very small concentrations. **Figure 4.10** in your text shows excellent comparisons for visualizing just how small these concentrations are. The key is to remember that in dilute aqueous solutions, we can express these small concentrations as follows:

$$\text{parts per million (ppm)} = \frac{\text{mg solute}}{\text{L soln.}}$$

$$\text{parts per billion (ppb)} = \frac{\mu\text{g solute}}{\text{L soln.}}$$

$$\text{parts per trillion (ppt)} = \frac{\text{ng solute}}{\text{L soln.}}$$

Example 6	**Converting Parts per Billion to Molarity**

Traces of Pb^{2+} are found in a solution in the range of 4.0 ppb. What does this concentration represent in units of molarity?

First Thoughts

First we want to represent the 4.0 ppb as a mass-to-volume ratio:

$$4.0 \text{ ppb} = \frac{4.0 \ \mu\text{g Pb}^{2+}}{\text{L soln.}}$$

Solution

Let's use a flowchart to guide us in making the conversion:

$$\frac{\mu\text{g Pb}^{2+}}{\text{L soln.}} \xrightarrow{\frac{1\text{g}}{10^6 \ \mu\text{g}}} \frac{\text{g Pb}^{2+}}{\text{L soln.}} \xrightarrow{\frac{\text{mol}}{\text{g}}} \frac{\text{mol Pb}^{2+}}{\text{L soln.}}$$

$$\left(\frac{4.0 \ \mu\text{g Pb}^{2+}}{\text{L soln.}}\right)\left(\frac{1 \text{ g Pb}^{2+}}{10^6 \ \mu\text{g Pb}^{2+}}\right)\left(\frac{1 \text{ mol Pb}^{2+}}{207.2 \text{ g Pb}^{2+}}\right) = 1.9 \times 10^{-8} \ M \ Pb^{2+}$$

Further Insight

Do you now see how parts per billion correspond to such small concentrations?

Example 7	**Converting Molarity to Parts per Million**

An acceptable mid-range value for fluoride ion (F^-) in drinking water is 1.0 ppm. You test a sample of water and find the concentration to be $3.95 \times 10^{-4} \ M \ F^-$. Is this water sample safe to drink?

Solution

Convert the concentration in mol/L to mg/L (to obtain ppm):

$$\frac{\text{mol F}^-}{\text{L soln.}} \xrightarrow{\frac{\text{g}}{\text{mol}}} \frac{\text{g F}^-}{\text{L soln.}} \xrightarrow{\frac{1000 \text{ mg}}{1 \text{ g}}} \frac{\text{mg F}^-}{\text{L soln.}}$$

$$\left(\frac{3.95 \times 10^{-4} \text{ mol } F^-}{L \text{ soln.}}\right)\left(\frac{19.00 \text{ g } F^-}{1 \text{ mol } F^-}\right)\left(\frac{10^3 \text{ mg } F^-}{1 \text{ g } F^-}\right) = 7.51 \text{ ppm}$$

Given that 7.51 ppm is much higher than 1.0 ppm, this water sample is unsafe to drink.

Another important aspect of solution stoichiometry is **dilution.** In the laboratory, chemicals are often stored in higher concentrations than what we need for a particular experiment. Therefore, we must dilute a certain volume of a known concentration with water to obtain the concentration we need for our experiment. The key point to remember is that *the number of moles in the concentrated solution must equal the number of moles in the dilute solution.* The number of moles of solute particles *does not change* just because we add water!

Example 8   Diluting a Solution

For a particular experiment, we need 25.0 mL of 0.500 *M* NaOH. We have only 3.00 *M* NaOH in the stock room. How many milliliters of the 3.00 *M* NaOH will we need to prepare our desired solution?

First Thoughts

Let's figure out how many moles of NaOH will be present in the desired solution:

$$\left(25.0 \text{ mL soln.}\right)\left(\frac{1 \text{ L soln.}}{1000 \text{ mL}}\right)\left(\frac{0.500 \text{ mol NaOH}}{1 \text{ L soln.}}\right) = 0.0125 \text{ mol NaOH in desired soln.}$$

⌐ Molarity of NaOH
  needed for experiment

Solution

The desired solution will contain 0.0125 mol of NaOH. We need to take this *same number* of moles from the 3.00 *M* NaOH stock solution. Remember, the number of moles of NaOH measured from the concentrated solution equals the number of moles needed in the dilute solution. So, what volume of the 3.00 *M* NaOH must we measure out so that we have 0.0125 mol of NaOH?

$$\left(0.0125 \text{ mol NaOH}\right)\left(\frac{1 \text{ L soln.}}{3.00 \text{ mol NaOH}}\right)\left(\frac{1000 \text{ mL}}{1 \text{ L}}\right) = 4.17 \text{ mL NaOH}$$

⌐ Molarity of NaOH
  stock solution

Therefore, you take 4.17 mL of the 3.00 *M* stock solution and add water until the solution reaches the 25.0-mL line (25.0 mL of total solution).

Further Insight

Your text simplifies the dilution concept into one equation:

$$C_{initial} \times V_{initial} = C_{final} \times V_{final}$$

(where C = concentration and V = volume)

Number of moles measured from initial solution = Number of moles needed in dilute solution

$$\left(3.00 \text{ } M\right) \times V_{initial} = \left(0.500 \text{ } M\right) \times \left(25.0 \text{ mL}\right)$$

$$V_{initial} = 4.17 \text{ mL NaOH}$$

We recommend that you use this equation with *caution!* If you do not understand how to use it properly, you will make many errors in your dilutions! See Example 9 for another demonstration of how to use the equation correctly.

Example 9   **How Much Water Must Be Added?**

You need 25.0 mL of 0.800 M HCl for an experiment. You have only 2.68 M HCl available. How much water must you add to the concentrated HCl solution to obtain the desired volume and concentration for your experiment? Use the most practical method to solve this problem.

Solution

Your first instinct might be to use the simplified equation to solve the entire problem:

$$C_{initial} \times V_{initial} = C_{final} \times V_{final}$$

$$(2.68\ M) \times V_{initial} = (0.800\ M) \times (25.0\ mL)$$

$$V_{initial} = 7.46\ mL$$

However, the answer 7.46 mL is *incorrect*. This quantity is the volume of HCl that must be *taken* to have the same number of moles needed in our dilute solution. That is, you must take 7.46 mL of the 2.68 M HCl solution and dilute it with water until you reach the 25.0 mL-line on your glassware. The amount of water added is approximately

$$
\begin{array}{r}
25.0\ mL \\
-\quad \underline{7.46\ mL} \\
17.54\ mL \quad = \quad \text{volume of } H_2O \text{ added}
\end{array}
$$

Here is a picture to help you visualize this process:

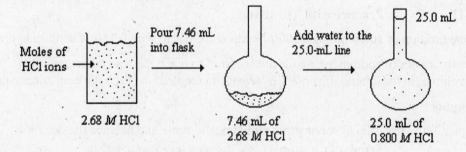

Moles of HCl ions → 2.68 M HCl

Pour 7.46 mL into flask → 7.46 mL of 2.68 M HCl

Add water to the 25.0-mL line → 25.0 mL / 25.0 mL of 0.800 M HCl

Example 10   **Determining Concentration**

Your lab partner gives you 30.0 mL of 0.10 M barium nitrate. She tells you that she started with 1.50 mL of the concentrated barium nitrate solution before diluting it with water. What was the concentration of the original barium nitrate solution?

Solution

Because the number of moles measured from the concentrated solution must equal the number of moles in the dilute solution, we first calculate the number of moles in the 0.10 M solution of barium nitrate.

$$(30.0\ mL)\left(\frac{1\ L}{1000\ mL}\right)\left(\frac{0.10\ mol\ Ba(NO_3)_2}{1\ L}\right) = 0.0030\ mol\ Ba(NO_3)_2$$

Therefore, your lab partner obtained 0.0030 mol $Ba(NO_3)_2$ when she took 1.50 mL of the concentrated solution. The molarity of the concentrated solution is

$$\text{Molarity} = \frac{\text{moles of solute}}{\text{L soln.}} = \frac{0.0030\ mol\ Ba(NO_3)_2}{\left(\dfrac{1.50\ mL}{1000\ mL/L}\right)} = 2.0\ M\ Ba(NO_3)_2$$

You can also solve this problem by using the simplified equation:

$$C_{initial} \times V_{initial} = C_{final} \times V_{final}$$

$$C_{initial} \times (1.50 \text{ mL}) = (0.10 \text{ } M) \times (30.0 \text{ mL})$$

$$C_{initial} = 2.0 \text{ } M$$

Using this equation is much faster, but make sure you understand when and how to use it!

4.3   Stoichiometric Analysis of Solutions

Using concentration (measured in molarity) and the principles we learned in Chapter 3 about chemical reactions, we can do solution stoichiometry. Here are some important points concerning solution stoichiometry:

- The key concept that links the concentration of a solution to a chemical equation is moles.
- One type of reaction that involves solutions is a **titration.** In a titration, one reactant of known concentration (called the standard solution) is added to an unknown amount of another reactant until the reaction is complete.
- The **equivalence point** of a titration is the point at which all of the reactants have been completely used up. During an actual experiment, we call the equivalence point the **end point** because these points are very close to each other. The end point is usually signaled by some type of color change in your reaction.

Example 11   **Practice with Titration**

Consider the titration of 10.00 mL of 0.500 M hydrochloric acid with 0.150 M sodium hydroxide.

a)   How many moles of sodium hydroxide are required to reach the equivalence point?
b)   What volume (in milliliters) of sodium hydroxide is required to reach the equivalence point?

First Thoughts

As in Chapter 3, the first step to solving this problem is to write and balance the equation:

$$HCl(aq) + NaOH(aq) \rightarrow H_2O(l) + NaCl(aq)$$

Next, we must make sure that we understand what the equivalence point of a titration means. In this case, the equivalence point is the point at which all of the moles of HCl and NaOH have been completely consumed (they react in a perfect stoichiometric ratio). Thus we must determine how many moles of HCl are present, and then use the mole ratio between HCl and NaOH to determine how many moles of NaOH are used up in the reaction.

Solution

a)   mL HCl $\xrightarrow{\frac{1 \text{ L}}{1000 \text{ mL}}}$ L HCl $\xrightarrow{\frac{\text{mol HCl}}{\text{L}}}$ mol HCl $\xrightarrow{\frac{\text{mol NaOH}}{\text{mol HCl}}}$ mol NaOH

$$(10.00 \text{ mL HCl})\left(\frac{1 \text{ L HCl}}{1000 \text{ mL HCl}}\right)\left(\frac{0.500 \text{ mol HCl}}{1 \text{ L}}\right)\left(\frac{1 \text{ mol NaOH}}{1 \text{ mol HCl}}\right) = 0.00500 \text{ mol NaOH}$$

$\uparrow$ Concentration of HCl $\uparrow$ Mole ratio in balanced equation

b)   Use the answer from part a and the given concentration of NaOH to solve for the volume required.

$$\text{mole NaOH} \xrightarrow[\text{mol NaOH}]{1\,L} \text{L NaOH} \xrightarrow[L]{1000\,mL} \text{mL NaOH}$$

$$\left(0.00500 \text{ mol NaOH}\right)\left(\frac{1\text{ L}}{0.150\text{ mol NaOH}}\right)\left(\frac{1000\text{ mL}}{1\text{ L}}\right) = 33.3\text{ mL NaOH}$$

Further Insight

You could have also determined how much water was produced from this reaction:

$$\left(0.00500 \text{ mol NaOH}\right)\left(\frac{1\text{ mol H}_2\text{O}}{1\text{ mol NaOH}}\right)\left(\frac{18.016\text{ g H}_2\text{O}}{1\text{ mol H}_2\text{O}}\right) = 0.0901\text{ g H}_2\text{O produced}$$

Example 12   **Titration of Sulfuric Acid with Lithium Hydroxide**

It was experimentally determined that 35.10 mL of 0.200 M sulfuric acid was added to 28.50 mL of lithium hydroxide to reach the end point. What was the molarity of the lithium hydroxide?

Solution

The balanced equation is

$$\text{H}_2\text{SO}_4(aq) + 2\text{LiOH}(aq) \rightarrow 2\text{H}_2\text{O}(l) + \text{Li}_2\text{SO}_4(aq)$$

The molarity of a solution is

$$\text{Molarity} = \frac{\text{moles of solute}}{\text{L soln.}}$$

We have already been given the volume. Now we need to determine the moles of LiOH present before the reaction occurred.

$$\text{mL H}_2\text{SO}_4 \xrightarrow[1000\,mL]{1\,L} \text{L H}_2\text{SO}_4 \xrightarrow[L]{\text{mol H}_2\text{SO}_4} \text{mol H}_2\text{SO}_4 \xrightarrow[\text{mol H}_2\text{SO}_4]{\text{mol LiOH}} \text{mol LiOH}$$

$$\left(35.10 \text{ mL H}_2\text{SO}_4\right)\left(\frac{1\text{ L}}{1000\text{ mL}}\right)\left(\frac{0.200\text{ mol H}_2\text{SO}_4}{1\text{ L}}\right)\left(\frac{2\text{ mol LiOH}}{1\text{ mol H}_2\text{SO}_4}\right) = 0.01404\text{ mol LiOH}$$

The volume of LiOH in liters is

$$\left(28.50 \text{ mL LiOH}\right)\left(\frac{1\text{ L}}{1000\text{ mL}}\right) = 0.02850\text{ L LiOH}$$

Therefore the concentration of LiOH to the appropriate number of significant figures is

$$\text{Molarity} = \frac{0.01404\text{ mol LiOH}}{0.02850\text{ L}} = 0.493\ M\text{ LiOH}$$

4.4   Types of Chemical Reactions

Three of the more important types of chemical reactions are

- Precipitation reactions (discussed in Section 4.5),
- Acid–base reactions (discussed in Section 4.6), and
- Oxidation–reduction reactions (discussed in Section 4.7).

Before we can analyze each of these types of reactions, we need to understand how to write chemical reactions involving aqueous solutions. Chemical reactions containing aqueous reactants or products can be written in three ways: as a molecular equation, as a complete ionic equation, or as a net ionic equation.

- A **molecular equation** includes the reactants and products written as compounds but does not show what is truly happening in solution.
- A **complete ionic equation** shows what is happening in solution, including the dissociation of ions where appropriate.
- A **net ionic equation** shows only those components that actually take part in the reaction.

Example 13   Working with Aqueous Reactions

Aqueous nitric acid and aqueous potassium hydroxide react to form aqueous potassium nitrate and water. Write the molecular, complete ionic, and net ionic equations for this reaction.

First Thoughts

Make sure you write the appropriate formulas for each reactant and product: nitric acid, HNO_3; potassium nitrate, KNO_3; potassium hydroxide, KOH; and water, H_2O.

Solution

Molecular equation:

$$HNO_3(aq) + KOH(aq) \rightarrow KNO_3(aq) + H_2O(l)$$

Nitric acid, potassium hydroxide, and potassium nitrate are strong electrolytes, so they dissociate into their ions in solution (which the molecular equation does not show).

Complete ionic equation:

$$H^+(aq) + NO_3^-(aq) + K^+(aq) + OH^-(aq) \rightarrow K^+(aq) + NO_3^-(aq) + H_2O(l)$$

The complete ionic equation shows every component in solution, whether or not it actually participates in the chemical reaction. Notice how the charges balance on both sides of the equation. Notice also how the NO_3^- ions and K^+ ions appear on both sides of the equation, yet do not actually take part in the chemical reaction. They are called *spectator ions* because they are not doing anything except balancing out the charges (they are just "watching").

Net ionic equation:

$$H^+(aq) + OH^-(aq) \rightarrow H_2O(l)$$

The H^+ ions and the OH^- ions react to form water. These ions are participating in the chemical reaction.

Further Insight

Make sure that all three equations always balance! This example all contained 1:1 ratios, but other ratios are possible. Remember the reaction between sulfuric acid and lithium hydroxide in Example 12?

4.5   Precipitation Reactions

Let's analyze precipitation reactions in more detail.

- A precipitation reaction produces at least one **precipitate** (a solid material that forms in a solution). The precipitate is *insoluble* because not very much of it dissolves into the solvent.
- The aqueous substances in a reaction are *soluble* because they dissolve into the solvent. The reactants of a precipitation reaction are soluble.
- You can predict when a precipitation reaction will occur by conducting experiments with metal ions in the laboratory *or* you can use **Table 4.3** in your text, which summarizes the results of several reactions for you.

Example 14   How Much Precipitate Is Produced?

Calculate the mass of solid formed when 1.50 L of 0.500 M silver nitrate and 1.85 L of 0.300 M ammonium carbonate are mixed.

First Thoughts

Identify the chemical reaction that occurs by first listing the ions present in solution and then using the solubility rules to determine which solid forms.

Ions present: $Ag^+, NO_3^-, NH_4^+, CO_3^{2-}$

Looking at Table 4.3, we find that Ag_2CO_3 is the precipitate. Therefore our balanced molecular, complete ionic, and net ionic equations are

$$2AgNO_3(aq) + (NH_4)_2CO_3(aq) \rightarrow Ag_2CO_3(s) + 2NH_4NO_3(aq)$$

$$2Ag^+(aq) + 2NO_3^-(aq) + 2NH_4^+(aq) + CO_3^{2-}(aq) \rightarrow Ag_2CO_3(s) + 2NH_4^+(aq) + 2NO_3^-(aq)$$

$$2Ag^+(aq) + CO_3^{2-}(aq) \rightarrow Ag_2CO_3(s)$$

Solution

Now that we have a balanced chemical equation, we can apply what we have learned about stoichiometry to calculate the mass of Ag_2CO_3 formed.

First, let's calculate the moles of reactants present. We know that 0.500 M $AgNO_3$ contains 0.500 M Ag^+ ions. Therefore the number of moles of Ag^+ present before the reaction occurs is

$$\left(1.50 \text{ L}\right)\left(\frac{0.500 \text{ mol } Ag^+}{1 \text{ L}}\right) = 0.750 \text{ mol } Ag^+$$

We also know that 0.300 M $(NH_4)_2CO_3$ contains 0.300 M CO_3^{2-} ions. Therefore the number of moles of CO_3^{2-} present before the reaction occurs is

$$\left(1.85 \text{ L}\right)\left(\frac{0.300 \text{ mol } CO_3^{2-}}{1 \text{ L}}\right) = 0.555 \text{ mol } CO_3^{2-}$$

Using the number of moles present and the net ionic equation, we can determine the limiting reactant and the smallest amount of Ag_2CO_3 that can be formed:

$$2Ag^+(aq) + CO_3^{2-}(aq) \rightarrow Ag_2CO_3(s)$$

$$\left(0.750 \text{ mol Ag}^+\right)\left(\frac{1 \text{ mol Ag}_2\text{CO}_3}{2 \text{ mol Ag}^+}\right) = 0.375 \text{ mol Ag}_2\text{CO}_3$$

$$\left(0.555 \text{ mol CO}_3^{2-}\right)\left(\frac{1 \text{ mol Ag}_2\text{CO}_3}{1 \text{ mol CO}_3^{2-}}\right) = 0.555 \text{ mol Ag}_2\text{CO}_3$$

Thus 0.375 mol of Ag_2CO_3 is formed and then the silver ion runs out (Ag^+ is the limiting reactant). Convert 0.375 mol to grams to calculate the mass of precipitate formed.

$$\left(0.375 \text{ mol Ag}_2\text{CO}_3\right)\left(\frac{275.81 \text{ g Ag}_2\text{CO}_3}{1 \text{ mol Ag}_2\text{CO}_3}\right) = 103 \text{ g Ag}_2\text{CO}_3 \text{ formed}$$

Further Insight

You can take these problems a step further by determining the concentration of ions remaining in solution after the reaction is complete. See Example 15 for a demonstration.

Example 15   Concentration of Ions Left in Solution

Using the information from Example 14, what is the concentration of ions left in solution after the reaction between silver nitrate and ammonium carbonate is complete?

First Thoughts

Possible ions left in solution: Ag^+, CO_3^{2-}, NO_3^-, NH_4^+

Solution

[Ag^+]: From our calculations in Example 14, we determined that Ag^+ was the limiting reactant and was therefore completely used up in the reaction; that is, all of the Ag^+ was used to make Ag_2CO_3. As a result, the concentration of Ag^+ must be 0 M because no moles of Ag^+ are left in solution.

[Ag^+] = 0 M

[CO_3^{2-}]: We started with 0.555 mol CO_3^{2-} but some was required to react with Ag^+ and form Ag_2CO_3.

$$\left(0.750 \text{ mol Ag}^+\right)\left(\frac{1 \text{ mol CO}_3^{2-}}{2 \text{ mol Ag}^+}\right) = 0.375 \text{ mol CO}_3^{2-} \text{ used up}$$

The excess CO_3^{2-} is

$$\begin{array}{l} 0.555 \text{ mol} \\ - \underline{0.375 \text{ mol}} \\ 0.180 \text{ mol} \quad CO_3^{2-} \text{ leftover} \end{array}$$

Remembering that

$$\text{Molarity} = \frac{\text{moles of solute}}{\text{L soln.}}$$

determine the *total* volume present after the reaction:

$$\begin{array}{ll} 1.50 \text{ L} & [\text{from the AgNO}_3] \\ + \underline{1.85 \text{ L}} & [\text{from the (NH}_4)_2\text{CO}_3] \\ 3.35 \text{ L} & \text{total volume after rxn.} \end{array}$$

Therefore the concentration of CO_3^{2-} is

$$\text{Molarity} = \frac{0.180 \text{ mol } CO_3^{2-}}{3.35 \text{ L}} = 0.0537 \, M \, CO_3^{2-}$$

$[CO_3^{2-}] = 0.0537 \, M$

$[NO_3^-]$: NO_3^- does not take part in the chemical reaction—it is a spectator ion. Therefore the number of moles of NO_3^- will not change after the reaction is complete. We know that $0.500 \, M$ $AgNO_3$ contains $0.500 \, M \, NO_3^-$ ions. The number of moles of NO_3^- ions present is

$$(1.50 \text{ L})\left(\frac{0.500 \text{ mol } NO_3^-}{1 \text{ L}}\right) = 0.750 \text{ mol } NO_3^-$$

The total volume of solution is again 3.35 L. The concentration of NO_3^- is

$$\text{Molarity} = \frac{0.750 \text{ mol } NO_3^-}{3.35 \text{ L}} = 0.224 \, M \, NO_3^-$$

Even though the number of moles of NO_3^- remains constant after the reaction, the concentration drops because the NO_3^- is present in a larger amount of solution.

$[NO_3^-] = 0.224 \, M$

$[NH_4^+]$: NH_4^+ is also a spectator ion, so the number of moles of NH_4^+ will not change. We know that $0.300 \, M \, (NH_4)_2CO_3$ contains $0.600 \, M \, NH_4^+$ ions: $0.300 \, M \, (NH_4)_2CO_3 \times \dfrac{2 \text{ mol } NH_4^+}{1 \text{ mol } (NH_4)_2CO_3}$. The number of moles of NH_4^+ ions present is

$$(1.85 \text{ L})\left(\frac{0.600 \text{ mol } NH_4^+}{1 \text{ L}}\right) = 1.11 \text{ mol } NH_4^+$$

The concentration of NH_4^+ is

$$\text{Molarity} = \frac{1.11 \text{ mol } NH_4^+}{3.35 \text{ L}} = 0.331 \, M \, NH_4^+$$

$[NH_4^+] = 0.331 \, M$

Further Insight

This kind of problem can be quite long and challenging, so make sure you lay everything out step-by-step and focus on the chemistry that is happening in solution. Ask yourself, "Do my answers make sense?"

4.6 Acid–Base Reactions

Another important type of reaction is an acid–base reaction. Here are some key points concerning acid–base reactions:

- An **acid** is a substance that releases hydrogen ions (H^+) in solution. We often call these H^+ ions protons.
- In Section 4.1, you learned that strong acids (and bases) are strong electrolytes: they completely dissociate into their ions in solution. One of the ions in a strong acid is always H^+.
- There are six strong acids: HCl, HBr, HI, HNO_3, $HClO_4$, and H_2SO_4. You should memorize these strong acids now because you will encounter acids again—in much more detail—in Chapters 17 and 18.
- Acids can be monoprotic, diprotic, or triprotic. HNO_3 is monoprotic: it produces just one H^+ ion (or one mole of H^+) in solution. H_2SO_4 is diprotic: it produces two H^+ ions (or two moles of H^+) in solution. Triprotic acids produce three H^+ ions (or three moles of H^+) in solution.
- The opposite of an acid is a **base,** a substance that releases hydroxide ions (OH^-) in solution.
- Strong bases completely dissociate into their ions in solution. One of the ions in a strong base is always OH^-. Most metals with the hydroxide ion are considered strong bases. **Table 4.4** in the text summarizes the strong acids and bases.
- When a strong acid and a strong base react, they can "neutralize" each other and produce water as one of the products. As an example, let's consider the reaction between HNO_3 and $LiOH$:

$$HNO_3(aq) + LiOH(aq) \rightarrow H_2O(l) + LiNO_3(aq)$$

To look at the *chemistry* that is occurring in solution, write the complete ionic and net ionic equations:

$$H^+(aq) + NO_3^-(aq) + Li^+(aq) + OH^-(aq) \rightarrow H_2O(l) + Li^+(aq) + NO_3^-(aq)$$

$$H^+(aq) + OH^-(aq) \rightarrow H_2O(l)$$

The H^+ from the strong acid and the OH^- from the strong base react to form water, which is neutral.

Example 16   **Stoichiometry of a Neutralization Reaction**

What volume of a 0.500 M HBr solution is needed to neutralize 45.0 mL of a 0.230 M KOH solution?

First Thoughts

Determine the chemical reaction that occurs by listing the ions present in solution.

Ions present: H^+, Br^-, K^+, OH^-

You should guess that H^+ and OH^- will react to form H_2O. Just to be sure, let's write the balanced molecular, complete ionic, and net ionic equations:

$$HBr(aq) + KOH(aq) \rightarrow H_2O(l) + KBr(aq)$$

$$H^+(aq) + Br^-(aq) + K^+(aq) + OH^-(aq) \rightarrow H_2O(l) + K^+(aq) + Br^-(aq)$$

$$H^+(aq) + OH^-(aq) \rightarrow H_2O(l)$$

Solution

Now that we have a balanced net ionic equation, let's apply what we know about solution stoichiometry to determine the volume of HBr needed to neutralize the KOH.

First, calculate the number of moles of OH⁻ present. We know that 0.230 M KOH contains 0.230 M OH⁻ ions. Therefore, the number of moles of OH⁻ present is

$$\left(45.0 \text{ mL}\right)\left(\frac{1 \text{ L}}{1000 \text{ mL}}\right)\left(\frac{0.230 \text{ mol OH}^-}{1 \text{ L}}\right) = 0.01035 \text{ mol OH}^-$$

To neutralize all of the OH⁻ in solution, the H⁺ from HBr must completely react with all of the OH⁻ ions.

$$\left(0.01035 \text{ mol OH}^-\right)\left(\frac{1 \text{ mol H}^+}{1 \text{ mol OH}^-}\right) = 0.01035 \text{ mol H}^+ \text{ needed to neutralize}$$

↑ Mole ratio from
balanced equation

Now we can solve for the volume of HBr needed using the moles of H⁺, the 1:1 ratio between H⁺ and HBR, and the concentration of HBr:

$$\left(0.01035 \text{ mol H}^+\right)\left(\frac{1 \text{ mol HBr}}{1 \text{ mol H}^+}\right)\left(\frac{1 \text{ L}}{0.500 \text{ mol HBr}}\right)\left(\frac{1000 \text{ mL}}{1 \text{ L}}\right) = 20.7 \text{ mL HBr}$$

Further Insight

In this example, there was no limiting reactant. The HBr and KOH reacted in a perfect stoichiometric ratio to exactly neutralize each other and form water.

Example 17   How Much Is in Excess?

Suppose 20.0 mL of a 0.887 M H_2SO_4 solution is mixed with 70.0 mL of a 0.330 M NaOH solution.

a)   How many moles of H_2SO_4 would be required to neutralize the NaOH solution?
b)   What excess volume of H_2SO_4 remains unreacted with the NaOH solution?

First Thoughts

Ions present: H^+, SO_4^{2-}, Na^+, OH^-

Water will form as one of the products. The balanced equations follow:

$$H_2SO_4(aq) + 2NaOH(aq) \rightarrow 2H_2O(l) + Na_2SO_4(aq)$$

$$2H^+(aq) + SO_4^{2-}(aq) + 2Na^+(aq) + 2OH^-(aq) \rightarrow 2H_2O(l) + 2Na^+(aq) + SO_4^{2-}(aq)$$

$$2H^+(aq) + 2OH^-(aq) \rightarrow 2H_2O(l)$$

Solution

a)   Calculate the number of moles of OH⁻ present in solution before the reaction occurs (0.330 M NaOH contains 0.330 M OH⁻ ions):

$$\left(70.0 \text{ mL}\right)\left(\frac{1 \text{ L}}{1000 \text{ mL}}\right)\left(\frac{0.330 \text{ mol OH}^-}{1 \text{ L}}\right) = 0.0231 \text{ mol OH}^-$$

To neutralize all of the OH⁻ in solution, the H⁺ ions must use up all of the moles of OH⁻ present.

$$\left(0.0231 \text{ mol OH}^-\right)\left(\frac{2 \text{ mol H}^+}{2 \text{ mol OH}^-}\right) = 0.0231 \text{ mol H}^+ \text{ needed}$$

To determine the moles of H_2SO_4 needed, remember that for every 1 mole of H_2SO_4, there are 2 moles of H^+ ions.

$$\left(0.0231 \text{ mol H}^+\right)\left(\frac{1 \text{ mol H}_2SO_4}{2 \text{ mol H}^+}\right) = 0.0116 \text{ mol H}_2SO_4 \text{ needed}$$

b) Using the moles of H_2SO_4 needed and the given concentration of H_2SO_4, we can calculate what volume of H_2SO_4 is necessary to neutralize all of the NaOH.

$$\left(0.0116 \text{ mol H}_2SO_4\right)\left(\frac{1 \text{ L}}{0.887 \text{ mol H}_2SO_4}\right)\left(\frac{1000 \text{ mL}}{1 \text{ L}}\right) = 13.1 \text{ mL H}_2SO_4 \text{ needed}$$

We started with 20.0 mL and used up 13.1 mL, so the volume of excess reactant is

$$\begin{array}{r} 20.0 \text{ mL} \\ - \quad 13.1 \text{ mL} \\ \hline 6.9 \text{ mL} \end{array} \quad H_2SO_4 \text{ not used to neutralize NaOH}$$

Further Insight

In this problem, NaOH was the limiting reactant and excess H_2SO_4 remains. Because extra H^+ ions are present in the solution after the reaction is complete, the solution is acidic.

4.7   Oxidation–Reduction Reactions

Here are some important points concerning **oxidation–reduction reactions**, also known as **redox reactions**:

- In a redox reaction, electrons are transferred between the reactants.
- Species that *lose* electrons in the reaction are **oxidized** (undergo the oxidation part of the reaction).
- Species that *gain* electrons in the reaction are **reduced** (undergo the reduction part of the reaction).
- To help us better understand which species are being reduced and which are being oxidized, we often write half-reactions. These half-reactions do not exist alone. Instead, the overall oxidation–reduction equation is the sum of the half-reactions.
- In addition to mass being conserved in a chemical reaction, charge must be conserved in oxidation–reduction reactions.
- One way to help us keep track of the electrons (and overall charge) in an oxidation–reduction reaction is to assign **oxidation numbers** (or **oxidation states**) to atoms in ionic or molecular forms. We assign oxidation numbers to atoms on the basis of where electrons in a bond are likely to be found.
- **Table 4.5** in the text contains rules for assigning oxidation numbers to atoms and ions. Use this table to assist you in understanding Example 18.

Example 18   Assigning Oxidation States

Assign oxidation states to all of the atoms in the following elements, compounds, or ions.

a) N_2 b) CO_2 c) Kr   d) $KMnO_4$ e) PO_4^{3-} f) Fe_3O_4

First Thoughts

With the exception of part e, these elements and compounds are electrically neutral. Therefore, the overall charge must add up to zero. Use Table 4.5 to assist you in assigning the oxidation numbers.

Solution

a) N_2 is an element, so the oxidation number is 0 (Rule 1 in Table 4.5).

   $N_2 = 0$

b) The oxidation number of oxygen in a covalent compound like CO_2 is -2 (Rule 6 in Table 4.5). There are two oxygen atoms in CO_2, so the overall charge from oxygen is -4. Therefore, the oxidation state of carbon must be +4.

   $+4 + 2(-2) = 0$

   $O = -2, C = +4$

c) Kr is an element so the oxidation number is 0 (Rule 1 in Table 4.5).

   $Kr = 0$

d) In this compound, a metal (potassium) is bonded to a polyatomic ion (MnO_4^-). The oxidation number on potassium must be +1 (Rule 3 in Table 4.5). Therefore the overall charge on the permanganate ion must be -1 to make the compound electrically neutral. Knowing that oxygen has a -2 oxidation number, the overall charge from oxygen in MnO_4^- is -8. The oxidation number of Mn must then be -7.

   $+7 + 4(-2) = -1$ (-1 is the charge on the permanganate ion needed to counterbalance the charge on potassium)

   $K = +1, O = -2, Mn = +7$

e) The oxidation state of oxygen is -2 (Rule 6 in Table 4.5), but the overall charge from oxygen is -8 because four oxygen atoms are present in PO_4^{3-}. The overall charge on the phosphate ion is -3, so the oxidation state of phosphorus must be +5.

   $+5 + 4(-2) = -3$

   $O = -2, P = +5$

f) The oxidation number of oxygen is again -2 (Rule 7 in Table 4.5), but now there are four oxygen atoms in Fe_3O_4. Thus the overall charge from oxygen is -8. To make Fe_3O_4 electrically neutral, the oxidation state of iron must be +8/3. (Yes! Oxidation states can be fractions!)

   $3(?) + 4(-2) = 0$

   $3(8/3) + 4(-2) = 0$

   $8 - 8 = 0$

   $O = -2, Fe = +8/3$

Example 19   Is It a Redox Reaction?

Which of the following reactions are redox reactions?

a) $2Mg(s) + O_2(g) \rightarrow 2MgO(s)$

b) $NaCl(aq) + AgNO_3(aq) \rightarrow AgCl(s) + NaNO_3(aq)$

c) $CH_4(g) + 2O_2(g) \rightarrow CO_2(g) + 2H_2O(g)$

First Thoughts

The first step in identifying whether an exchange of electrons occurs is to assign oxidation numbers to each element. If a change in oxidation number occurs, check whether electrons have been lost or gained.

Solution

a) Based on the oxidation number rules in Table 4.5, we can make the following assignments:

$$2Mg(s) + O_2(g) \rightarrow 2MgO(s)$$

          0   0   +2 -2

Magnesium goes from 0 to +2 (loses electrons = oxidation).
Oxygen goes from 0 to -2 (gains electrons = reduction).

When the oxidation state *decreases* in number, the substance is being *reduced*. Likewise, when the oxidation state *increases* in number, the substance is being *oxidized*.

Because electrons have been exchanged, this is a redox reaction.

b)

$$NaCl(aq) + AgNO_3(aq) \rightarrow AgCl(s) + NaNO_3(aq)$$

+1,-1   +1,+5,-2   +1, -1   +1,+5,-2

None of these atoms or ions changes its oxidation number, so this is not a redox reaction (no electrons have been lost or gained).

c)

$$CH_4(g) + 2O_2(g) \rightarrow CO_2(g) + 2H_2O(g)$$

-4,+1   0   +4,-2   +1, -2

Carbon goes from -4 to +4.
Oxygen goes from 0 to -2.
Hydrogen does not change its oxidation number.

Electrons have been exchanged, so this is a redox reaction.

Further Insight

These redox reactions can also be balanced. Because of the added electron exchange, however, this task is much more complex than balancing precipitation and acid–base reactions. You will learn how to balance redox reactions in Chapter 19.

In Chapter 4, you learned about solution stoichiometry and saw how it applies to different kinds of reactions, such as precipitation and acid–base reactions. Furthermore, you learned how to assign oxidation numbers to atoms and ions and explored how this concept relates to redox reactions. Try solving the exercises that follow to test your understanding of Chapter 4.

Exercises

Section 4.1

1. What is an electrolyte? Give an example of a strong electrolyte, a weak electrolyte, and a nonelectrolyte.

2. Classify the following compounds as strong electrolytes, weak electrolytes, or nonelectrolytes.
    a) $CaCl_2$
    b) HCN
    c) HBr
    d) KOH
    e) C_6H_{14}

3. The following aqueous solutions are connected to two electrodes, which are themselves attached to a light bulb. Which of the following solutions would cause the bulb to light up?
    a) HNO_2
    b) HClO
    c) NH_4Cl
    d) $MgBr_2$

4. Show how each of the following strong electrolytes dissociates ("breaks up") into its ions when dissolved in water. For example, $NaCl \rightarrow Na^+ + Cl^-$.
    a) NH_4Cl
    b) $Al(NO_3)_3$
    c) HI
    d) LiBr
    e) $HClO_4$

Section 4.2

5. How many moles of $HC_2H_3O_2$ are contained in 3.50 L of a 0.500 M solution?

6. How many grams of NaCl are contained in 45.0 mL of a 0.250 M solution?

7. If a 75.2-g sample of ammonium nitrate is dissolved in enough water to make 315 mL of solution, what will be the molarity of this solution?

8. What volume of solution (in milliliters) is required to make a 1.40 M solution of sodium carbonate? The mass of the sodium carbonate in solution is 25.0 g.

9. What volume (in milliliters) of a 0.600 M solution of HF contains 0.320 g of HF?

10. How many grams of LiOH are contained in 115 mL of a 0.900 M solution?

11. A 171.2-mg sample of $Na_2Cr_2O_7$ is dissolved in water to make 400.0 mL of solution. What is the molarity of this solution?

12. Determine the concentrations of all ions present in each of the following solutions.
    a) 1.3 M NaCl
    b) 0.42 M CaBr$_2$
    c) 0.15 M Fe(NO$_3$)$_3$
    d) 7.10 × 10^{-3} M (NH$_4$)$_2$SO$_4$

13. How many moles of SO$_4^{2-}$ ions will be present in 60.0 mL of 0.085 M Na$_2$SO$_4$? How many moles of Na$^+$ ions will be present? What is the total number of moles of ions present in the solution?

14. Which of the following solutions contains the *smallest* number of ions?
    a) 500.0 mL of 1.0 M lithium nitrate
    b) 300.0 mL of 2.0 M potassium hydroxide
    c) 150.0 mL of 2.0 M iron(III) chloride
    d) 300.0 mL of 1.0 M sodium sulfate
    e) At least two of these solutions contain the smallest number of ions.

15. Traces of Cu$^+$ are found in a solution in the range of 5.2 ppb. What does this concentration represent in units of molarity?

16. A solution contains 1.7 ppm of Ca^{2+} ion. What is the concentration of Ca^{2+} in units of molarity?

17. You test a sample of water and find the concentration of Fe^{3+} to be 4.66 × 10^{-7} M. What is this concentration in units of
    a) ppm?
    b) ppb?
    c) ppt?

18. You wish to prepare 0.170 M HCl from a 12.5 M stock solution of hydrochloric acid. How many milliliters of the stock solution do you require to make up 1.00 L of 0.170 M HCl?

19. You have a 16.0-oz (473-mL) glass of lemonade with a concentration of 2.00 M. The lemonade sits out on your counter for a couple of days and 150. mL of water evaporates from the glass. What is the new concentration of the lemonade?

    a) 1.52 M b) 2.00 M c) 2.93 M d) 6.31 M e) None of these

20. A chemist wants to prepare 0.580 M HNO$_3$. The stock solution in the laboratory is 10.0 M. How many milliliters of the stock solution does the chemist require to make up 2.00 L of the dilute HNO$_3$ acid?

21. Describe how you would prepare 500. mL of 0.850 M K$_3$PO$_4$ from its solid form.

22. Describe how you would prepare 2.00 L of each of the following solutions.
    a) 0.100 M NaOH from 1.00 M stock solution
    b) 0.100 M NaOH from solid NaOH
    c) 0.550 M Fe(NO$_3$)$_3$ from 2.00 M stock solution
    d) 0.550 M Fe(NO$_3$)$_3$ from solid Fe(NO$_3$)$_3$

23. True or false? To prepare 20.0 mL of a 2.00 M solution of hydrochloric acid from a 10.0 M stock solution, you should take 4.00 mL of the stock solution and add 20.0 mL of water. If true, verify this result with calculations. If false, explain why (including calculations) and state how to correctly make the desired solution.

Section 4.3

24. Consider the titration of 25.00 mL of 1.50 M $HClO_4$ with 0.860 M lithium hydroxide.
    a) How many moles of lithium hydroxide are required to reach the equivalence point?
    b) What volume (in milliliters) of lithium hydroxide is required to reach the equivalence point?

25. It was determined that 15.2 mL of 0.950 M sulfuric acid was added to 6.40 mL of potassium hydroxide to reach the end point of the titration. What was the molarity of the potassium hydroxide?

26. What volume (in milliliters) of each of the following acids is required to titrate 50.00 mL of 0.300 M NaOH to the equivalence point?
    a) 0.100 M HCl
    b) 0.250 M HBr
    c) 0.030 M HNO_3

27. What volume (in milliliters) of each of the following bases is required to titrate 75.00 mL of 0.100 M HCl to the equivalence point?
    a) 0.050 M NaOH
    b) 0.350 M LiOH
    c) 0.100 M $Ca(OH)_2$

Section 4.4

28. Write the balanced molecular, complete ionic, and net ionic equations for each of the following acid–base reactions.
    a) $HCl(aq) + NaOH(aq) \rightarrow$
    b) $HNO_3(aq) + Ca(OH)_2(s) \rightarrow$
    c) $HBr(aq) + KOH(aq) \rightarrow$
    d) $HClO_4(aq) + Al(OH)_3(s) \rightarrow$

29. Write net ionic equations for the following balanced molecular equations.
    a) $HI(aq) + KOH(aq) \rightarrow H_2O(l) + KI(aq)$
    b) $CaS(aq) + 2HBr(aq) \rightarrow CaBr_2(aq) + H_2S(g)$
    c) $NaOH(aq) + NH_4Br(aq) \rightarrow NaBr(aq) + NH_3(g) + H_2O(l)$
    d) $MgCO_3(s) + H_2SO_4(aq) \rightarrow MgSO_4(aq) + H_2O(l) + CO_2(g)$

Section 4.5

30. When the following solutions are mixed together, what precipitate (if any) will form?
    a) $CaCl_2(aq) + Na_2SO_4(aq)$
    b) $AgNO_3(aq) + Na_3PO_4(aq)$
    c) $K_2S(aq) + Ni(NO_3)_2(aq)$
    d) $MgSO_4(aq) + NaOH(aq)$

31. Which drawing *best* represents the mixing of aqueous calcium chloride with aqueous potassium sulfate when the two are mixed in stoichiometric amounts (neither reactant is limiting)?

a)

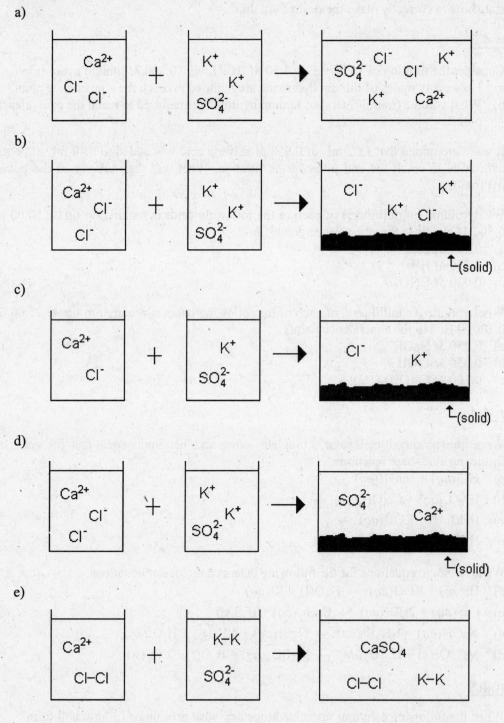

b)

c)

d)

e)

32. Write the molecular equation and the net ionic equation for each of the following aqueous reactions. If no reaction occurs, write NR after the arrow.
    a) $FeSO_4(aq) + NaCl(aq) \rightarrow$
    b) $Pb(NO_3)_2(aq) + KCl(aq) \rightarrow$
    c) $Na_2CO_3(aq) + CaCl_2(aq) \rightarrow$
    d) $BaCl_2(aq) + Na_2SO_4(aq) \rightarrow$

33. True or false? When solutions of barium hydroxide and sulfuric acid are mixed, the net ionic equation is $Ba^{2+}(aq) + SO_4^{2-}(aq) \rightarrow BaSO_4(s)$ because only the species involved in making the precipitate are included. Write a balanced molecular equation and complete ionic equation for the reaction between barium hydroxide and sulfuric acid to support your answer.

34. Calculate the mass of solid formed when 0.750 L of 1.30 M calcium chloride and 1.25 L of 0.500 M potassium sulfate are mixed.

35. What volume of 0.500 M sodium hydroxide is required to react with 21.0 mL of 0.890 M magnesium sulfate if 1.09 g of magnesium hydroxide is produced?

36. Determine the mass of solid formed when 20.0 mL of 0.900 M lead(II) nitrate is mixed with 25.0 mL of 1.10 M potassium chloride.

37. Using the information from Exercise 36, what is the concentration of the ions left in solution after the reaction between lead(II) nitrate and potassium chloride is complete?

38. For the reaction between 200.0 mL of 0.100 M silver nitrate and 95.00 mL of 0.100 M sodium chloride, what is the concentration of *silver* ions in solution *after* the reaction is complete?
    a) 0.00 M; all of the silver ions are used up to make the precipitate.
    b) 0.100 M; the silver ions do not participate in the chemical reaction and are considered spectator ions.
    c) Greater than 0.100 M; for every 1.0 mol of silver nitrate, there are 2.0 mol of silver ions in solution.
    d) Less than 0.100 M; even though silver ions do not react, they are diluted when mixed with the sodium chloride solution.
    e) Less than 0.100 M; some, but not all, of the silver ions are used to make the precipitate.

Section 4.6

39. What volume of a 0.261 M HCl solution is needed to neutralize 30.0 mL of 0.455 M NaOH?

40. In an experiment, 45.0 mL of KOH is neutralized by 23.0 mL of 1.04 M HBr. What was the original concentration of KOH?

41. Consider the reaction between 18.0 mL of 0.334 M NaOH and 15.0 mL of 0.540 M HClO₄.
    a) Calculate the mass of water formed.
    b) What is the concentration of H^+ or OH^- ions in excess after the reaction is complete?

Section 4.7

42. Determine the oxidation number for the element noted in each of the following:
    a) Ga in $GaPO_4$
    b) P in P_4O_6
    c) Hg in Hg_2Cl_2
    d) S in $CaSO_3$

43. Assign oxidation states for all atoms in each of the following compounds, elements, or ions.

    a) NiO_2 b) SF_4 c) CO d) S_8 e) $Mg(NO_3)_2$ f) Fe^{3+}

44. Assign the oxidation state for nitrogen in each of the following compounds.

    a) NH_3 b) NO_3^- c) N_2 d) N_2O

45. For the following reaction, determine which substance is being oxidized and which substance is being reduced.
$$Zn(s) + 2HCl(aq) \rightarrow ZnCl_2(aq) + H_2(g)$$

46. For the reaction $Cu(s) + 2Ag^+(aq) \rightarrow 2Ag(s) + Cu^{2+}(aq)$, state which substance is being oxidized and which substance is being reduced. Also identify the oxidizing agent and the reducing agent. (The oxidizing agent is the substance that causes another substance to be oxidized; the reducing agent is the substance that causes another substance to be reduced.)

47. The following oxidation–reduction reaction occurs:
$$SiCl_4(l) + 2Mg(s) \rightarrow 2MgCl_2(s) + Si(s)$$

    Determine which substance is being oxidized, which substance is being reduced, the oxidizing agent, and the reducing agent.

48. How many electrons are transferred in the following redox reaction?
$$2Al(s) + 3I_2(s) \rightarrow 2AlI_3(s)$$

49. Determine which substance is being oxidized, which substance is being reduced, and how many electrons are transferred in the following reaction:
$$2CuCl(aq) \rightarrow CuCl_2(aq) + Cu(s)$$

50. Which of the following reactions are redox reactions?
    a) $HCl(g) + NH_3(g) \rightarrow NH_4Cl(s)$
    b) $N_2(g) + 3Br_2(l) \rightarrow 2NBr_3(g)$
    c) $KCl(aq) + Mg(NO_3)_2(aq) \rightarrow MgCl_2(aq) + KNO_3(aq)$
    d) $BaCl_2(aq) + Na_2SO_4(aq) \rightarrow BaSO_4(s) + NaCl(aq)$

Answers to Exercises

1. An electrolyte is a compound that dissolves in a solvent to release free ions; these ions may then conduct electricity.

   Strong electrolyte: sodium chloride (NaCl)
   Weak electrolyte: acetic acid ($HC_2H_3O_2$)
   Nonelectrolyte: glucose ($C_6H_{12}O_6$)

2. a) strong electrolyte
   b) weak electrolyte
   c) strong electrolyte
   d) strong electrolyte
   e) nonelectrolyte

3. All of the compounds would light the bulb. NH_4Cl and $MgBr_2$ (strong electrolytes) would produce a brighter light than HNO_2 and $HClO$ (weak electrolytes).

4. a) $NH_4Cl \rightarrow NH_4^+ + Cl^-$
   b) $Al(NO_3)_3 \rightarrow Al^{3+} + 3NO_3^-$
   c) $HI \rightarrow H^+ + I^-$
   d) $LiBr \rightarrow Li^+ + Br^-$
   e) $HClO_4 \rightarrow H^+ + ClO_4^-$

5. 1.75 mol $HC_2H_3O_2$

6. 0.657 g NaCl

7. 2.98 M NH_4NO_3

8. 168 mL

9. 26.7 mL

10. 2.48 g LiOH

11. 1.634×10^{-3} M $Na_2Cr_2O_7$

12. a) $[Na^+] = 1.3$ M; $[Cl^-] = 1.3$ M
    b) $[Ca^{2+}] = 0.42$ M; $[Br^-] = 0.84$ M
    c) $[Fe^{3+}] = 0.15$ M; $[NO_3^-] = 0.45$ M
    d) $[NH_4^+] = 1.42 \times 10^{-2}$ M; $[SO_4^{2-}] = 7.10 \times 10^{-3}$ M

13. 5.1×10^{-3} mol SO_4^{2-}; 1.0×10^{-2} mol Na^+; 1.5×10^{-2} mol total ions

14. d

15. 8.2×10^{-8} M Cu^+

16. $4.2 \times 10^{-5}\ M\ Ca^{2+}$

17. a) 2.60×10^{-2} ppm
    b) 26.0 ppb
    c) 2.60×10^{4} ppt

18. 13.6 mL of stock solution

19. c

20. 116 mL of stock solution

21. Measure out 90.2 g of K_3PO_4 into a volumetric flask and slowly dissolve with H_2O to the 500.-mL line.

22. a) Pour out 0.200 L (or 200. mL) of the 1.00 M NaOH stock solution into a volumetric flask and dilute it with water to the 2.00-L line.
    b) Measure out 8.00 g of solid NaOH into a volumetric flask and slowly dissolve it with water until the solution reaches the 2.00-L line.
    c) Pour out 0.550 L (or 550. mL) of the 2.00 M $Fe(NO_3)_3$ stock solution into a volumetric flask and dilute with water to the 2.00-L line.
    d) Measure out 266 g of solid $Fe(NO_3)_3$ into a volumetric flask and slowly dissolve it with water until the solution reaches the 2.00-L line.

23. False. By adding 20.0 mL of water to the stock solution, you will make the desired solution too dilute and you will have the incorrect volume. See the calculations below:

$$10.0\ M = \frac{x \text{ moles}}{0.00400 \text{ mL}} = 0.0400 \text{ mol of stock soln.}$$

$$\text{New concentration} = \frac{0.0400 \text{ mol}}{(0.00400 \text{ L} + 0.0200 \text{ L})} = 1.67\ M$$

You want 20.0 mL of a 2.00 M solution, *not* 24 mL of a 1.67 M solution. To make the desired solution, take 4.00 mL of stock solution and *add* 16.0 mL of water.

$$\text{New concentration} = \frac{0.0400 \text{ mol}}{(0.00400 \text{ L} + 0.0160 \text{ L})} = 2.00\ M$$

24. a) 0.0375 mol LiOH
    b) 43.6 mL LiOH

25. 4.51 M KOH

26. a) 150. mL HCl
    b) 60.0 mL HBr
    c) 5.0×10^{2} mL HNO_3

27. a) 150 mL NaOH
    b) 21.4 mL LiOH
    c) 37.5 mL $Ca(OH)_2$

28. a) $HCl(aq) + NaOH(aq) \rightarrow H_2O(l) + NaCl(aq)$

$H^+(aq) + Cl^-(aq) + Na^+(aq) + OH^-(aq) \rightarrow H_2O(l) + Na^+(aq) + Cl^-(aq)$

$H^+(aq) + OH^-(aq) \rightarrow H_2O(l)$

b) $2HNO_3(aq) + Ca(OH)_2(s) \rightarrow 2H_2O(l) + Ca(NO_3)_2(aq)$

$2H^+(aq) + 2NO_3^-(aq) + Ca(OH)_2(s) \rightarrow 2H_2O(l) + Ca^{2+}(aq) + 2NO_3^-(aq)$

$2H^+(aq) + Ca(OH)_2(s) \rightarrow 2H_2O(l) + Ca^{2+}(aq)$

c) $HBr(aq) + KOH(aq) \rightarrow H_2O(l) + KBr(aq)$

$H^+(aq) + Br^-(aq) + K^+(aq) + OH^-(aq) \rightarrow H_2O(l) + K^+(aq) + Br^-(aq)$

$H^+(aq) + OH^-(aq) \rightarrow H_2O(l)$

d) $3HClO_4(aq) + Al(OH)_3(s) \rightarrow 3H_2O(l) + Al(ClO_4)_3(aq)$

$3H^+(aq) + 3ClO_4^-(aq) + Al(OH)_3(s) \rightarrow 3H_2O(l) + Al^{3+}(aq) + 3ClO_4^-(aq)$

$3H^+(aq) + Al(OH)_3(s) \rightarrow 3H_2O(l) + Al^{3+}(aq)$

29. a) $H^+(aq) + OH^-(aq) \rightarrow H_2O(l)$

b) $S^{2-}(aq) + 2H^+(aq) \rightarrow H_2S(g)$

c) $OH^-(aq) + NH_4^+(aq) \rightarrow NH_3(g) + H_2O(l)$

d) $MgCO_3(s) + 2H^+(aq) \rightarrow Mg^{2+}(aq) + H_2O(l) + CO_2(g)$

30. a) $CaSO_4$
    b) Ag_3PO_4
    c) NiS
    d) $Mg(OH)_2$

31. b

32. a)   NR

b) $Pb(NO_3)_2(aq) + 2KCl(aq) \rightarrow 2KNO_3(aq) + PbCl_2(s)$

$Pb^{2+}(aq) + 2Cl^-(aq) \rightarrow PbCl_2(s)$

c) $Na_2CO_3(aq) + CaCl_2(aq) \rightarrow CaCO_3(s) + 2NaCl(aq)$

$CO_3^{2-}(aq) + Ca^{2+}(aq) \rightarrow CaCO_3(s)$

d) $BaCl_2(aq) + Na_2SO_4(aq) \rightarrow BaSO_4(s) + 2NaCl(aq)$

$Ba^{2+}(aq) + SO_4^{2-}(aq) \rightarrow BaSO_4(s)$

33. False. The balanced molecular equation is

$$Ba(OH)_2(aq) + H_2SO_4(aq) \rightarrow BaSO_4(s) + 2\,H_2O(l)$$

The complete ionic equation is

$$Ba^{2+}(aq) + 2OH^-(aq) + 2H^+(aq) + SO_4^{2-}(aq) \rightarrow BaSO_4(s) + 2\,H_2O(l)$$

The net ionic equation includes all species that take part in the chemical reaction. The OH^- and H^+ ions form water, so they are also included in the net ionic equation. Thus the complete ionic equation and net ionic equation are the same.

34. 85.1 g $CaSO_4$

35. 74.8 mL NaOH

36. 3.82 g $PbCl_2$

37. $[Cl^-] = 0\ M;\ [Pb^{2+}] = 0.0944\ M;\ [K^+] = 0.611\ M;\ [NO_3^-] = 0.800\ M$

38. e

39. 52.3 mL HCl

40. 0.532 M KOH

41. a) 0.108 g H_2O formed
    b) 0.0633 $M\ H^+$ in excess

42. a) Ga = +3
    b) P = +3
    c) Hg = +1
    d) S = +4

43. a) Ni = +4; O = -2
    b) S = +4; F = -1
    c) C = +2; O = -2
    d) $S_8 = 0$
    e) Mg = +2; N = +5; O = -2
    f) Fe = +3

44. a) -3
    b) +5
    c) 0
    d) +1

45. HCl (H) is reduced; Zn is oxidized

46. Cu is being oxidized and is the reducing agent. Ag^+ is being reduced and is the oxidizing agent.

47. Mg is being oxidized and is the reducing agent. $SiCl_4$ (Si) is being reduced and is the oxidizing agent.

48. 3 electrons

49. CuCl (Cu) is being both reduced and oxidized; 1 electron is transferred

50. b only

Chapter 5

Energy

The Bottom Line

When you finish this chapter, you will be able to:

- Explain that chemical reactions are accompanied by the gain or release of energy.
- Comprehend that thermodynamics is the study of energy changes.
- Recall that there are different types of energy such as kinetic, potential, and heat.
- Understand that energy is neither created nor destroyed in a chemical reaction.
- Describe and solve problems involving work, heat, internal energy, and enthalpy.
- Describe and solve problems involving heat capacity, calorimetry, and Hess's law.

5.1 The Concept of Energy

There are several important concepts in this section:

- There are three natural forces: gravitational, electroweak, and strong nuclear.

- Potential energy is the energy of position. It is energy stored within a system.

- Kinetic energy is the energy of motion. $KE = \frac{1}{2} mv^2$.

- Law of conservation of energy: Energy can be neither created nor destroyed.

- The system is that which is being studied.

- The surroundings are everything outside the system.

- Work and heat are two ways that energy can move from the system to the surroundings, or vice versa.

- Heat energy flowing from the system to the surroundings represents an **exothermic** process. Heat flowing from the surroundings to the system represents an **endothermic** process.

- Energy changes in a reaction can be studied by examining a reaction profile diagram. (See **Figure 5.9** in your text for an example.)

Work (w) is defined as the product of force (f) times distance (d):

$$w = f \times d$$

In chemical reactions, work is usually associated with the expansion or compression of gases. The system would be the gas in the chemical reaction. In this case, the definition of work is the product of the external pressure on the gas (P) times the change in volume (ΔV):

$$w = -P \times \Delta V$$

Why the negative sign? The sign conventions for work are important:

- If the **system expands,** it is doing positive work on the surroundings; therefore it is doing **negative work on the system.**

- If the **system contracts,** the surroundings have done work on the system; therefore there is **positive work done on the system.**

Example 1   Work

Calculate the work in L · atm and joules associated with the expansion of a gas from 23.0 L to 60.0 L at a constant external pressure of 2.50 atm. (1 L · atm = 101.3 J.)

First Thoughts

Remember that expansion of a gas results in negative work and that $\Delta V = V_{final} - V_{initial}$.

Solution

$$w = -P \times \Delta V$$

$$w = -2.50 \text{ atm } (60.0 \text{ L} - 23.0 \text{ L}) = -92.5 \text{ L} \cdot \text{atm}$$

$$w = -92.5 \text{ L} \cdot \text{atm} \left(\frac{101.3 \text{ J}}{\text{L} \cdot \text{atm}} \right) = -9.37 \times 10^3 \text{ J}$$

The **law of conservation of energy** is also called the **first law of thermodynamics.** In a chemical reaction, chemical bonds are made and broken. The energy is converted between the potential energy stored in bonds and the thermal energy (kinetic energy) released as heat.

The change in **internal energy** of the system (ΔU) is the sum of the **heat** (q) and the **work** (w):

$$\Delta U = q + w$$

and

$$\Delta U_{system} = \Delta U_{surroundings}$$

ΔU_{system} has a positive (+) sign if the system loses energy; ΔU_{system} has a negative (–) sign if the system gains energy.

Example 2   First Law of Thermodynamics

What is the change in the internal energy of a system if 92.6 J of work is done by the system and 21.3 J of heat is lost?

First Thoughts

You must pay close attention to the sign conventions for heat and work. If the system does work it is negative. If the system loses heat it is also negative.

Solution

$$\Delta U = q + w$$

$$\Delta U = -21.3 \text{ J} + (-92.6 \text{ J}) = -113.9 \text{ J}$$

Therefore 113.9 J of energy is transferred from the system to the surroundings.

5.2   Keeping Track of Energy

In this section, it is important to know the units of energy and how to convert between units.

$$1 \text{ joule (J)} = 1 \ \frac{\text{kg m}^2}{\text{s}^2}$$

Energy can also be expressed in calories:

$$1 \text{ cal} = 4.184 \text{ J}$$

Do not confuse an energy calorie (cal) with a nutritional or food calorie (Cal). Notice that the difference between the two is the small "c" versus the big "C".

$$1 \text{ Cal} = 1000 \text{ cal} = 1 \text{ kcal}$$

So a 300-Calorie ice cream cone is equal to 300,000 energy calories!

Example 3   **Energy Units**

Premium vanilla ice cream contains 230. Cal per ½-cup serving (103 g). How many joules and kilojoules are in this serving?

Solution

$$\frac{\text{joules}}{\text{g}} = \left(\frac{230. \text{ Cal}}{103 \text{ g}} \right) \left(\frac{1000 \text{ cal}}{\text{Cal}} \right) \left(\frac{4.184 \text{ J}}{\text{cal}} \right) = 9.34 \times 10^3 \text{ J/g}$$

$$\frac{\text{kilojoules}}{\text{g}} = 9.34 \times 10^3 \text{ J} \left(\frac{1 \text{ kJ}}{1000 \text{ J}} \right) = 9.34 \text{ kJ/g}$$

5.3   Specific Heat and Heat Capacity

The following are important definitions from this section of your text:

1. **Specific heat capacity (c):** the amount of heat needed to raise the temperature of one gram of a substance by 1°C (or 1 K). The units are J/g · °C or J/g · K.

2. **Heat capacity (C):** the heat required to raise the temperature by a given amount. The units are J/°C or J/ K.

3. **Molar heat capacity:** the heat required to heat one mole of a substance by 1 °C (or 1 K). The units are J/mol · °C or J/mol · K.

To calculate the amount of heat necessary to change the temperature of a substance, you can use the equation:

$$q = m \times c \times \Delta T$$

q = heat, m = mass, ΔT = change in temperature in °C or K.

Example 4   **Specific Heat Capacity I**

How much heat is required to heat 26.5 g of ethanol, whose specific heat is 2.46 J/g · °C, from 23.0°C to 90.0°C?

First Thoughts

The mass of ethanol is given. The specific heat is also given (or we could have looked it up in a table). Remember that $\Delta T = T_{\text{final}} - T_{\text{initial}}$.

Solution

$$q = (26.5 \text{ g}) \left(2.46 \ \frac{J}{g \cdot {}^{\circ}C} \right) (90.0^{\circ}C - 23.0^{\circ}C) = 4.37 \times 10^{3} \text{ J}$$

Example 5   **Specific Heat Capacity II**

How much heat must be removed from 125 g of water to lower the water's temperature from 25.0°C to 4.0°C?

Solution

$$q = (125 \text{ g}) \left(4.184 \ \frac{J}{g \cdot {}^{\circ}C} \right) (4.0^{\circ}C - 25.0^{\circ}C) = -1.10 \times 10^{4} \text{J}$$

Further Insight

Notice that the answer is negative! The amount of heat is 1.10×10^{4} J. The negative sign indicates that the heat is *removed*.

   Calorimetry is the experimental method used to measure the amount of heat transfer between the system and the surroundings. As described in your textbook, calorimeters can be simple, like Styrofoam coffee cups, or more sophisticated, like the bomb calorimeter. Using a Styrofoam calorimeter, the heat lost by a sample is related to the heat gained by the water in the calorimeter. The heat can be calculated from $q = m \times c \times \Delta T$.

   The bomb calorimeter uses a term for the heat capacity of the calorimeter itself since it is constructed from steel, which absorbs some of the heat. In this case, the heat is calculated from $q_{rxn} = -c_{cal} \times \Delta T$, where c_{cal} is the heat capacity of the bomb calorimeter. The following examples will show each of the techniques.

Example 6   **Styrofoam Calorimeter**

A student uses a "coffee-cup calorimeter" to determine the specific heat of a metal sample. She heats 37.0 g of the metal to 100.0°C and then places the hot metal into a coffee cup containing 97.0 g of water at 21.0°C. The temperature of the water and the metal equilibrates to 27.0°C. Calculate the specific heat of the metal, given that the specific heat of water is 4.184 J/°C.

First Thoughts

The heat lost by the metal $(-q_m)$ equals the heat gained by the water (q_w). In each case, $q = m \times c \times \Delta T$.

Solution

$$-q_m = q_w$$

$$-(m_m)(c_m)(\Delta T_m) = (m_w)(c_w)(\Delta T_w)$$

$$-(37.0\text{g})(c_m)(27.0 \ ^{\circ}C - 100.0^{\circ}C) = (97.0\text{g}) \left(4.184 \ \frac{J}{g \ ^{\circ}C} \right) (27.0^{\circ}C - 21.0 \ ^{\circ}C)$$

$$c = 0.90 \ \frac{J}{g \ ^{\circ}C}$$

Example 7   **Bomb Calorimeter**

A 3.393 g sample of methylhydrazine, CH_6N_2, was reacted with excess oxygen in a bomb calorimeter whose heat capacity was 6.643 kJ/°C. The temperature rose from 25.00°C to 38.25°C. Calculate the energy released in kJ/mol for this reaction.

First Thoughts

The heat capacity of the bomb calorimeter is given, so we can calculate the energy released by this reaction (q_{rxn}) from $q_{rxn} = -c_{cal} \times \Delta T$. That will be the amount of heat energy released by the sample (3.393 g). We then have to express the amount of heat in kJ/mol.

Solution

$$q_{rxn} = -c_{cal} \times \Delta T$$

$$= -\left(6.643 \frac{kJ}{°C}\right)\left(38.25°C - 25.00°C\right)$$

$$= -88.02 \text{ kJ}$$

$$\left(\frac{-88.02 \text{ kJ}}{3.393 \text{ g}}\right)\left(\frac{46.10 \text{ g}}{1 \text{ mol } CH_6N_2}\right) = -1196 \text{ kJ/mol } CH_6N_2$$

Molar mass of CH_6N_2

5.4   Enthalpy

The term **enthalpy (H)** is discussed and derived in your textbook. For a chemical reaction:

$$\Delta H = H_{products} - H_{reactants}$$

The change in enthalpy, **under constant pressure conditions,** is equal to the energy flow:

$$\Delta H = q_p$$

If $\Delta H > 0$, the reaction is endothermic: heat is absorbed by the system.
If $\Delta H < 0$, the reaction is exothermic: heat is evolved by the system.

Example 8   **Enthalpy**

Consider the following reaction:

$$HCl(aq) + NaOH(aq) \rightarrow H_2O(l) + NaCl(aq)$$

For this reaction under constant pressure, $\Delta H = -1.7$ kJ/mol. At the end of the reaction, does the beaker get warmer or colder, and is the reaction endothermic or exothermic?

Solution

The sign of ΔH indicates that the reaction is exothermic; heat is given from the system to the surroundings, making the beaker warmer.

The **standard enthalpy of formation** ($\Delta_f H^o$) is the enthalpy change for the formation of one mole of the substance in its standard state from its elements in their reference forms. Here are the **standard states** for substances:

- For a pure solid, liquid, or gas, it is the state of the substance at a pressure of exactly 1 atmosphere.

- For any substance in solution, it is a concentration of exactly 1 M.

The normal thermodynamic reference temperature is 298 K. An example of a formation reaction is

$$\tfrac{1}{2}H_2(g) + \tfrac{1}{2}Cl_2(g) \rightarrow HCl(g) \qquad \Delta_f H^o = -92.3 \text{ kJ}$$

The reference states for hydrogen and chlorine, respectively, are H_2 and Cl_2, not H and Cl. Notice that, according to the definition, we have to write only **one mole** of product. To do so, we have to use fractional coefficients for the reactants.

By definition, $\Delta_f H^o$ of the most stable form of any element is **zero.** For example, Cu(s), $O_2(g)$, and Hg(l) all have a standard enthalpy of formation of zero.

Example 9   Using $\Delta_f H^o$

Using the equation above for the formation of HCl, how much heat would be released if 250. g of HCl was formed?

First Thoughts

Remember that the definition of a formation reaction requires that one mole of product be formed. In this case, 92.3 kJ is released for every mole of HCl formed.

Solution

$$(250. \text{ g HCl})\left(\frac{1 \text{ mol HCl}}{36.5 \text{ g HCl}}\right) = 6.85 \text{ mol HCl}$$

$$(6.85 \text{ mol HCl})\left(\frac{92.3 \text{ kJ}}{1 \text{ mol HCl}}\right) = 632 \text{ kJ}$$

5.5   Hess's Law

Enthalpy is a **state function,** which means that the enthalpy change for a chemical reaction does not depend on the path by which the products were obtained. What is important are the initial and final states. Hess realized that it does not matter if the ΔH for a reaction is calculated from one step or from a series of steps.

If we have a series of reactions with known ΔH values, we can manipulate them algebraically to find the value of a reaction whose ΔH is unknown:

1. We can reverse the entire equation—the ΔH then has to be multiplied by -1.

2. We can multiply the entire equation by a factor like 2, 3, or ½— the same thing has to be done to ΔH.

3. We can add the individual equations to get the overall equation—the same thing has to be done to ΔH.

An example will make this procedure clearer.

Example 10   Hess's Law

Calculate ΔH for the reaction

$$C_2H_2(g) + 5/2O_2(g) \rightarrow 2CO_2(g) + H_2O(l)$$

given the following information:

$2C(s) + H_2(g) \rightarrow C_2H_2(g)$	$\Delta H = 226.8$ kJ
$C(s) + O_2(g) \rightarrow CO_2(g)$	$\Delta H = -393.5$ kJ
$H_2(g) + \frac{1}{2}O_2(g) \rightarrow H_2O(l)$	$\Delta H = -285.8$ kJ

First Thoughts

Looking at the equation that we want, we realize that we have to reverse the first equation to get the C_2H_2 over to the reactant side. We must multiply the second equation by 2 to get $2CO_2$. Finally, we have to add the third equation to the other two equations. Remember to do the same things to the ΔH's.

Solution

$C_2H_2(g) \rightarrow 2C(s) + H_2(g)$	$\Delta H = -226.8$ kJ
$2C(s) + 2O_2(g) \rightarrow 2CO_2(g)$	$\Delta H = 2(-393.5$ kJ$)$
$H_2(g) + \frac{1}{2}O_2(g) \rightarrow H_2O(l)$	$\Delta H = -285.8$ kJ

$C_2H_2(g) + 5/2O_2(g) \rightarrow 2CO_2(g) + H_2O(l)$	$\Delta H = -1299.6$ kJ

If table values of standard enthalpies of formation are available, the **enthalpy of reaction,** $\Delta H°_{reaction}$ can be calculated by taking the **sum** of the $\Delta_f H°$ values of the products and subtracting the **sum** of the $\Delta_f H°$ values of the reactants.

$$\Delta H°\text{reaction} = \sum n_p \Delta_f H°\text{ (products)} - \sum n_r \Delta_f H° \text{ (reactants)}$$

n_p = the number of moles of each product in the reaction equation
n_r = the number of moles of each reactant in the reaction equation

Example 11 $\Delta H°_{reaction}$

Calculate $\Delta H°_{reaction}$ for the reaction:

$$C_4H_{10}(l) + 9/2O_2(g) \rightarrow 2CO_2(g) + 5H_2O(l)$$

using the table of standard enthalpies of formation in your text.

First Thoughts

The table gives $\Delta_f H°$ in units of kJ/mol. The value for each compound must be multiplied by the number of moles of either reactant or product. The value for O_2 is zero, as it is an element in its reference state.

Solution

$$\Delta H°\text{reaction} = \sum n_p \Delta_f H° (\text{products}) - \sum n_r \Delta_f H° (\text{reactants})$$

$$\Delta H°\text{reaction} = \left[2(-393.5 \text{ kJ}) + 5(-285.8) \right] - \left[-147.6 + 9/2(0) \right]$$

$$= 2068.4 \text{ kJ}$$

5.6 Energy Choices

This last section of the chapter raises some important issues that you should think about.

1. Fossil fuels account for about 70% of our energy needs. Will they last forever?

2. What effect does burning fossil fuel have on the problem of "global warming"?

3. What is the impact of burning fossil fuels on the environment?

4. Renewable sources are an attractive option. They include solar cells, solar thermal systems, biomass conversion, hydroelectric systems, wind power, and geothermal systems. Will they make an impact in our future energy needs?

Exercises

Section 5.1

1. Define kinetic and potential energy.

2. If a basic solution is added to a beaker containing an acidic solution, what is the system and what are the surroundings?

3. In Exercise 2, the acid was neutralized by the base. The beaker feels hot to the touch. In terms of the system and the surroundings, which way is the heat flowing?

4. The reaction of aluminum powder with iron oxide gives off heat and light and forms molten iron. Is this reaction exothermic or endothermic?

5. People often use a "cool pack" on an injury. This plastic bag contains a liquid and a solid that are mixed when the seal is broken. Once the solid dissolves, the bag becomes cold. Is this an exothermic or endothermic process?

6. An electron has a mass of 9.11×10^{-31} kg. Calculate the kinetic energy (in joules) of an electron traveling at the speed of light, 3.00×10^8 m/s.

7. If a pitcher throws a baseball weighing 143 g at 75.0 mph (33.5 m/s), what is the kinetic energy of the baseball (in joules)?

8. Calculate the amount of work (in joules) necessary to compress a gas from 75.0 L to 13.0 L at a constant external pressure of 10.5 atm.

9. How much work (in joules) is done by the expansion of a gas from 0.50 L to 20.00 L at an external pressure of 0.85 atm?

10. What is the change in internal energy of a system if 163.5 J of work is done on the system as it gains 99.4 J of heat?

Section 5.2

11. If your diet allows for 2200. Cal per day, how many joules is this?

12. If a 5.00 kg cannon ball is traveling with a kinetic energy of 1250 J, how fast is it traveling?

13. If a gaseous system has 1.25×10^3 J of energy, how many kilocalories is this?

14. You exercise and burn 120. Cal/h. How many joules per second is this?

Section 5.3

15. Explain the difference between specific heat capacity and molar heat capacity.

16. The specific heat of $CO_2(g)$ is 0.843 J/g · °C and that of $Cl_2(g)$ is 0.478 J/g · °C. If the 10.0 J of heat is transferred to 1.00 g of each gas, and both are at the same initial temperature of 20.0°C, which will have the greater final temperature?

17. How much heat is needed to raise 28.00 g of water (specific heat, 4.184 J/g · °C) from 25.00°C to 26.48°C?

18. The specific heat of a metal is 0.467 J/g · °C. How much energy is released when a 35.0 g piece of this metal drops in temperature from 353.3°C to 297.4°C?

19. If 2.17 kJ of heat is used to warm 2.0×10^2 g of air initially at 20.0°C, calculate the final temperature of the air (the specific heat of air is 1.0 J/g · °C).

20. A 47.00 g sample of iron absorbs 914 J of heat, causing a temperature rise from 28.34°C to 71.75°C. Determine the specific heat of iron.

21. The specific heat of gold is 0.132 J/g · °C. A sample of gold absorbs 3.59 kJ of heat. The sample temperature rises from 24.50°C to 100.0°C. How many grams of gold are in the sample?

22. A student heated 119 g of an unknown metal to a temperature of 86.6°C., then dropped the metal into a Styrofoam calorimeter containing 390 g of water at 23.2°C. The final temperature in the calorimeter was 24.3°C. What is the specific heat of the metal?

23. The reaction of 43.60 g of a compound in a bomb calorimeter whose heat capacity is 952.0 J/°C causes a temperature rise from 21.34°C to 32.75°C. What is the heat of this reaction in units of J/g?

24. A bomb calorimeter has a heat capacity of 32.5 kJ/°C. Combustion of 20.0 g of graphite in this calorimeter resulted in a temperature change from 25.2°C to 57.2°C. Determine the heat of combustion of the sample in units of kJ/mol.

Section 5.4

25. What are the "standard states" used to define the enthalpies of formation?

26. What is the value of the standard enthalpy of formation of the stable form of any element?

27. In an endothermic reaction, will the system gain or lose heat energy?

28. If the enthalpy of a forward reaction is +10 kJ, what will the enthalpy of the reverse reaction be?

29. The $\Delta_f H°$ for $CO_2(g)$ is −393.5 kJ/mol. What is the formation reaction for CO_2?

    a) $C(s) + O_2(g) \rightarrow CO_2(g)$

    b) $2C(s) + 2O_2(g) \rightarrow CO_2(g)$

    c) $CO(g) + \frac{1}{2}O_2(g) \rightarrow CO_2(g)$

    d) $C(g) + \frac{1}{2}O_2(g) \rightarrow CO_2(g)$

30. The standard enthalpy of formation for Fe_2O_3 is -824 kJ/mol. How much heat is evolved when 332 g of Fe_2O_3 is formed?

31. Natural gas (methane, CH_4) has $\Delta_f H° = -74.8$ kJ/mol. How much heat is evolved from burning 1.00 g of methane?

Section 5.5

32. The ΔH for the reaction $C(s) + 2H_2(g) \rightarrow CH_4(g)$ is -74.8 kJ. What is ΔH for the following reaction:

$$3CH_4(g) \rightarrow 6H_2(g) + 3C(s)$$

33. Calculate the enthalpy for the reaction $P_4O_6(s) + 2O_2(g) \rightarrow P_4O_{10}(s)$, given the following information:

$P_4(s) + 3O_2(g) \rightarrow P_4O_6(s)$	$\Delta H = -1640.1$ kJ
$P_4(s) + 5O_2(g) \rightarrow P_4O_{10}(s)$	$\Delta H = -2940.1$ kJ

34. Calculate the enthalpy for the reaction $C_2H_4(g) + 6F_2(g) \rightarrow 2CF_4(g) + 4HF(g)$, given the following information:

$H_2(g) + F_2(g) \rightarrow 2HF(g)$	$\Delta H = -537$ kJ
$C(s) + 2F_2(g) \rightarrow CF_4(g)$	$\Delta H = -680.$ kJ
$2C(s) + 2H_2(g) \rightarrow C_2H_4(g)$	$\Delta H = -52.3$ kJ

35. Calculate the enthalpy for the reaction $N_2O(g) + NO_2(g) \rightarrow 3NO(g)$, given the following information:

$N_2(g) + O_2(g) \rightarrow 2NO(g)$	$\Delta H = -180.7$ kJ
$2NO(g) + O_2(g) \rightarrow 2NO_2(g)$	$\Delta H = -113.1$ kJ
$2N_2O(g) \rightarrow 2N_2(g) + O_2(g)$	$\Delta H = -163.2$ kJ

In Exercises 36–40, calculate $\Delta H°_{rxn}$ for the reactions using the standard heats of formation from Appendix A-3 in the textbook.

36. $2PbS(s) + 3O_2(g) \rightarrow 2SO_2(g) + 2PbO(s, yellow)$

37. $N_2O_4(g) + 4H_2(g) \rightarrow N_2(g) + 4H_2O(g)$

38. $CS_2(l) + 3O_2(g) \rightarrow CO_2(g) + 2SO_2(g)$

39. $4NH_3(g) + 5O_2(g) \rightarrow 4NO(g) + 6H_2O(g)$

40. $Fe_2O_3(s) + 3CO(g) \rightarrow 2Fe(s) + 3CO_2(g)$

41. Given the reaction $2Cu_2O(s) + O_2(g) \rightarrow 4CuO(s)$; $\Delta H°_{rxn} = -284$ kJ. Calculate $\Delta_f H°$ for $Cu_2O(s)$.

42. Given the reaction $CaC_2(s) + 2H_2O(l) \rightarrow Ca(OH)_2(s) + C_2H_2(g)$; $\Delta H^\circ_{rxn} = -125$ kJ. Calculate $\Delta_f H^\circ$ for $CaC_2(s)$.

Section 5.6

43. What are fossil fuels? Give some examples.

44. Why is the development of renewable resources important to future energy needs?

45. What would be the economic impact of fossil fuels becoming even more scarce?

Answers to Exercises

1. Kinetic energy is the energy of motion; it is equal to $\frac{1}{2}mv^2$. Potential energy is the energy of position.

2. The solutions and the beaker are the system. The surroundings are everything outside the beaker.

3. The heat (neutralization reactions evolve heat) flows from the system to the surroundings.

4. exothermic

5. endothermic

6. 4.10×10^{-14} J

7. 80.2 J

8. 6.59×10^4 J (compression results in positive work)

9. -1.7×10^3 J (expansion results in negative work)

10. 262.9 J

11. 9.205×10^6 J

12. 22.4 m/s

13. 0.299 kcal

14. 139 J/s

15. Specific heat capacity is the amount of heat necessary to raise the temperature of one **gram** of substance by one degree Celsius. Molar heat capacity is the amount of heat necessary to raise the temperature of one **mole** of substance by one degree Celsius.

16. The final temperature of Cl_2 will be greater because it has the smaller specific heat.

17. 173 J

18. -914 J

19. 31°C

20. 0.448 J/g $\cdot$ $^\circ$C

21. 360. g

22. 0.24 J/g $\cdot$ $^\circ$C

23. 249.1 J/g

24. 625 kJ/mol

25. Standard states: (a) for a pure solid, liquid, or gas—the state of the substance at a pressure of 1 atm; (b) for any substance in solution—a concentration of 1 M.

26. zero

27. An endothermic reaction will gain energy.

28. −10 kJ

29. d

30. −1.71 × 10^3 kJ

31. −55.6 kJ

32. 224.4 kJ

33. −1300.0 kJ

34. −2382 kJ

35. −205.8 kJ

36. −828 kJ

37. −977 kJ

38. −1078 kJ

39. −908 kJ

40. −23 kJ

41. −170 kJ

42. −63 kJ

43. Fossil fuels result when plants and animals that died millions of years ago form deposits of oil, natural gas, and coal.

44. Eventually fossil fuels will be used up and will not be replaced. Renewable resources can be used and replaced. Wood is an example.

45. Our economy depends greatly on abundant energy. If fuels become scarce, then prices will rise dramatically, forcing the price of goods and services to increase as well.

Chapter 6

Quantum Chemistry

The Bottom Line

When you finish this chapter, you will be able to:

- Apply and relate the concepts of wavelength, frequency, and energy of electromagnetic radiation.
- Explain and solve problems involving the concept of wave–particle duality, where atoms demonstrate both wave and particle-like behavior at the atomic scale.
- Predict and justify orbital diagrams, quantum numbers, and electron configurations.

6.1   Introducing Quantum Chemistry

Quantum chemistry allows us to probe the behavior of atoms and molecules. Studying the electromagnetic spectrum gives us information about the electronic structure of atoms. From this information, we can understand why atoms are found in groups in the periodic table and predict chemical properties such as size and ionization energy.

6.2   Electromagnetic Radiation

In this section you will learn to convert between wavelength, frequency, and the energy of electromagnetic radiation. The definitions for frequency, wavelength, and the energy associated with electromagnetic radiation are given in **Section 6.2** of your text. Make sure that you learn them. You should also know the different regions of the electromagnetic spectrum and the wavelengths covered by them. For example, visible light is composed of light colored red, orange, yellow, green, blue, indigo, and violet, whose wavelengths range from about 400 to 700 nm. The relationship between frequency, ν, and wavelength, λ is:

$$\nu = c/\lambda$$

Wavelength (λ) is in meters.
Frequency (ν) is in s^{-1} or Hz.
Speed of light (c) is 3.00×10^8 m/s.

Example 1   **Wavelength–Frequency Conversion**

What is the frequency of a green light having a wavelength of 520 nm?

First Thoughts

The relationship of frequency to wavelength is $\nu = c/\lambda$. Remember that the speed of light (c) is 3.00×10^8 m/s, so the wavelength in nanometers must be converted to meters so that the meter units cancel and you are left with s^{-1}, the unit for frequency. This is how you convert the wavelength from nanometers to meters:

$$520\,\text{nm}\left(\frac{1\,\text{m}}{1\times10^9\,\text{nm}}\right) = 5.20\times10^{-7}\,\text{m}$$

Solution

Using the speed of light and the wavelength in meters, calculate the frequency:

$$\nu = c/\lambda$$

$$\nu = \frac{3.00 \times 10^8 \, \text{m/s}}{5.20 \times 10^{-7} \, \text{m}}$$

$$= 5.77 \times 10^{14} \, \text{s}^{-1}$$

Further Insights

You could have also solved the problem by leaving the wavelength in nanometers (as given in the problem) and converting the speed of light from m/s to nm/s. You would then again use the relationship $\nu = c/\lambda$ to get your final answer. Either way, the units properly cancel so that the frequency is in s^{-1}. Furthermore, these data can be used to solve for the energy of a given wavelength as discussed below.

The energy associated with a particular wavelength of light or a certain frequency is given by

$$E = \frac{hc}{\lambda} = h\nu$$

E is the energy in joules of each photon of light, and h is Planck's constant, 6.626×10^{-34} J·s.

Example 2   The Energy of Electromagnetic Radiation

Ultraviolet radiation that contributes to sunburn has a wavelength of 1.0×10^{-8} m. What is the energy of a photon of ultraviolet radiation?

First Thoughts

We can quickly relate the energy to the wavelength by using the equation given above. Given that the wavelength is already in meters, we can substitute the value directly into the energy equation, looking up Planck's constant, 6.626×10^{-34} J·s (for h), and the speed of light, 3.00×10^8 m/s (for c).

Solution

$$E = \frac{hc}{\lambda} = \frac{\left(6.626 \times 10^{-34} \, \text{J} \cdot \text{s}\right)\left(3.00 \times 10^8 \, \text{m/s}\right)}{1.0 \times 10^{-8} \, \text{m}}$$

$$E = 2.0 \times 10^{-17} \, \text{J}$$

Further Insights

You could have solved the problem by first solving for the frequency of the ultraviolet wave, using the equation $\nu = c/\lambda$, and then using $E = h\nu$ to solve for the energy. This method requires two steps, instead of only one.

$$\nu = c/\lambda$$

$$\nu = \frac{3.00 \times 10^8 \, \text{m/s}}{1.0 \times 10^{-8} \, \text{m}}$$

$$\nu = 3.0 \times 10^{16} \, \text{s}^{-1}$$

Using the frequency, solve for the energy.

$$E = h\nu$$
$$E = (6.626 \times 10^{-34}\ \text{J} \cdot \text{s})(3.0 \times 10^{16}\ \text{s}^{-1})$$
$$E = 2.0 \times 10^{-17}\ \text{J}$$

6.3 Atomic Emission Spectroscopy and Absorption Spectroscopy, Chemical Analysis, and the Quantum Number

Emission spectroscopy results from the emission of electromagnetic radiation by hydrogen atoms that are excited by energy. The light emitted by the hydrogen atoms when passed through a prism is separated into distinct wavelengths.

The emission spectrum of the hydrogen atom is separated into four series as summarized in the table below. Notice that n_f determines which series the wavelengths fall into. The wavelengths of the hydrogen spectrum in the various regions of the electromagnetic spectrum can be calculated from

$$\frac{1}{\lambda} = 1.0968 \times 10^{-2} \left[\frac{1}{n_f^2} - \frac{1}{n_i^2} \right] \text{nm}^{-1}$$

Spectral Series	n_f	n_i	Wavelength range
Lyman	1	2, 3, 4, …	X-ray and UV
Balmer	2	3, 4, 5, …	Visible
Ritz-Paschen	3	4, 5, 6, …	Short-wave infrared
Pfund	4	5, 6, 7, …	Long-wave infrared

Example 3   Using the Balmer Equation

The shortest wavelength for the Balmer series is found when $n = 6$:

$$\frac{1}{\lambda} = 1.0968 \times 10^{-2} \left[\frac{1}{2^2} - \frac{1}{6^2} \right] \text{nm}^{-1}$$

$$= 410.2\ \text{nm}$$

(Note that this wavelength falls within the visible range on the electromagnetic spectrum, which corresponds to the table above.)

What is the energy of this radiation?

Solution

The wavelength is calculated in nanometers, so you must convert to meters using $1\ \text{m} = 1 \times 10^9\ \text{nm}$ and then substitute this value into the energy equation:

$$E = \frac{hc}{\lambda} = \frac{\left(6.626 \times 10^{-34} \text{J} \cdot \text{s}\right)\left(3.00 \times 10^8 \text{m/s}\right)}{4.102 \times 10^{-7} \text{m}}$$

$$E = 4.85 \times 10^{-19}\ \text{J}$$

6.4 The Bohr Model of Atomic Structure

In this section, you will learn to calculate the energy, wavelengths, and frequencies of the electronic transitions in the hydrogen atom. Bohr postulated that the energy that an electron has when it occupies a particular orbit is given by this equation:

$$E_n = -\frac{2.1786 \times 10^{-18} \text{ J}}{n^2}$$

where the quantum number $n = 1, 2, 3, 4, \ldots$.

For hydrogen, the orbit closest to the nucleus is $n = 1$. This lowest energy level is called the **ground state.** If the electron moves to a higher energy level, it occupies an **excited state.** To make this electronic transition from a lower energy level to a higher energy level, the electron must **absorb** energy. If the electron moves from a higher energy level to a lower energy level, it **emits** energy. The change in energy, ΔE, is given by

$$\Delta E = E_f - E_i$$

where E_f is the final energy of the electron and E_i is the initial energy of the electron. To calculate ΔE in joules, we can use the following equation:

$$\Delta E = -2.1786 \times 10^{-18} \text{ J} \left[\frac{1}{n_f^2} - \frac{1}{n_i^2} \right]$$

Remembering that $\Delta E = hc/\lambda$, we can use the equation stated in Section 6.3, which allows us to calculate the wavelength in nanometers directly:

$$\frac{1}{\lambda} = 1.097 \times 10^{-2} \left[\frac{1}{n_f^2} - \frac{1}{n_i^2} \right] \text{nm}^{-1}$$

Example 4 **Calculating Energy, Wavelengths and Frequencies of Transitions**

What is the wavelength of the light emitted from the hydrogen atom when the electron undergoes a transition from $n = 4$ to $n = 1$? Calculate the change in energy for this transition.

First Thoughts

Since the electron is moving from an excited state to the ground state, energy is released. Therefore, the calculated change in energy should be a negative value. Before calculating the energy, though, you must first determine the wavelength emitted.

Solution

We can calculate the wavelength in nanometers directly:

$$\frac{1}{\lambda} = 1.097 \times 10^{-2} \left[\frac{1}{n_f^2} - \frac{1}{n_i^2} \right] \text{nm}^{-1}$$

$$\frac{1}{\lambda} = 1.097 \times 10^{-2} \left[\frac{1}{1^2} - \frac{1}{4^2} \right] \text{nm}^{-1}$$

$$\lambda = 97.23 \text{ nm}$$

Once we know the wavelength, we can calculate the change in energy:

$$E = \frac{hc}{\lambda}$$

$$E = \frac{(6.626 \times 10^{-34} \text{ J} \cdot \text{s})(3.00 \times 10^8 \text{ m/s})}{97.23 \text{ nm} \left(\dfrac{1 \text{ m}}{1 \times 10^9 \text{nm}} \right)}$$

$$E_{photon} = 2.044 \times 10^{-18} \text{ J}$$

$$E_{photon} = -\Delta E_{electron}$$

$$\Delta E_{electron} = -2.04 \times 10^{-18} \text{ J}$$

Further Insights

Once we have calculated the wavelength or energy, we can also determine the frequency:

$$\nu = \frac{c}{\lambda}$$

$$\nu = \frac{3.00 \times 10^8 \text{ m/s}}{97.23 \text{ nm} \left(\dfrac{1 \text{ m}}{1 \times 10^9 \text{ nm}} \right)}$$

$$= 3.09 \times 10^{15} \text{ s}^{-1}$$

6.5   Wave–Particle Duality

Read the discussion in your text about classical mechanics, and study how it views particles and waves differently. According to classical mechanics, particles have exact locations and the positions can be defined precisely in space. A particle has momentum, p, given by the product of its mass, m, and velocity, v:

$$p = m \times v$$

The idea behind wave–particle duality is that very small, light-weight particles, such as electrons, protons, and photons of electromagnetic radiation, exhibit wave-like behavior.

6.6   Why Treating Things as "Waves" Allows Us to Quantize Their Behavior

If electrons behave like waves, we can find their wavelength by using the de Broglie equation:

$$\lambda = \frac{h}{mv}$$

where λ is the wavelength, h is Planck's constant, m is the mass, and v is the velocity. This equation relates the particle-like properties of an electron (due to the fact that the electron has mass) to its wave-like properties.

Example 5   **Calculating the de Broglie Wavelength**

What is the wavelength of an electron traveling at a velocity of 5.31×10^6 m/s?

Solution

We look up the mass of the electron (9.11×10^{-31} kg) and the Planck's constant (6.626×10^{-34} J·s). Remembering that 1 J = 1 kg m^2/s^2, we can substitute these values into the de Broglie wavelength equation:

$$\lambda = \frac{h}{mv}$$

$$\lambda = \frac{6.626 \times 10^{-34} \text{ J·s}}{9.11 \times 10^{-31} \text{kg} \left(5.31 \times 10^{6} \text{m/s}\right)}$$

$$= 1.37 \times 10^{-10} \text{ m or } 0.137 \text{ nm}$$

6.7 The Heisenberg Uncertainty Principle

The Heisenberg uncertainty principle states that there is a limit to just how precisely one can know both the position and momentum of a particle at a particular time. Expressed mathematically:

$$\Delta x \Delta p_x \geq \frac{h}{4\pi}$$

where Δx is the uncertainty in the position, Δp_x is the uncertainty in the momentum in the x direction, and h is Planck's constant. *This principle tells us that we cannot know exactly where an electron is around an atom at any given time.*

6.8 More About the Photon: The de Broglie and Heisenberg Discussions

Light is electromagnetic radiation that is a wave. The smallest unit of light is the photon, which can have momentum (a particle-like characteristic), illustrating again the wave–particle duality. Again using the de Broglie equation, we can calculate the momentum of a wave.

Example 6   Photons and Momentum

What is the momentum of a mole of photons with a wavelength of 420 nm?

Solution

Use the de Broglie equation, $\lambda = h/p$, to calculate the momentum of one photon:

$$p = \frac{h}{\lambda} = \frac{6.626 \times 10^{-34} \text{ J·s}}{4.20 \times 10^{-7} \text{ m}} = 1.58 \times 10^{-27} \text{ kg·m·s}^{-1}$$

Then, for a mole of photons, multiply by Avogadro's number:

$$p = \frac{1.58 \times 10^{-27} \text{ kg·m}}{\text{s·photon}} \times \frac{6.022 \times 10^{23} \text{ photons}}{\text{mol}} = \frac{9.51 \times 10^{-4} \text{ kg·m}}{\text{s·mol}}$$

6.9 The Mathematical Language of Quantum Chemistry

Quantum chemistry uses mathematics to describe the wave properties of submicroscopic particles. You should no longer think about an electron as having a precise orbit about the nucleus of an atom, because we cannot know its exact position or momentum as it moves about the nucleus. The

Schrödinger equation allows us to calculate the energy that an electron would have available to it in the atom. The wave function, Ψ, describes the path and position of the electron in its particular energy level. More importantly, to describe where we might find the electron in the space about the nucleus, we use the square of the wave function, Ψ^2 (or $\Psi^*\Psi$), which is the **probability** of finding the electron in a region of space. So, the electron does not exist in a single point in space but rather is more likely to be found in some places (high probability) than in others (low probability). We usually consider 90% of the probability of finding an electron in a certain region of space to be an **orbital.**

6.10   Atomic Orbitals: Quantum Numbers

Four quantum numbers are used to describe the electron in the hydrogen atom:

n: principal quantum number—it describes the size and energy of the orbital.
$\quad\quad n = 1, 2, 3, 4, \ldots$ (only integers)

l: angular momentum quantum number—describes the shape of the orbital.
$\quad\quad l = 0$ to $n - 1$ (only integers)

m_l: magnetic quantum number—describes the spatial orientation of the orbital.
$\quad\quad m_l = -l$ to 0 to $+l$ (only integers)

m_s: spin quantum number—describes the direction of spin of the electron.
$\quad\quad m_s = +\frac{1}{2}$ or $-\frac{1}{2}$ (only two possible values)

Example 7   **Quantum Numbers**

If $n = 2$, what are the possible values of l, m_l, and m_s? How many different quantum numbers are possible?

First Thoughts

If $n = 2$, then there are two values for l: $l = 0$ and $l = 1$.
For $l = 0$, $m_l = 0$.
For $l = 1$, $m_l = -1, 0$, or $+1$.
For each m_l value, $m_s = +\frac{1}{2}$ and $m_s = -\frac{1}{2}$.

Solution

Let's summarize the possibilities for the position of the electron in a table:

n	l	m_l	m_s
2	0	0	$+\frac{1}{2}$
2	0	0	$-\frac{1}{2}$
2	1	1	$+\frac{1}{2}$
2	1	1	$-\frac{1}{2}$
2	1	0	$+\frac{1}{2}$
2	1	0	$-\frac{1}{2}$
2	1	-1	$+\frac{1}{2}$
2	1	-1	$-\frac{1}{2}$

Therefore, eight different quantum numbers are possible.

Further Insights

Try $n = 3$ and see if you can get the 18 quantum numbers. The best approach is to create a table like the one above to determine all of the possibilities.

Example 8

Which of the following sets of quantum numbers are allowed. If not, why not?

	n	l	m_l	m_s
a)	3	3	0	$-\frac{1}{2}$
b)	4	1	1	$+\frac{1}{2}$
c)	2	1	1	$-\frac{1}{2}$
d)	5	-4	2	$+\frac{1}{2}$
e)	2	0	0	1

Solution

a)   Not allowed. l cannot equal n.
b)   Allowed.
c)   Allowed.
d)   Not allowed. l cannot be negative.
e)   Not allowed. m_s must be $+\frac{1}{2}$ or $-\frac{1}{2}$.

The numerical values of l are also given letter symbols. You should memorize the following:

l Value	Orbital Name
0	*s*
1	*p*
2	*d*
3	*f*

The $l = 0$ value represents an *s* orbital; since $m_l = 0$, it can have only one orientation in space, that of a sphere. The $l = 1$ value represents a *p* orbital; since $m_l = +1, 0, -1$, three *p* orbitals are possible (oriented along the *x*-, *y*-, and *z*-axes, respectively). For *d* orbitals ($l = 2$), $m_l = +2, +1, 0, -1, -2$; thus there are 5 *d* orbitals. Refer to your text for visual representations of the *s, p, d,* and *f* orbitals.

6.11   Electron Spin and the Pauli Exclusion Principle

An electron spins about an axis. Because there are only two possible spin directions, clockwise and counterclockwise, there are only two possible values of m_s, $+\frac{1}{2}$ and $-\frac{1}{2}$. The **Pauli exclusion principle** states that no two electrons can have the same four quantum numbers. (Verify this in the table above for $n = 2$.) As a result, an electron can have the same three quantum numbers, n, l, and m_l, but it must have a different m_s. In other words, only two electrons can occupy an orbital and only if they have opposite spins.

6.12   Orbitals and Energy Levels in Multielectron Atoms

Atoms with more than one electron are complicated by the fact that electrons, having the same charge, repel one another. In such cases, the energies and the orbital wave functions can no longer be solved exactly. Using approximation techniques, however, we can derive shapes for the orbitals in multielectron atoms.

Due to electron shielding, the energy levels for multielectron atoms are quite different from those for the hydrogen atom. **Figure 6.39** in the textbook provides a diagram of the energy levels for multielectron atoms.

6.13   Electron Configurations and the Aufbau Principle

Review the section of your text explaining the aufbau principle. The text illustrates the electron configurations for the first ten elements, hydrogen through neon. One method of finding the order in which the electrons fill the orbitals as the atomic number increases is shown in **Figure 6.47** of the textbook. To determine the order of filling, draw arrows from the upper right to the lower left.

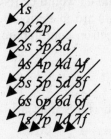

The resulting pattern is $1s\ 2s\ 2p\ 3s\ 3p\ 4s\ 3d\ 4p\ 5s\ 4d\ 5p\ 6s\ 4f\ 5d\ 6p\ 7s$. . . . Now recall the maximum number of electrons that each orbital can hold (s can hold 2, p can hold 6, d can hold 10, f can hold 14). Alternatively, carefully review the periodic table in **Figure 6.48** of the text, which shows how the table can be broken down into regions. Practice writing electron configurations using the table so that you can gain an appreciation of how the elements are arranged and see the patterns. This is the preferred (and easier way) of writing these configurations.

Example 9   Using the Aufbau Principle

Try using one of the two methods described above to determine the electron configurations for Na ($Z = 11$) and C ($Z = 6$).

First Thoughts

Locate each element on the periodic table. Sodium (Na) is in Period 3, Group 1A, and is an alkali metal. Carbon (C) is in Period 2, Group 4A. Based on each element's location on the periodic table, determine its electron configuration.

Solution

$$\text{Na:} \quad 1s^22s^22p^63s^1 \qquad\qquad \text{C: } 1s^22s^22p^2$$

Further Insights

An orbital diagram shows the electron configuration pictorially for sodium:

Na: $(\uparrow\downarrow)\ (\uparrow\downarrow)\quad (\uparrow\downarrow)\ (\uparrow\downarrow)\ (\uparrow\downarrow)\quad (\uparrow\ )$
$\quad\ \ \ 1s \quad\ \ 2s \qquad\qquad 2p \qquad\qquad 3s$

For carbon:

C: $(\uparrow\downarrow)\ (\uparrow\downarrow)\quad (\uparrow\ )\ (\uparrow\ )\ (\ \ )$
$\quad\ 1s \quad\ \ 2s \qquad\quad 2p$

   Notice that the two $2p$ electrons are placed in different orbitals. This is an example of **Hund's rule,** which states that you want to maximize the number of unpaired electrons as long as the orbitals have *equal energy*. In this case, the three different $2p$ orbital orientations have the same energy, so the electrons will spread out between the three orbitals instead of pairing up. Also notice that the two electrons are designated by two up arrows, meaning that they have parallel spins. This combination of spreading the electrons out with the same spins provides the lowest-energy configuration.

Another way of representing an electron configuration is to use the configuration of the noble gases to represent the core electrons. Here are examples of the "cores":

$[He] = 1s^2$
$[Ne] = 1s^2 2s^2 2p^6$
$[Ar] = 1s^2 2s^2 2p^6 3s^2 3p^6$

Thus Na would be $[Ne]3s^1$, C would be $[He]2s^2 2p^2$, and Cl would be $[Ne]3s^2 3p^5$. Notice that Na is in Group 1A and Cl is in Group 7A, and both elements are in Period 3. All of the members of the representative groups have the same outer orbital electron configuration: s^1 for members of Group 1A, s^2 for members of Group 2A, and so on.

Example 10

Use the periodic table to write the electron configuration for S and Ca.

Solution

S ($Z = 16$) is in Period 3, Group 6A: $[Ne]3s^2 3p^4$. We use the noble gas before the element—in this case, neon, which counts for 10 electrons, followed by 6 electrons (2 in the s orbital, 4 in the p orbital).

Ca ($Z = 20$) is in Period 4, Group 2A: $[Ar]4s^2$.

But what about the transition metals in the nonrepresentative "B" groups? Transition metals are "d fillers." They fill the $3d$ and $4d$ orbitals. The $3d$'s are in the fourth period, while the $4d$'s are in the fifth period. The transition metals fill in between the s orbitals (Groups 1A and 2A) and the p orbitals (Groups 3A to 8A). The transition metals also fill the f orbitals on up once you get to Period 6 on the periodic table.

Example 11

Write the electron configuration for vanadium, rhenium, and the stable iodine ion, I^-.

Solution

V ($Z = 25$) is a transition metal in the fourth period: $[Ar]4s^2 3d^3$. Argon counts for 18 electrons, then 2 electrons in the $4s$ orbital, and the final 3 electrons in the $3d$ orbital (remember that the d-orbital coefficient is always one less than the period number).

Re ($Z = 75$) is a transition metal in the sixth period: $[Xe]6s^2 4f^{14} 5d^5$. Xenon counts for 54 electrons, then 2 electrons in the $6s$ orbital, 14 electrons in the $4f$ orbital, and the final 5 electrons in the $5d$ orbital (remember that the d-orbital coefficient is always one less than the period number, and the f-orbital coefficient is always two less than the period number).

I ($Z = 53$) is in the fifth period: $[Kr]5s^2 4d^{10} 5p^5$ (notice that the $4d$ electrons come between the $5s$ and $5p$ orbitals). But the ion I^- means that there is one extra electron added to the neutral iodine atom, so you have to add one more electron to your electron configuration: $[Kr]5s^2 4d^{10} 5p^6$.

Exercises

Section 6.2

1. Convert 332 nanometers to meters.

2. An electromagnetic wave has a wavelength of 1.042×10^{-9} km. What is this wavelength in nanometers?

3. What are the differences between amplitude, wavelength, and frequency?

4. Calculate the frequency of light of wavelength 525 nm.

5. An FM radio station broadcasts its signal at 88.5 MHz. What is the wavelength of this radio signal?

6. X rays have a wavelength range of 10 nm to 10 pm. What is the frequency range?

7. Calculate the wavelength of blue light (in nanometers) of frequency 6.70×10^{14} Hz.

8. Which form of visible light has a higher frequency—red light with a wavelength of 700 nm or green light with a wavelength of 500 nm?

   a) Red light has a higher frequency because of the direct relationship between wavelength and frequency.
   b) Green light has a higher frequency because the length between waves is smaller, so more wave cycles pass by a given point each second.
   c) Green light has a higher frequency because it must be traveling faster due to its smaller wavelength.
   d) Red light has a higher frequency because it has a higher amount of energy.
   e) This question is impossible to answer because we do not know the speed at which each light wave is traveling.

9. What is the energy of a photon from a radio signal with a frequency of 101.3 MHz?

10. What is the wavelength (in nanometers) of the radio signal from Exercise 9?

11. Blue light has a wavelength of 450 nm. What is the energy of this light per photon?

12. Green light has a frequency of 5.70×10^{14} Hz. What is the energy of this light per mole of photons?

13. Calculate the energy of a photon of red light having a wavelength of 7.00×10^{2} nm.

14. The energy required to ionize gaseous C atoms is 1.08×10^{3} kJ/mol. What is the maximum wavelength of electromagnetic radiation (in nanometers) that can cause the ionization of one atom?

15. When examining the electromagnetic spectrum, why is it more harmful to be exposed to ultraviolet rays than to radio waves over a period of time? Be sure to use the concepts of frequency, waves, and energy in your answer.

Section 6.3

16. The Balmer equation is based on experiments conducted with which atom?

17. In which wavelength region of the spectrum would you look for the radiation associated with the $n = 6$ to $n = 3$ transition of the hydrogen atom?

18. In which wavelength region of the spectrum would you look for the radiation associated with the $n = 5$ to $n = 1$ transition of the hydrogen atom?

Section 6.4

19. How much energy will an electron absorb, if it undergoes an $n = 2$ to $n = 3$ transition?

20. Calculate the wavelength (in nanometers) of light emitted by the electron transition of $n = 5$ to $n = 3$ in the hydrogen atom.

21. Calculate the wavelength (in nanometers) of an electromagnetic wave emitted by the electronic transition of $n = 4$ to $n = 2$ in the hydrogen atom. Why is it different than the wavelength in Exercise 20, even though the electron falls two energy levels in both cases?

22. Which of the following statements are true for a hydrogen electron in the $n = 1$ energy level?

    a) The electron must emit energy to occupy an excited state.
    b) The electron is in the energy level closest to the nucleus.
    c) The electron is in the ground state.
    d) Statements b and c are both true.
    e) Statements a, b, and c are all true.

23. What is the energy and frequency of the electronic transition $n = 5$ to $n = 1$? Is energy absorbed or released? Which is more stable, the electron in the $n = 5$ energy level or in the $n = 1$ energy level? Why?

24. Which emits shorter wavelengths—the electronic transition from $n = 4$ to $n = 3$ or the electronic transition from $n = 4$ to $n = 1$? Explain your answer.

25. An excited hydrogen atom emits radiation with a wavelength of 410.3 nm to reach the energy level $n = 6$. In which principal quantum level did the electron begin?

Section 6.6

26. What is the purpose of the de Broglie equation?

27. What is the wavelength (in nanometers) of a neutron ($m = 1.67 \times 10^{-27}$ kg) traveling at a speed of 4.30 km/s?

28. What is the wavelength (in meters) of a 1.0-lb baseball traveling at a speed of 100. mi/h?

29. Calculate the wavelength (in nanometers) of a car with a mass of 2200. kg moving with a velocity of 1.11×10^2 km/h.

Section 6.10

30. Orbitals that have equal energies are said to be _____ .

31. Which of the following sets of quantum numbers are not allowed? If not, why not?

	n	l	m_l	m_s
a)	1	0	0	1
b)	1	3	+3	½
c)	3	2	+3	−½
d)	1	0	0	−½
e)	2	2	0	½

32. You may recall a typical diagram for the model of the atom showing electrons moving around the nucleus in circular orbits, like planets orbiting the Sun. Why is this model of the atom fundamentally incorrect?

33. How many electrons can occupy

    a)   3d orbitals?
    b)   4f orbitals?
    c)   2p orbitals?

34. Consider the following representation of a 3d orbital:

Which of the following statements best describes the movement of electrons in a 3d orbital?

a)   The electrons move within the four lobes of the d orbital, but never beyond the outside surface of the orbital.
b)   The electrons mostly stay at the center of the four lobes (the node).
c)   The electrons move along the outside surface of the d orbital, similar to a "flower-like" type of movement.
d)   The electron movement cannot be exactly determined.
e)   The electrons are moving in only one lobe at any given time.

35. Transition metals in the fourth and fifth rows of the periodic table are d-orbital fillers. What is the relationship between the row number and the coefficient (energy level) of the d orbitals?

36. Which orbitals are filled by the actinides and the lanthanides?

Section 6.13

37. Write the electron configurations for Si, Co, and Se.

38. Write the electron configurations for Ca, P, and Cl.

39. Choose the best orbital diagram for nitrogen, justify your answer, and explain why the other orbital diagrams are incorrect.

   a)   N: ($\uparrow\downarrow$) ($\uparrow\downarrow$) ($\uparrow$) ($\uparrow$) ($\uparrow$)
          1s   2s   2p

   b)   N: ($\uparrow\downarrow$) ($\uparrow\downarrow$) ($\uparrow\downarrow$) ($\uparrow$) ( )
          1s   2s   2p

   c)   N: ($\uparrow\downarrow$) ($\uparrow$) ($\uparrow\downarrow$) ($\uparrow\downarrow$) ( )
          1s   2s   2p

   d)   N: ($\uparrow\downarrow$) ($\uparrow\downarrow$) ($\uparrow\downarrow$) ($\uparrow$) ($\uparrow$)
          1s   2s   2p

   e)   N: ($\uparrow$) ( ) ($\uparrow\downarrow$) ($\uparrow\downarrow$) ($\uparrow\downarrow$)
          1s   2s   2p

40. Why is the lowest-energy configuration for an atom the one having the maximum number of unpaired electrons within a particular set of degenerate orbitals? In other words, justify Hund's rule.

41. How many unpaired electrons does Co have?

42. How many unpaired electrons does Si have?

43. Use the periodic table to write the electron configurations for the following atoms:

   a) K
   b) Te
   c) P

44. Use the periodic table to write the electron configurations for the following ions:

   a) O^{2-}
   b) Cs^+
   c) P^{3+}
   d) Fe^{2+}

Use the following list of atoms for Exercises 45 and 46:

   Na, Zn, Nb, Cs, O, I, Ga, Pd

45. Which atoms contain electrons that occupy $4d$ orbitals?

46. Which atoms contain at least one unpaired electron?

Use the following list of ions to answer Exercises 47–50:

$$K^+, Cl^-, Ru^+, Mg^{2+}, N^{3-}, Ge^{2+}$$

47. Which ions lost electrons to get from their neutral state to the ionic state?

48. Which ions have a complete valence shell?

49. Which ions contain at least one electron with a spin quantum number of $+\frac{1}{2}$?

50. Which ions contain six $3p$ electrons?

Answers to Exercises

1. 3.32×10^{-7} m

2. 1042 nm

3. The amplitude of a wave is the distance between its highest point and zero. The wavelength is the distance between two consecutive peaks in a wave. The frequency is the number of waves that pass a given point each second.

4. 5.71×10^{14} Hz

5. 3.39 m

6. The range is 3×10^{16} Hz to 3×10^{19} Hz.

7. 448 nm

8. b

9. 6.712×10^{-26} J

10. 2.960×10^{9} nm

11. 4.4×10^{-19} J

12. 2.27×10^{5} J

13. 2.84×10^{-19} J

14. 1.11×10^{2} nm

15. Ultraviolet rays have a shorter wavelength (in the 10^{-8} range) and therefore a higher frequency (since $c = \lambda \nu$). This means that the number of waves that penetrate your body over a period of time will be greater than the number of radio waves that hit your body in that same period of time (since radio waves have a longer wavelength and lower frequency). Because more ultraviolet waves penetrate your body, you are being exposed to more energy, which can then burn your skin.

16. hydrogen

17. short-wave infrared

18. X-ray and ultraviolet

19. 3.03×10^{-19} J

20. 1282 nm

21. 486 nm; The wavelength is different because the wavelength is proportional to the *square* of the principal energy levels.

22. d,

23. $\Delta E = -2.0934 \times 10^{-18}$ J. $\upsilon = 3.159 \times 10^{15}$ Hz. Energy is released. The electron in the $n = 1$ energy level is more stable because it is closer to the nucleus.

24. The transition from $n = 4$ to $n = 1$ emits shorter wavelengths because more energy is released in this process ($\Delta E = hc/\lambda$).

25. $n = 2$

26. It illustrates the basic principle of the wave–particle duality: moving particles exhibit definite wave-like characteristics.

27. 0.0923 nm

28. 3.3×10^{-35} m

29. 9.77×10^{-30} nm (wave properties are too small to observe.)

30. degenerate

31. a) Not allowed. m_s must be $+\frac{1}{2}$ or $-\frac{1}{2}$.
    b) Not allowed. l cannot be equal to or larger than n.
    c) Not allowed. m_l cannot be equal to $+3$; the largest it can be is $+2$.
    d) Allowed.
    e) Not allowed. l cannot equal n.

32. This type of diagram is representative of the Bohr model. We now use the wave mechanical model. Although we still believe that atoms exist in quantized energy levels, we do not know exactly how the electrons move in an atom. We can only predict the probability of finding an electron in a certain location around the nucleus.

33. a) 10
    b) 14
    c) 6

34. d

35. The coefficient is one less ($n - 1$) than the row number. Thus the fourth-row transition metals fill the $3d$ orbitals.

36. They fill the f orbitals.

37. Si: $[Ne]3s^2 3p^2$
    Co: $[Ar]4s^2 3d^7$
    Se: $[Ar]4s^2 3d^{10} 4p^4$

38. Ca: $[Ar]4s^2$

P: $[Ne]3s^23p^3$
Cl: $[Ne]3s^23p^5$

39. Answer a is correct because the electrons fill the s orbitals first and then the p orbitals. It also follows Hund's rule, therefore creating the lowest-energy configuration. Answer b is incorrect because it does not follow Hund's rule. Answers c and e are incorrect because the electrons fill the $1s$ and $2s$ orbitals first before filling up the $2p$ orbitals (lower energy configuration). Answer d has too many electrons (eight instead of seven).

40. Unpaired electrons are favored within a particular set of degenerate orbitals because all electrons have negative charges, so their like charges repel one another. The farther apart the electrons are within a set of degenerate orbitals, the lower the energy.

41. 3

42. 2

43. a) $[Ar]4s^1$
    b) $[Kr]5s^24d^{10}5p^4$
    c) $[Ne]2s^22p^3$

44. a) $[Ne]$
    b) $[Xe]$
    c) $[Ne]3s^2$
    d) $[Ar]3d^6$ (This electron configuration is trickier. The *ions* of the transition metal elements will fill the d orbitals first before filling the outer-level s orbital.)

45. Nb, I, Cs, Pd

46. Na, Nb, Cs, O, I, Ga, Pd

47. K^+, Ru^+, Mg^{2+}, Ge^{2+}

48. K^+, Cl^-, Mg^{2+}, N^{3-}

49. All of them (all have at least two electrons)

50. K^+, Cl^-, Ru^+, Ge^{2+}

Chapter 7

Periodic Properties of the Elements

The Bottom Line

When you finish this chapter, you will be able to:

- Understand how the structure of the periodic table was developed.
- Appreciate how elements fall into "blocks" that are defined by the type of orbital that the element's electrons have filled.
- Comprehend the structure of periods and groups in the periodic table.
- Explain the differences between metals, metalloids, and nonmetals.
- Apply the concept of "periodicity" to atomic size, ionization energy, electron affinity, and electronegativity.
- Appreciate the concept of reactivity and the elemental makeup of the Earth's crust and atmosphere.

7.1   The Big Picture: Building the Periodic Table

The historic arrangement of the elements in the periodic table was based on the properties and reactivities of the elements as determined experimentally. The structure of the periodic table includes several "blocks" of electron orbitals: the *s*-block, *p*-block, *d*-block, and *f*-block orbitals. See **Figure 7.1** in the text.

Of equal importance are the following observations:

- The row or period indicates the highest energy level for the electrons in each element.
- Elements in the same column or group have the same number of electrons in the highest energy level or valence level.
- Elements in the same group have similar properties.

Example 1   **Identifying Elements**

Identify the block, period, and group number for each of the following elements:

Li, Ca, Fe, S, I

Solution

Lithium is in an *s*-block, Period 2, Group IA.

Calcium is in an *s*-block, Period 4, Group IIA.

Iron is in a *d*-block, Period 4, and Group VIIIB.

Sulfur is in a *p*-block, Period 3, Group VIA.

Iodine is in a *p*-block, Period 5, Group VIIA.

7.2 The First Level of Structure: Metals, Nonmetals, and Metalloids

The three main sections of the periodic table are metals, nonmetals and metalloids as shown in **Figure 7.1** in your text.

The characteristics of **metals** are:

- Shiny
- Solids at room temperature and pressure (except for mercury)
- Good conductors of electricity
- Malleable
- Likely to form positively charged ions when forming ionic compounds

"Heavy" metals include metals whose atomic masses are greater than or equal to 63.546 g/mol (copper).

The characteristics of **nonmetals** are:

- Dull, brittle solids, or gases at room temperature and pressure
- Poor conductors of electricity (except for graphite)
- Likely to form negatively charged ions when forming ionic compounds

Metalloids have characteristics between metals and nonmetals. They do not fit well into either classification. The metalloids are boron, silicon, germanium, arsenic, antimony, tellurium, and astatine.

Example 2   **Metals, Nonmetals, and Metalloids**

Classify each of the following as a metal, nonmetal, or metalloid:

Rb, Ga, Mn, Se, As, P

Solution

Rb: rubidium is a metal.
Ga: gallium is a metal.
Mn: manganese is a metal.
Se: selenium is a nonmetal.
As: arsenic is a metalloid.
P: phosphorus is a nonmetal.

7.3 The Next Level of Structure: Groups in the Periodic Table

Elements in a group exhibit similar chemical properties. The reason for this similarity is the arrangement of the valence electrons, the electrons in the outermost energy levels. Elements in the main groups, IA–VIIIA, all have the same number of electrons in their outer energy level, corresponding to the highest principal quantum number. For transition elements (B groups), the valence electrons include two energy levels.

Group IA—Hydrogen and the Alkali Metals: These highly reactive metals form +1 ions. They react violently with water, resulting in a basic solution. Hydrogen is placed in this group because it also forms a +1 ion. When reacting with alkali metals, hydrogen is capable of forming a −1 ion, known as a hydride.

Group IIA—Alkaline Earth Metals: These very reactive metals form +2 ions. They react readily with oxygen, forming oxides that are basic when dissolved in water.

Group IIIA: These much less reactive metals form +3 ions. Notice the first evidence of a "trend" in the periodic table—the reactivity of a metal within a period decreases with increasing ionic charge.

Group IVA: The carbon group contains atoms that have four valence electrons, two in the s orbital and two in the p orbitals. Carbon forms four bonds that are covalent rather than ionic. Covalent bonds are formed by sharing electrons. Carbon also concatenates (bonds to itself), forming compounds with chains of carbon atoms. These compounds are essential to life. Silicon is also in this group. This atom is very important to the semiconductor industry.

Group VA: Atoms in this group have five valence electrons, two in the s orbital and three in the p orbitals. They form ions having a −3 charge by accepting three electrons; they also form covalent bonds. Nitrogen is in this group. It is extremely important to the fertilizer industry.

Group VIA—Chalcogens: These atoms have six valence electrons, two in the s orbital and four in the p orbitals. This group contains oxygen, which is extremely important to our respiration, and sulfur, which is found in important compounds such as proteins, fertilizers, and sulfuric acid.

Group VIIA—Halogens: These very reactive nonmetals form −1 ions. They react with metals, forming compounds known as "salts." Halogens have seven valence electrons, two in the s orbital and five in the p orbitals.

Group VIIIA—Noble Gases: These atoms are largely unreactive. They have eight valence electrons, two in the s orbital and six filling the p orbitals. These very stable monatomic gases are used in a variety of applications by the lighting industry.

Transition Elements: The middle of the periodic table contains the transition elements in Groups IB–VIIIB. These elements fill the d orbitals. Notice that between Sc and Zn there are 10 elements corresponding to the maximum occupancy of the d orbitals. Notice also a gap in numbering between La ($Z = 57$) and Hf ($Z = 72$). The missing 14 elements listed at the bottom of the table are known as the **lanthanides.** There are also 14 elements missing between Ac ($Z = 89$) and Rf ($Z = 104$), which are known as the **actinides.** These two rows are often referred to as the inner transition elements. The lanthanides and the actinides fill the f orbitals. Unlike the A group elements, the number of valence electrons of the B group elements is not easy to predict. The placement of the valence electrons often involves the s orbital and either the d orbitals or the f orbitals. This accounts for the fact that transition elements form ions with different charges. For example, cobalt forms both +2 and +3 ions, while copper forms both +1 and +2 ions.

Example 3   The Periodic Table

Give the name, group name, and number of valence electrons for the following elements:

Ca, Cs, S, I, Rn

Solution

Ca: calcium, alkaline earth metal, 2 electrons in the $4s$ orbital.
Cs: cesium, alkali metal, 1 electron in the $6s$ orbital.
S: sulfur, chalcogen, 2 electrons in the $3s$ orbital and 4 electrons in the $4p$ orbitals.
I: iodine, halogen, 2 electrons in the $5s$ orbital and 5 electrons in the $5p$ orbitals.
Rn: radon, noble gas, 2 electrons in the $6s$ orbital and 6 electrons in the $6p$ orbitals.

Example 4   **Transition Elements**

Locate the following elements in the periodic table, give their symbol, and indicate whether they are a transition metal, an actinide, or a lanthanide.

uranium, europium, cobalt, zirconium, titanium

Solution

U is an actinide, Eu is a lanthanide, Co is a transition metal, Zr is a transition metal, and Ti is a transition metal.

7.4   The Concept of Periodicity

Why are the elements arranged in the order that is found in the periodic table? The elements are arranged in order of increasing atomic number. When organized in this way, the elements demonstrate periodic variations in their physical and chemical behavior. If you examine the chemical behavior of the alkali metals, you will see that they all form +1 ions in ionic compounds and they all react violently with water. In the same manner, alkaline earth metals all form +2 ions and also react with water but much less violently. Halogens, at the other end of the table, all form –1 ions and react with alkali metals and alkaline metals to form salts.

The physical properties of atomic size, first ionization energy, electron affinity, and electronegativity also show this periodic behavior, as we shall see in the following sections.

Example 5   **Periodicity**

What common electron configuration do the elements in Group IA have that explains their position in the periodic table?

Solution

All of the elements in Group IA have one electron in the outermost s orbital. For example, Li is $2s^1$, Na is $3s^1$, K is $4s^1$, and so on. As we will see, it is the removal of this electron that allows these elements to form +1 ions.

7.5   Atomic Size

As explained in your textbook, atomic size refers to the atomic radius of an atom. The periodic trend is that **atomic size decreases going across a period from left to right and increases going down a group.** The reason that atomic size increases going down a group is that the atoms have an increasing number of electrons occupying orbitals with a higher principal quantum number.

Why does the atomic size decrease going across a period? In this case, the electrons that are being added as the atomic number increases occupy the same orbitals as in the previous elements. In addition, as the nucleus adds more protons it increases the pull on the electrons, causing them to come closer to the nucleus and resulting in a smaller atom.

Example 6   **Atomic Size**

Using the periodic trends for atomic size, order the following elements from smallest to largest:

V, Ba, S

Solution

S, V, Ba: The atomic size increases going down a group and going from right to left across a period.

7.6 Ionization Energies

Ionization energy is the amount of energy necessary to remove one or more electrons from an atom. As discussed in the textbook, the first ionization energy is the energy required to remove the outermost electron from the atom:

$$X(g) \rightarrow X^+(g) + e^-$$

The second ionization energy is the energy necessary to remove the next electron:

$$X^+(g) \rightarrow X^{2+}(g) + e^-$$

First ionization energies generally decrease down a group and generally increase across a period from left to right. This periodic trend is the opposite of the trend seen with atomic size. There are some anomalies in this trend, however (see **Figure 7.11** in the text).

What is the explanation for these trends? As we move down a group, the outermost electron is farther and farther from the nucleus—hence the force of attraction diminishes. In addition, the larger number of electrons of a larger atom shields the outermost electron from the nuclear pull. In going across a period, extra energy is needed to remove the electron from the pull of increasing nuclear charge. Remember that in going across a period, the nuclear charge increases with atomic number.

Example 7   **Ionization Energies**

Using the periodic trends for first ionization energies, order the following elements in terms of increasing ionization energy:

Mg, Cs, Cl

Solution

Cs, Mg, Cl: Ionization energies increase going up a group and across a period from left to right.

7.7 Electron Affinity

Electron affinity is the change in energy associated with the addition of an electron to an atom in the gaseous state:

$$Y(g) + e^- \rightarrow Y^-(g)$$

The general trend for electron affinities is that as you move down a group, the electron affinities become less negative (more positive), which means less energy is being released. Electron affinities also become more negative as you move from left to right across a period. As is the case for ionization energies, there are many exceptions to this general rule (see **Figure 7.12** in the text).

7.8 Electronegativity

The electronegativity of an atom is the ability of that atom in a molecule to attract shared electrons to it. The Pauling electronegativity scale (see **Figure 7.13** in the text) assigns an electronegativity value to each atom in the periodic table on a scale of 0 to 4. Fluorine is the most electronegative element, with a value of 4.0. **Electronegativities generally increase as we move across the periodic table from left to right. Electronegativities also increase going up a group.** If you draw arrows pointing across to fluorine and up to fluorine, you will have an easy way of remembering these trends.

Another important trend that may be deduced from looking at ionization energies and electronegativity values is that atoms on the extreme left of the periodic table, especially the lower elements on the extreme left, will most readily form positive ions. Atoms on the extreme right, especially those on the upper extreme right, will most readily form negative ions. These trends do not include Group VIIIA, the noble gases.

Example 8   Electronegativity

Using the periodic trends rather than the Pauling electronegativity scale, predict which of the elements in the following pairs will be more electronegative:

$$B \text{ or } N \quad Br \text{ or } I$$

Solution

N is more electronegative; it is closer to F than is B. Br is more electronegative than I; it is closer to F than is I.

7.9   Reactivity

An element is considered to be highly reactive if it reacts readily to form compounds. An element is considered to be unreactive if it does not react readily to form compounds. Highly reactive elements are found in the environment combined with other elements as compounds. Metals such as sodium and potassium are never found uncombined but rather form salts found in the Earth. Gold, by contrast, is found as a pure metal in an uncombined form. In the text, **Table 7.14** lists a reactivity series for some important metals.

7.10 The Elements and the Environment

Elements are distributed throughout the Earth, most importantly in its crust, hydrosphere, and atmosphere. The Earth's crust provides the elements that are used for industrial applications and energy. **Table 7.15** of the text lists the abundances of the elements in the Earth's crust. The hydrosphere, which is made up of the oceans, seas, lakes, rivers, and underground aquifers, contains hydrogen and oxygen in the form of water and a variety of dissolved salts. The gases found in the atmosphere are of great importance to the environment and the health of the Earth's inhabitants. **Table 7.16** of the text lists the low-density gases that are held to the planet by the Earth's gravity.

Exercises

Section 7.1

1. The periodic table is organized into blocks. List them.

2. Give the blocks in which you would find the following elements: Ba, Tc, I, Cs.

3. Give the blocks in which you would find the following elements: V, As, Ce, U.

4. Give the blocks in which you would find the following elements: Cr, Mg, C, Np.

5. Identify the period number and group number for Zr, K, Pb, and F.

6. Identify the period number and group number for Sn, O, Cu, and Ca.

7. Identify the period number and group number for In, Br, W, and Rb.

Section 7.2

8. Classify each of the following as a metal, metalloid, or nonmetal: B, Sc, S, Kr.

9. Classify each of the following as a metal, metalloid, or nonmetal: Ca, Co, N, Te.

10. Classify each of the following as a metal, metalloid, or nonmetal: Zr, Pd, Ar, Si.

11. At room temperature, mercury and bromine are liquids. What would be the classification of each element?

12. Classify the elements (metal, metalloid, or nonmetal) in Group IVA from top to bottom.

13. Classify the elements (metal, metalloid, or nonmetal) in Group VA from top to bottom.

Section 7.3

14. List the name of the element, the name of the group, and the number of valence electrons for Li, Be, O, and Ne.

15. List the name of the element, the name of the group, and the number of valence electrons for Ra, Se, Br, and Rn.

16. List the name of the element, the name of the group, and the number of valence electrons for Rb, Ca, S, and F.

17. Give the chemical symbol and identify the following elements as a transition metal, lanthanide, or actinide: cerium, molybdenum, zinc.

18. Give the chemical symbol and identify the following elements as a transition metal, lanthanide, or actinide: scandium, niobium, plutonium.

19. Give the chemical symbol and identify the following elements as a transition metal, lanthanide, or actinide: gold, thorium, lutetium, platinum.

20. Write the formulas for the compounds that would most likely be formed from the combination of K, Ca, and C with Cl.

21. Write the formulas for the compounds that would most likely be formed from the combination of Na, Mg, and Ge with S.

Section 7.4

22. What is the valence-electron orbital configuration for the alkaline earth metals?

23. What is the valence-electron orbital configuration for the chalcogens?

24. What is the valence-electron orbital configuration for the halogens?

25. What is the valence-electron orbital configuration for the noble gases?

Section 7.5

26. Using the positions of the atoms in the periodic table, order the following elements in terms of atomic size from smallest to largest: Ca, Ge, Br.

27. Using the positions of the atoms in the periodic table, order the following elements in terms of atomic size from smallest to largest: Al, P, Na.

28. Which is the largest atom in Group IVA?

29. Which is the smallest atom in Group VIIA?

30. Why is the sodium atom so much larger than the sulfur atom?

31. Why is the barium atom so much larger than the beryllium atom?

Section 7.6

32. Using the periodic trends for first ionization energy, rank the following elements from lowest to highest in terms of ionization energy: Se, Na, Ge.

33. Using the periodic trends for ionization energy, rank the following elements from lowest to highest in terms of ionization energy: Ca, Sr, Ba.

34. Using the periodic trends for first ionization energy, rank the following elements from lowest to highest in terms of ionization energy: Ag, S, Rb.

35. Which alkali metal would have the lowest first ionization energy?

36. Which Group VA element would have the highest first ionization energy?

37. Two elements have electron configurations of $[He]2s^2$ and $[Ar]4s^2$, respectively. One of them has a first ionization energy of 590 kJ/mol; the other has a first ionization energy of 899 kJ/mol. Match the ionization energy with the appropriate electron configuration and justify your choice.

Section 7.7

38. Define electron affinity.

39. Using the periodic trends, arrange the following elements in order of increasing electron affinity: Li, Na, K.

40. Using the periodic trends, arrange the following elements in order of increasing electron affinity: Cl, Rb, P.

41. Which of the halogens would have the most positive electron affinity?

Section 7.8

42. Define electronegativity. How does it differ from electron affinity?

43. Using the periodic trends, arrange the following elements in order of increasing electronegativity: Br, K, P.

44. Using the periodic trends, arrange the following elements in order of increasing electronegativity: As, Cl, Sr.

45. Consider the bond in C—O. Which atom would have a stronger attraction for the electrons in the bond?

46. Consider the bond in Si—Cl. Which atom would have a stronger attraction for the electrons in the bond?

47. Which of the chalcogens is the most electronegative?

Section 7.9

48. Order the following metals from generally least reactive to most reactive: Al, Fe, Na, Cu, Ag.

Section 7.10

49. Which metal is the most abundant metal in the Earth's crust?

50. Why does the Earth have an atmosphere? That is, why don't the gases in the atmosphere escape into space?

Answers to Exercises

1. *s*-block, *p*-block, *d*-block, *f*-block

2. Ba: *s*-block; Tc: *d*-block; I: *p*-block; Cs: *s*-block

3. V: *d*-block; As: *p*-block; Ce: *f*-block; U: *f*-block

4. Cr: *d*-block; Mg: *s*-block; C: *p*-block; Np: *f*-block

5. Zr: Period 5, Group IVB; K: Period 4, Group IA; Pb: Period 6, Group IVA; F: Period 2, Group VIIA

6. Sn: Period 5, Group IVA; O: Period 2, Group VIA; Cu: Period 4, Group IB; Ca: Period 4, Group IIA

7. In: Period 5, Group IIIA; Br: Period 4, Group VIIA; W: Period 6, Group VIB; Rb: Period 5, Group IA

8. B: metalloid; Sc: metal; S: nonmetal; Kr: nonmetal

9. Ca: metal; Co: metal; N: nonmetal; Te: metalloid

10. Zr: metal; Pd: metal; Ar: nonmetal; Si: metalloid

11. Hg: metal; Br: nonmetal

12. C: nonmetal; Si: metalloid; Ge: metalloid; Sn: metal; Pb: metal

13. N: nonmetal; P: nonmetal; As: metalloid; Sb: metalloid; Bi: metal

14. lithium, alkali metal, 1 electron; beryllium, alkaline earth metal, 2 electrons; oxygen, chalcogen, 6 electrons; neon, noble gas, 8 electrons

15. radium, alkaline earth metal, 2 electrons; selenium, chalcogen, 6 electrons; bromine, halogen, 7 electrons; radon, noble gas, 8 electrons

16. rubidium, alkali metal, 1 electron; calcium, alkaline earth metal, 2 electrons; sulfur, chalcogen, 6 electrons; fluorine, halogen, 7 electrons

17. Ce: lanthanide; Mo: transition metal; Zn: transition metal

18. Sc: transition metal; Nb: transition metal; Pu: actinide

19. Au: transition metal; Th: actinide; Lu: lanthanide; Pt: transition metal

20. KCl, $CaCl_2$, CCl_4

21. Na_2S, MgS, GeS_2

22. ns^2

23. ns^2np^4

24. ns^2np^5

25. ns^2np^6

26. Br, Ge, Ca

27. P, Al, Na

28. Pb

29. F

30. The outermost electron in Na is in the $3s$ orbital, whereas the outermost electrons of S are in the $3p$ orbitals. While the p orbitals are slightly higher in energy than the s orbital, the nucleus of sulfur contains more positively charged protons that pull the electrons in closer to the nucleus, making S smaller than Na.

31. Barium's outermost electrons are in the $6s$ orbital as compared to beryllium's $2s$-orbital electrons.

32. Na, Ge, Se

33. Ba, Sr, Ca

34. Rb, Ag, S

35. Fr

36. N

37. The element $[Ar]4s^2$ would have 590 kJ/mol for its first ionization energy. Because it is larger than the $[He]2s^2$ atom, its outermost electron is farther from the pull of the nucleus. It will take less energy to remove the outermost electron from the atom. Therefore, 899 kJ/mol belongs to $[He]2s^2$.

38. Electron affinity is the energy change associated with the addition of an electron to an atom in the gaseous state.

39. K, Na, Li

40. Rb, P, Cl

41. At

42. Electronegativity is the ability of an atom in a molecule to attract a shared electron to it. Electronegativity is a relative scale and deals with attraction of electrons. Electron affinity measures the energy realized by forming a negative gaseous ion.

43. K, P, Br

44. Sr, As, Cl

45. O; it is closer to F than C, making it more electronegative.

46. Cl; it is closer to F than Si, making it more electronegative.

47. oxygen

48. Ag, Cu, Fe, Al, Na

49. Al

50. The low-density gases in the Earth's atmosphere are held there by gravity.

Chapter 8

Bonding Basics

The Bottom Line

When you finish this chapter, you will be able to:

- Differentiate between an ionic bond and a covalent bond.
- Predict the sizes of ions in relationship to one another.
- Predict the lattice enthalpies of compounds in relationship to one another.
- Draw Lewis dot structures.
- Calculate the formal charge on atoms.
- Determine when resonance occurs in a molecule.
- Calculate the enthalpy change of a reaction based on bond dissociation energies.
- Predict the shape, bond angle, and polarity of various molecules using the VSEPR model.

8.1   Modeling Bonds

In this chapter, we move beyond writing formulas and into modeling the different arrangements of electrons to help determine the shape and polarity of compounds. There are three major types of chemical bonds:

- **Ionic bond**—the atoms are held together by the force of attraction of opposite charges
- **Covalent bond**—electrons are shared between the atoms (but not necessarily equally)
- **Metallic bond**—metal cations are spaced throughout a sea of mobile electrons

When ionic and covalent bonding occurs, one of the driving forces is the ability of each atom to reach a stable electron configuration. The main-group atoms typically react by changing their number of electrons in an attempt to attain the more stable electron configuration of a noble gas. We refer to this as the **octet rule** because the most stable arrangement features eight valence electrons. The element hydrogen follows the **duet rule** because it needs only two electrons to achieve a noble gas configuration.

Example 1   Arrangement of Electrons

Write the electron configuration and the Lewis dot symbol for O^{2-}. With which element is this isoelectronic?

First Thoughts

You should recall how to write electron configurations from Chapter 6. To draw a Lewis dot symbol, count the number of valence electrons for oxygen, including two extra electrons because oxygen is in the 2– state. Isoelectronic means "the same number of electrons" or "the same electron configuration."

Solution

Total number of valence electrons in $O^{2-} = 6e^- + 2e^- = 8e^-$

Electron configuration for O^{2-}:

$1s^2 2s^2 2p^6$ (valence: $2s^2 2p^6$)

Lewis dot symbol:

$$\left[\ddot{\underset{\cdot\cdot}{\text{O}}} \right]^{2-}$$

O^{2-} is isoelectronic with the element neon (Ne).

8.2   Ionic Bonding

Some examples of ionic bonding are shown in **Table 8.1** and **Figure 8.3** in your text. How does the ionic bonding in these compounds occur?

- The metal loses electrons to become a cation (thus attaining a noble gas configuration).
- The nonmetal gains electrons to become an anion (also attaining a noble gas configuration).
- The two oppositely charged atoms are attracted to each other and form an **ion pair** (such as K^+Cl^-).
- Eventually a collection of these ions aggregate into a **crystalline lattice.**

Your text describes the ionic bonding between sodium and chloride in detail. Keep in mind that even though energy must be added to the system to remove electrons from a metal atom, even more energy is released when the ionic bond forms. In addition, the anions and cations that form an ionic bond make the compound neutral overall.

Example 2   **Ionic Compound Formation**

Using Lewis dot symbols show the transfer of electrons to form the following compounds:

a) $MgBr_2$ from magnesium metal and bromine atoms
b) CaS from calcium metal and a sulfur atom

First Thoughts

The metals (Mg and Ca) will lose electrons. The nonmetals (Br and S) will gain electrons.

Solution

$MgBr_2$:

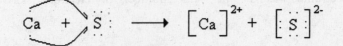

CaS:

$$\text{Ca} + \ddot{\underset{\cdot\cdot}{\text{S}}} \longrightarrow \left[\text{Ca} \right]^{2+} + \left[\ddot{\underset{\cdot\cdot}{:\text{S}}} \right]^{2-}$$

Further Insight

Notice that each ion ends up with a noble gas configuration.

Because these metal and nonmetal atoms become ions before they bond, the sizes of the ions are important. The size helps determine the structure of the ionic crystal and the strength of the ionic bond. The size of an ion is determined by its electrons. Here is the bottom line:

- The cation for an element is smaller than the atom from which it was formed due to its higher effective nuclear charge. For example, Na^+ is smaller in size than Na. **Table 8.2** lists the sizes of some common atoms and ions.
- The anion for an element is larger than the atom from which it was formed due to its decreased effective nuclear charge. For example, Cl^- is larger in size than Cl. **Figure 8.7** compares the sizes of a series of ions.
- As you go down a group on the periodic table, the distance of the valence electrons from the nucleus increases because the principal quantum number increases. For example, Ba^{2+} is larger in size than Mg^{2+}.

Example 3   **Ions and Their Sizes I**

List the following four ions in order of increasing size: Cl^-, Br^-, F^-, I^-.

First Thoughts

Each of these ions has a noble gas configuration but has valence electrons in different principal quantum levels. The higher the principal quantum number, the larger the ion.

Solution

$$F^- < Cl^- < Br^- < I^-$$

Example 4   **Ions and Their Sizes II**

List four ions that are isoelectronic with neon, and arrange them in order of decreasing size.

First Thoughts

Ions that are isoelectronic with neon must contain 10 electrons. The ion with the greatest number of protons will be the smallest in size due to its higher effective nuclear charge.

Solution

We selected four ions that are close in atomic number to neon: Na^+, Mg^{2+}, O^{2-}, F^-. Oxygen has the smallest number of protons, so it will be the largest in size. Magnesium has the largest number of protons, so it will be the smallest in size.

$$O^{2-} > F^- > Na^+ > Mg^{2+}$$

The strength of an ionic bond is measured by the **lattice enthalpy** of the ionic solid. Lattice enthalpy is defined as the amount of energy required to separate one mole of a solid ionic crystalline compound into its gaseous ions. **Table 8.3** in your text lists the lattice enthalpies for some common ionic solids. Calculating lattice enthalpies requires you to make use of the Born-Haber cycle, which is discussed in detail in your text. In general, lattice enthalpies are greatest for ionic compounds that are made up of *small, highly charged* particles.

Example 5   **Predicting Lattice Energies**

Use the relationship of ionic size and charge to predict which of the following would have the smaller lattice enthalpy:

a) magnesium chloride or magnesium bromide
b) sodium fluoride or barium fluoride

Solution

a) The bromine ion in magnesium bromide is larger than the chlorine ion in magnesium chloride; therefore magnesium bromide has a smaller lattice enthalpy.

b) The charge on the sodium ion is +1 in sodium fluoride; the charge on the barium ion is +2 charge in barium fluoride. Even though the sodium ion is smaller in size, sodium fluoride has a smaller lattice enthalpy because the sodium ion is not as highly charged as the barium ion. Charge has a larger effect on lattice enthalpy than does size.

Further Insight

These answers can be verified by looking at Table 8.3.

a)   magnesium bromide   =   2097 kJ/mol
      magnesium chloride   =   2326 kJ/mol

b)   sodium fluoride   =   923 kJ/mol
      barium fluoride   =   2341 kJ/mol

8.3   Covalent Bonding

Covalent bonding is the second major type of bonding in compounds. Here are some key points about covalent bonding:

- The valence electrons are shared between the atoms. Where there are different electronegativities between the atoms, the sharing is unequal.
- When these electrons are shared, each atom attains a noble gas configuration.
- Due to the similarities in electronegativity, the majority of covalent bonds exist between nonmetals.
- Unequal electron sharing results in a **polar covalent bond.** The charge separation can be represented as

$$\overset{\delta^+}{H} - \overset{\delta^-}{Cl}: \quad \text{(Chlorine is more electronegative than hydrogen.)}$$

- Equal electron sharing results in a **nonpolar covalent bond.** An example of this is in nitrogen gas, N_2.
- Bonds can range from nonpolar covalent bonding, to polar covalent bonding, to ionic bonding.

Example 6   **Bond Type**

Predict the type of bond—ionic, polar covalent, or nonpolar covalent—in each of the following compounds.

   a) NaBr
   b) O_2
   c) NH_3

Solution

a) ionic bonding   (bond between a metal and nonmetal; a large electronegativity difference between the two atoms)

b) nonpolar covalent bonding (two nonmetals; no electronegativity difference between the two atoms; electrons shared equally)

c) polar covalent bonding (two nonmetals; electronegativity difference between the two atoms; electrons shared unequally)

Further Insight

Remember that when you predict the type of bond between two atoms, you are often making a generalization. Most bonds have some ionic and some covalent character.

One way to model how atoms are attached in covalent bonds is to draw Lewis dot structures. **Table 8.5** in your text gives the rules for drawing Lewis dot structures. These rules include drawing a skeletal picture of a molecule, then placing extra valence electrons in the skeleton until the model of the compound is complete.

Example 7   Drawing a Lewis Dot Structure

Draw the Lewis dot structure for ammonia, NH_3.

First Thoughts

The first step is to determine the total number of valence electrons in ammonia.

Nitrogen = 5 valence electrons

Hydrogen = $1 \times 3 = 3$ valence electrons

Total number of valence electrons = $5e^- + 3e^- = 8e^-$

Solution

According to Table 8.5, the next step is to draw a skeletal structure of ammonia. Because hydrogen atoms are on the edges of the molecule (to satisfy the duet rule), nitrogen must be the central atom.

$$H \!-\! N \!-\! H$$
$$|$$
$$H$$

So far, the structure contains six valence electrons. We place the two remaining valence electrons on nitrogen because it needs to satisfy the octet rule.

Further Insight

The lines between N and H are called **bonding pairs** because the electrons are involved in the bonding. The two dots are considered a **lone pair** because the electrons are not involved in the bonding.

It is not uncommon for more than one Lewis dot structure to exist for a molecule. When this occurs, it is useful to assign **formal charges** to each atom in the molecule. Formal charge is calculated using the following formula:

Formal charge = valence electrons – number of bonds – number of nonbonded electrons

The best structure meets these criteria:

- It has the smallest magnitude for all of the formal charges.
- It places negative formal charges on electronegative atoms.
- It has the smallest number of nonzero formal charges.

Your text shows an excellent example of how formal charge is important for choosing the best Lewis dot structure for methanol (**Figure 8.13**).

Example 8   **Formal Charges**

Draw the Lewis dot structure for PO_4^{3-} and assign formal charges on each atom.

First Thoughts

The first step is to determine the total number of valence electrons and then draw a skeletal structure for the phosphate ion.

> Phosphorus = 5 × 1 = 5 valence electrons
>
> Oxygen = 6 × 4 = 24 valence electrons
>
> 3⁻ charge = 3 electrons
>
> Total number of valence electrons = 5e⁻ + 24e⁻ + 3e⁻ = 32e⁻

Phosphorus has a lower electronegativity than oxygen, so it is the central atom.

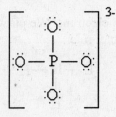

Solution

So far we have used eight valence electrons to model the bonds between P and O. We now position the remaining valence electrons so that the octet rule for each oxygen atom is satisfied. We also put the structure in brackets with the charge because it is an ion.

$$\left[\begin{array}{c} :\overset{..}{O}: \\ | \\ :\overset{..}{O} - P - \overset{..}{O}: \\ | \\ :\overset{..}{O}: \end{array} \right]^{3-}$$

We can now assign formal charges to each atom:

> Formal charge = valence electrons – number of bonds – number of nonbonded electrons
>
> Formal charge (P) = 5 – 4 – 0 = +1
>
> Formal charge (O) = 6 – 1 – 6 = –1
> (The formal charge on each oxygen atom is therefore –1.)

Further Insight

Notice how the formal charges on the atoms of the phosphate ion add up to the –3 charge on the ion overall.

$$+1 + (-1)(4) = -3$$

For some Lewis dot structures, the atoms must share more electrons, which is represented by forming **double** or **triple bonds.** Double bonds contain four bonding electrons between two atoms. Triple bonds contain six bonding electrons between two atoms.

Example 9   **Multiple Bonds**

Draw the Lewis dot structure for HCN.

Solution

Total number of valence electrons = $1e^- + 4e^- + 5e^- = 10e^-$

Skeletal structure:

$$H - C - N$$ (H is on end; C is less electronegative
than N so it's the central atom)

To satisfy the octet rule for both carbon and nitrogen, six electrons must be shared (forming a triple bond). If you did not use a triple bond, you would exceed the total number of valence electrons allowed.

<u>Correct:</u>   <u>Incorrect:</u>

$$H - C \equiv N:$$ $$H - \overset{..}{C} - \overset{..}{N}:$$

(10 electrons total)   (14 electrons total)

When drawing Lewis dot structures involving double or triple bonds, be aware that **resonance** sometimes occurs. A resonance structure is a model of a molecule in which the positions of the electrons have changed, but the positions of the atoms remain fixed. For example, a double or triple bond could be positioned in more than one place between various atoms, yet all of the structures be valid Lewis dot structures. When resonance occurs, the best model combines all of the resonance structures to form a **resonance hybrid**.

Example 10   **Resonance Structures**

Draw a Lewis dot structure for NO_2^- and show all resonance structures for this ion.

First Thoughts

Determine the total valence electrons and draw a skeletal structure.

Total number of valence electrons = $5e^- + (2)(6e^-) + 1e^- = 18e^-$

Skeletal structure:

$$O - N - O$$ (N is the central atom because
it is less electronegative than O.)

Solution

Position the remaining electrons to satisfy the octet rule for each atom.

$$\left[:\overset{..}{O} - \overset{..}{N} = \overset{..}{O}: \right]^-$$

You could also place the double bond on the left side of nitrogen to make another valid Lewis structure.

$$\left[:\overset{..}{O} = \overset{..}{N} - \overset{..}{O}: \right]^-$$

Although there are two valid resonance structures for this ion, keep in mind that the real structure is actually a hybrid of the two resonance structures.

Further Insight

You may have drawn a Lewis dot structure for this ion in which oxygen was the central atom instead of nitrogen. This structure is not favored, especially when you take the formal charges into account.

Because all models are simplifications of reality, the Lewis dot structure model has some problems. In particular, some molecules exceed the octet rule. This can generally occur for atoms in the third row or higher on the periodic table. The higher-energy unfilled set of d orbitals can be used to hold extra electrons if needed.

Example 11   **Exceeding the Octet Rule**

Draw the Lewis dot structure for PF_5.

Solution

Total number of valence electrons = $5e^- + (5)(7e^-) = 40e^-$

P is less electronegative than F, so it is the central atom. However, P needs to exceed the octet rule to be in the center. P contains empty $3d$ orbitals that hold the extra electrons.

Now that you have learned a little about modeling covalent bonds, it is important to consider the strength of the covalent bond as well. **Table 8.6** in your text lists some average bond dissociation energies for various types of covalent bonds. We can calculate the overall enthalpy change of a chemical reaction (ΔH_{rxn}) by using these bond dissociation energies, drawing Lewis dot structures, and utilizing the following equation:

$$\Delta H_{rxn} = \sum (\Delta H_{diss})_{Breaking\ Bonds} - \sum (\Delta H_{diss})_{Making\ Bonds}$$

Example 12   **Bond Energies**

Use the bond energies in Table 8.6 to calculate ΔH_{rxn} for the following reaction:

$$N_2 + 3H_2 \rightarrow 2NH_3$$

First Thoughts

To determine the bond energy values needed, we need to draw the Lewis dot structures for each substance. We can then see which bonds are being broken and which are being formed.

$$:N \equiv N: \ + \ \begin{matrix} H—H \\ H—H \\ H—H \end{matrix} \ \longrightarrow \ \begin{matrix} H—\overset{..}{N}—H \\ | \\ H \end{matrix} \quad \begin{matrix} H—\overset{..}{N}—H \\ | \\ H \end{matrix}$$

Solution

Bonds Broken:

1 mol N_2 bond × 941 kJ/mol	=	+941 kJ/mol
3 mol H_2 bonds × 432 kJ/mol	=	+1296 kJ/mol
Total bonds broken	=	+2237 kJ/mol

Bonds Formed:

6 mol N—H bonds × 391 kJ/mol		= +2346 kJ/mol
Total bonds formed		= +2346 kJ/mol

ΔH_{rxn} = +2237 kJ/mol – 2346 kJ/mol = –109 kJ/mol

Further Insight

Since ΔH_{rxn} is a negative value, the reaction of N_2 and H_2 to produce ammonia is exothermic.

8.4   VSEPR: A Better Model

The next step in understanding why covalent compounds have certain properties is to determine the shape of a molecule. Lewis dot structures are helpful when thinking about these kinds of compounds, but they are two-dimensional drawings, not three-dimensional. We use the **valence shell electron-pair repulsion** (VSEPR) model to help us predict the shapes of molecules. The key assumption when using the VSEPR model is that bonding pairs and lone pairs of electrons move away from each other and orient themselves in three-dimensional space so as to give minimal repulsions (lowest energy configurations). **Table 8.7** illustrates the shapes of the electron-group geometries predicted by the VSEPR model. Your text also gives several examples of different shapes and explains how to visualize them. Be sure you understand the difference between **electron-group geometry** (which includes lone pairs and bonding electrons) and **molecular geometry** (which is the actual shape of the molecule and does not include lone pairs). Also keep in mind that lone pairs require the most space compared with triple, double, and single bonds.

Example 13   **Predicting Shapes I**

Predict the molecular geometry (shape) and bond angles for each of the following molecules:

$$BH_3, NH_3, CHCl_3, N_2, SeF_2$$

First Thoughts

First, draw the Lewis dot structure for each molecule.

Solution

Next, count the electron pairs and arrange them in a way that minimizes electron repulsion (put the pairs as far apart as possible). Then, determine the shape and bond angles of each molecule by looking at the positions of the atoms and using Table 8.7.

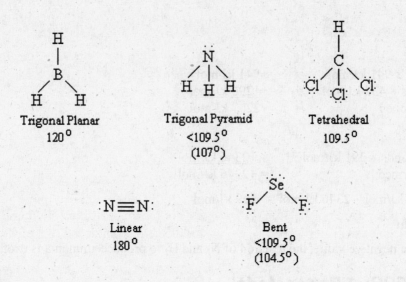

Trigonal Planar 120°   Trigonal Pyramid <109.5° (107°)   Tetrahedral 109.5°

Linear 180°   Bent <109.5° (104.5°)

Further Insight

If you are having difficulty visualizing these shapes, we recommend building these molecules using a model set or tying balloons together to "see" the molecules in three dimensions.

Example 14   **Predicting Shapes II**

Predict the molecular geometry (shape) and bond angles for each of the following molecules; note that these molecules do not obey the octet rule:

$$PF_5, BeF_2, SF_4, XeF_4, ClF_5$$

Solution

Lewis dot structures:

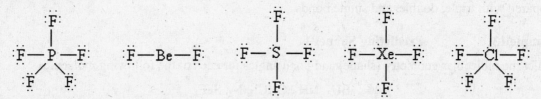

Arrange the electron pairs as far apart as possible to minimize repulsions and then look at the positions of the atoms.

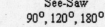

Trigonal Bipyramid 90°, 120°, 180°   Linear 180°   See-Saw 90°, 120°, 180°   Square Planar 90°, 180°   Square Pyramid 90°, 180°

8.5   Properties of Ionic and Molecular Compounds

Once you can determine the shape of molecules, the next step to predicting the properties of compounds is to understand polarization. Here are some key points about polarity:

- The polarization of electrons in a bond is commonly referred to as a **bond dipole.** It is caused by a difference in electronegativity between the atoms in the bond.
- A bond dipole could be represented two ways:

$$\overset{\delta^+}{H} - \overset{\delta^-}{\ddot{C}l}\colon \quad \text{or} \quad H \longleftrightarrow \ddot{C}l\colon$$

- The net polarization of electrons in a molecule is known as a **dipole moment.**
- For molecules with more than two atoms, the dipole moment is the sum of all of the bond dipoles in the molecule.
- If the bond dipoles in a molecule cancel each other out, the result is a *nonpolar* molecule.
- If the bond dipoles in a molecule do not cancel each other out, the result is a *polar* molecule.

Example 15   **Determining the Polarity of a Molecule**

Which of the following molecules are polar (have dipole moments)?

$$BF_3 \qquad SeO_2$$

First Thoughts

Before you can determine whether these molecules have dipole moments, you need to draw their Lewis structures and determine their shapes.

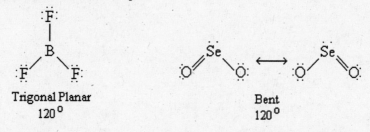

Trigonal Planar
120°

Bent
120°

Solution

Now draw the individual bond dipoles in each molecule and see if they cancel. You must be able to visualize these molecules three-dimensionally to do this.

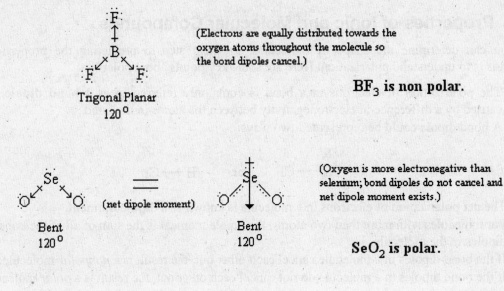

Trigonal Planar
120°

(Electrons are equally distributed towards the
oxygen atoms throughout the molecule so
the bond dipoles cancel.)

BF$_3$ is non polar.

(net dipole moment)

Bent
120°

Bent
120°

(Oxygen is more electronegative than
selenium; bond dipoles do not cancel and
net dipole moment exists.)

SeO$_2$ is polar.

8.6   A Look Ahead to the Next Steps

Lewis dot structures and the VSEPR model do a good job of addressing the shape and polarity of molecules. More models need to be studied, though, to better understand bond lengths, bond strengths, and reactivities. These models are discussed in Chapter 9.

Exercises

Section 8.1

1. What are the three main types of chemical bonds?

2. What is the significance of the octet rule?

3. Give three examples of elements or ions that do not follow the octet rule, but instead follow the duet rule. Justify your examples.

4. Write electron configurations and Lewis dot symbols for N^{3-} and Ca^{2+}. With which element is each ion isoelectronic?

5. Determine the element that is isoelectronic with each of the following ions:
   a) I^- b) Sr^{2+} c) Fe^{3+}

Section 8.2

6. Using Lewis dot symbols, show the transfer of electrons to form KCl from potassium metal and a chlorine atom.

7. Using Lewis dot symbols, show the transfer of electrons to form Na_2S from sodium metal and a sulfur atom.

8. Which of the following pairs of elements are likely to form an ionic bond?
   a) sodium and bromine
   b) lithium and oxygen
   c) potassium and rubidium
   d) nitrogen and oxygen
   e) iron and chlorine

9. Arrange the following ions in order of decreasing size and explain your answer.

$$Li^+, Na^+, K^+, Cs^+$$

10. Arrange the following ions in order of increasing size and explain your answer.

$$Mn^{2+}, Mn^{3+}, Mn^{4+}, Mn^{7+}$$

11. List four ions that are isoelectronic with krypton and arrange them in order of increasing size. Justify your answer.

12. Use the relationship of ionic sizes to predict which would have the greater lattice enthalpy: calcium oxide or calcium sulfide.

13. Use the relationship of ionic sizes and charge to predict which would have the smaller lattice enthalpy: potassium chloride or iron(III) chloride.

14. Which compound in each of the following pairs has the most exothermic lattice energy? Justify your answer.
    a) LiF or NaF
    b) $MgCl_2$ or MgO
    c) CuOH or $Cu(OH)_2$

Section 8.3

15. Predict the type of bond—ionic, polar covalent, or nonpolar covalent—in each of the following compounds.
    a) H_2 b) $CaCl_2$ c) H_2O

16. Identify the errors in each of the following Lewis dot structures and draw the correct structure:
    a)   b)   c)

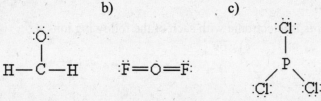

17. Identify the errors in each of the following Lewis dot structures and draw the correct structure:
    a)   b)   c)

18. Write Lewis dot structures for each of the following molecules or ions:
    a) NH_4^+ b) O_2 c) SeF_2

19. Write Lewis dot structures for each of the following molecules or ions:
    a) NF_3 b) SCl_2 c) ClO_3^-

20. Write Lewis dot structures for each of the following molecules or ions; these molecules and ions have central atoms that do not obey the octet rule.
    a) PCl_5 b) SF_4 c) I_3^-

21. Write Lewis dot structures for each of the following molecules; these molecules have central atoms that do not obey the octet rule.
    a) BH_3 b) $XeCl_4$ c) ClF_5

22. Write Lewis dot structures for each of the following molecules; these molecules may or may not obey the octet rule.
    a) HBr   b) ClF_3 c) SF_6

23. Write Lewis dot structures for each of the following molecules or ions. Show all resonance structures where applicable.

    a) N_2 b) NO_2^- c) N_3^-

24. Which of the following molecules exhibit resonance?

$$NO_3^-, SO_2, BeH_2$$

25. Draw Lewis dot structures for the following molecules or ions. Assign the formal charge for each central atom.

    a) XeO_4 b) NO_4^{3-} c) SO_2Cl_2

26. N_2O could have the following Lewis dot structures:

$$\ddot{N}-N\equiv O: \qquad \ddot{N}=O=\ddot{N}: \qquad N\equiv N-\ddot{O}:$$

Which Lewis dot structure is likely to be the most favored (stable)? Justify your answer by assigning formal charges.

27. Use the bond energy values in Table 8.6 of your text to calculate ΔH_{rxn} (in kilojoules) for the following reaction: $2H_2(g) + O_2(g) \rightarrow 2H_2O(g)$

28. Use bond energy values to determine ΔH_{rxn} (in kilojoules) for the combustion of methane:
$$CH_4(g) + 2O_2(g) \rightarrow CO_2(g) + 2H_2O(g)$$

29. Consider the following reaction: $A_2 + B_2 \rightarrow 2AB$ $\Delta H_{rxn} = -315$ kJ

The bond energy for B_2 is half the amount of AB. The bond energy of $A_2 = 422$ kJ/mol. What is the bond energy of B_2?

    a)   107 kJ/mol
    b)   246 kJ/mol
    c)   491 kJ/mol
    d)   806 kJ/mol
    e)   982 kJ/mol

Section 8.4

30. Predict the electron-group geometry and molecular geometry (shape) for each of the molecules or ions in Exercise 18.

31. Predict the electron-group geometry and molecular geometry (shape) for each of the molecules or ions in Exercise 19.

32. Predict the electron-group geometry and molecular geometry (shape) for each of the molecules or ions in Exercise 20.

33. Predict the electron-group geometry and molecular geometry (shape) for each of the molecules in Exercise 21.

34. Predict the electron-group geometry and molecular geometry (shape) for each of the molecules in Exercise 22.

35. Predict the electron-group geometry and molecular geometry (shape) for each of the molecules or ions in Exercise 23.

36. Predict the bond angles around the central atom for each of the molecules or ions in Exercise 18.

37. Predict the bond angles around the central atom for each of the molecules or ions in Exercise 19.

38. Predict the bond angles around the central atom for each of the molecules or ions in Exercise 20.

39. Predict the bond angles around the central atom for each of the molecules in Exercise 21.

40. Predict the bond angles around the central atom for each of the molecules in Exercise 22.

41. Predict the bond angles around the central atom for each of the molecules or ions in Exercise 23.

Section 8.5

42. How does a bond dipole occur?

43. Rank the following bonds from least polar to most polar:

    Si—Cl, P—Cl, Mg—Cl, S—Cl

44. Draw the bond dipole in two different ways for the following bond: C—O.

45. Predict whether each molecule in Exercise 18 will be polar or nonpolar (ignore part a, which is an ion).

46. Predict whether each molecule in Exercise 19 will be polar or nonpolar (ignore part c, which is an ion).

47. Predict whether each molecule in Exercise 20 will be polar or nonpolar (ignore part c, which is an ion).

48. Predict whether each molecule in Exercise 21 will be polar or nonpolar.

49. Predict whether each molecule in Exercise 22 will be polar or nonpolar.

50. Predict whether each molecule in Exercise 24 will be polar or nonpolar (ignore NO_3^-, which is an ion).

Answers to Exercises

1. ionic bond, covalent bond, metallic bond

2. One of the driving forces for ionic and covalent bonding is the ability of each atom to achieve a stable electron configuration. For most atoms, the stable arrangement results when there are eight valence electrons (an octet).

3. H, Li^+, Be^{2+}; They need only two electrons to achieve a noble gas configuration.

4. N^{3-}: $1s^2 2s^2 2p^6$ Ca^{2+}: $1s^2 2s^2 2p^6 3s^2 3p^6$

   $$\left[:\ddot{N}: \right]^{3-}$$ $$\left[Ca \right]^{2+}$$

   N^{3-} is isoelectronic with neon (Ne).
   Ca^{2+} is isoelectronic with argon (Ar).

5. a) Xe   b) Kr   c) V

6.

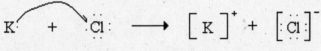

7.

8. a, b, e

9. $Cs^+ > K^+ > Na^+ > Li^+$; Each ion has valence electrons in different principal quantum numbers. The higher the principal quantum number, the farther the valence electrons are from the nucleus, and thus the larger the size of the ion.

10. $Mn^{7+} < Mn^{4+} < Mn^{3+} < Mn^{2+}$; They all contain the same number of protons but a different number of electrons. As more and more electrons are removed, the effective nuclear charge increases and pulls the remaining electrons in closer, thus making the ions smaller in size.

11. $Sr^{2+} < Rb^+ < Br^- < Se^{2-}$; All four ions contain 36 electrons, which is identical to the number of electrons in krypton. Sr^{2+} contains the largest number of protons (38), so it has the highest effective nuclear charge and pulls the electrons in closer (thus making it the smallest in size). Rb^+ follows with 37 protons, then Br^- with 35 protons, and finally Se with 34 protons (and therefore the weakest nuclear charge of the four).

12. Calcium oxide has the greater lattice enthalpy.

13. Potassium chloride has the smaller lattice enthalpy.

14. a) LiF; Li^+ is smaller in size than Na^+
    b) MgO; O^{2-} has a greater charge than Cl^-
    c) $Cu(OH)_2$; Cu^{2+} has a greater charge than Cu^+

15. a) nonpolar covalent
    b) ionic
    c) polar covalent

16. a) Carbon does not satisfy the octet rule.

    b) There are supposed to be 20 valence electrons, but only 16 are present. It is also very unlikely that there would be a double bond on F.

    :F̈—Ö—F̈:

    c) There are supposed to be 26 valence electrons, but only 24 are present.

    (structure of PCl_3 with central P bonded to three Cl atoms)

17. a) There are too many valence electrons. There are supposed to be only 16 valence electrons, not 20.

    :Ö=C=Ö:

    b) Hydrogen needs only two valence electrons to attain a noble gas configuration (H follows the duet rule). Carbon should be the central atom.

    (structure of $CHCl_3$ with central C bonded to three Cl and one H)

    c) Three electrons need to be added to include the 3– charge from the phosphate ion. This way, every atom completes an octet.

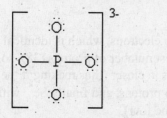

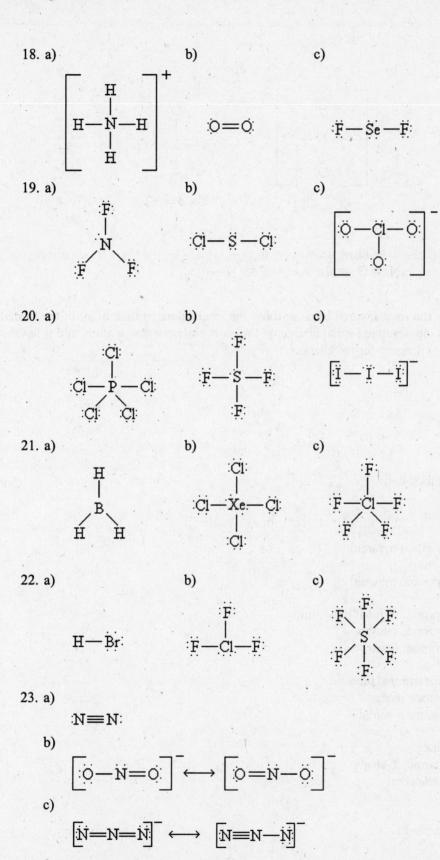

18. a) b) c)

19. a) b) c)

20. a) b) c)

21. a) b) c)

22. a) b) c)

23. a)

b)

c)

24. NO_3^-, SO_2

25. a)   b)   c)

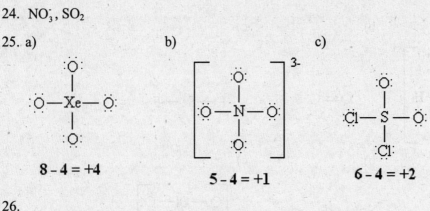

        8 - 4 = +4   5 - 4 = +1   6 - 4 = +2

26.

   -2 +1 +1   -1 +2 -1   0   +1 -1

:N—N≡O: :N=O=N: :N≡N—O:

:N≡N—O: is the most favored because it has the smallest magnitude of all of the formal charges, it places the negative formal charge on the most electronegative atom, and it has the smallest number of nonzero formal charges.

27. -509 kJ

28. -824 kJ

29. b

30. a)   tetrahedral; tetrahedral
    b)   linear; linear
    c)   tetrahedral; bent

31. a)   tetrahedral; trigonal pyramid
    b)   tetrahedral; bent
    c)   tetrahedral; trigonal pyramid

32. a)   trigonal bipyramid; trigonal bipyramid
    b)   trigonal bipyramid; see-saw
    c)   trigonal bipyramid; linear

33. a)   trigonal planar; trigonal planar
    b)   octahedral; square planar
    c)   octahedral; square pyramid

34. a)   tetrahedral; linear
    b)   trigonal bipyramid; T-shape
    c)   octahedral; octahedral

35. a)   linear; linear
    b)   trigonal planar; bent
    c)   linear; linear

36. a) 109.5°
    b) 180°
    c) 104.5°

37. a) 107°
    b) 104.5°
    c) 107°

38. a) 120°; 90°; 180°
    b) 120°; 90°; 180°
    c) 180°

39. a) 120°
    b) 90°; 180°
    c) 90°; 180°

40. a) 180°
    b) 90°; 180°
    c) 90°; 180°

41. a) 180°
    b) 120°
    c) 180°

42. A bond dipole occurs when there is a difference in electronegativity between two atoms in a bond.

43. S—Cl, P—Cl, Si—Cl, Mg—Cl

44.

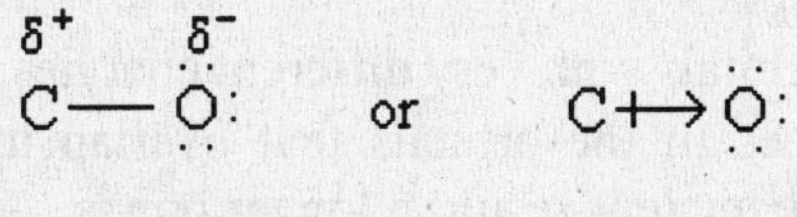

$\overset{\delta^+}{C} — \overset{\delta^-}{\ddot{O}:}$ or $C \longleftrightarrow \ddot{O}:$

45. b) nonpolar
    c) polar

46. a) polar
    b) polar

47. a) nonpolar
    b) polar

48. a) nonpolar
    b) nonpolar
    c) polar

49. a) polar
    b) polar
    c) nonpolar

50. SO_2 is polar. BeH_2 is nonpolar.

Chapter 9

Advanced Models of Bonding

The Bottom Line

When you finish this chapter, you will be able to:

- Differentiate between the VSEPR theory and valence bond theory.
- Predict the hybridization of atoms in a molecule or ion.
- Distinguish between a sigma (σ) bond and a pi (π) bond, and determine how many of these bonds exist in a molecule.
- Create molecular orbital diagrams and use them to predict the bond order of a molecule.
- Predict whether a species will be paramagnetic or diamagnetic.

9.1 Valence Bond Theory

While the VSEPR model and Lewis dot structures are useful for determining the shape and polarity of molecules, they do not adequately predict many of their properties. As a result, we need to understand better models, starting with the valence bond theory. Here are the key points about the valence bond model:

- A bond is the overlap of atomic orbitals on adjacent atoms.
- Orbitals that exhibit more overlap result in more electron sharing and stronger covalent bonds.
- Smaller orbitals overlap more than larger orbitals.
- Orbitals with similar sizes overlap more than orbitals with mismatched sizes. **Table 9.2** in your text shows how orbital size affects bond strength.
- Orbitals with similar energies overlap more than orbitals with very different energies.
- Differences in bond length result from differences in the orbitals that overlap to form a covalent bond. Orbitals that extend farther from the nucleus result in longer bonds.

Example 1   **Bond Strength**

Arrange the following compounds in order from strongest to weakest bond strength.

HBr, HI, HCl

First Thoughts

Think about the type of orbital overlap that occurs in each bond. The more overlap, the stronger the covalent bond.

Solution

HBr contains $1s$–$4p$ orbital overlapping.
HI contains $1s$–$5p$ orbital overlapping.
HCl contains $1s$–$3p$ orbital overlapping.

Orbitals with similar energies overlap more than orbitals with different energies. HCl contains the overlapping orbitals that are closest in energy, so it contains the strongest bond. HI contains the overlapping orbitals with the greatest difference in energy, so it contains the weakest bond.

146

$$HCl > HBr > HI$$

Further Insight

You can verify these strengths by looking at the bond dissociation energies in Table 9.2.

$$HCl = 431.9 \text{ kJ/mol}$$
$$HBr = 366.3 \text{ kJ/mol}$$
$$HI = 298.4 \text{ kJ/mol}$$

9.2   Hybridization

The valence bond theory does not properly explain bond angles in molecules, so it was modified to include **hybridization.** Through the hybridization process, we can mathematically combine two or more orbitals to provide new orbitals of equal energy. Here are some important facts about hybridization:

- Typically, only orbitals of the same subshell are hybridized (such as $2s$ and $2p$), resulting in a set of new orbitals that have the properties of all the orbitals from which they were made.
- The number of orbitals that are hybridized determines the number of new orbitals that are made.
- The new orbitals that are created are degenerate (have equal energy) and have the same shape, but are oriented in different directions.
- The energy of the resulting hybridized orbitals should be the weighted average energy of the atomic orbitals that were mixed.
- **Table 9.3** in your text explains the steps needed to convert atomic orbitals into hybridized orbitals.
- The bond angles between the resulting hybridized orbitals agree with what we expect from the VSEPR model.

Example 2 sp, sp^2, and sp^3 **Hybridization**

Determine the shape and hybridization of the central atom in each of the following molecules.

a) $CHCl_3$ b) BH_3 c) CO_2

First Thoughts

Use Table 9.3 to assist you in determining the hybridization. Start by drawing the Lewis structure for each molecule. Note the number of attached atoms and lone pairs, and then select the orbitals to be hybridized.

Solution

a) $CHCl_3$

The four electron pairs around the carbon atom require a tetrahedral arrangement according to the VSEPR model. The hybridization is obtained by mixing the $2s$ orbital and the three $2p$ orbitals. The hybridization on carbon is therefore sp^3. (The four sp^3 orbitals are used to form bonds to the three chlorine atoms and one hydrogen atom.)

b) BH_3

The three electron pairs around the boron atom require a trigonal planar arrangement according to the VSEPR model. The hybridization is obtained by mixing the $2s$ orbital and the two $2p$ orbitals. The hybridization on boron is therefore sp^2. (The three sp^2 orbitals are used to form bonds to the three hydrogen atoms.)

c) CO_2

The two effective electron pairs around the carbon atom require a linear arrangement according to the VSEPR model. The hybridization is obtained by mixing the $2s$ orbital and the $2p$ orbital. The hybridization on carbon is therefore sp. (The sp orbitals are used to form a single bond with each oxygen atom. We will discuss how the double bond forms later in this section.)

Example 3 sp^3d and sp^3d^2 **Hybridization**

Determine the shape and hybridization of the central atom in each of the following molecules.

                    a) PF_5 b) SF_6

Solution

a) PF_5

The five electron pairs around the phosphorus atom require a trigonal bipyramidal arrangement according to the VSEPR model. The hybridization is obtained by mixing one s orbital, three p orbitals, and one d orbital. The hybridization on phosphorus is therefore sp^3d. (The five sp^3d orbitals are used to form bonds to the five fluorine atoms.)

b) SF_6

The six electron pairs around the sulfur atom require an octahedral arrangement according to the VSEPR model. The hybridization is obtained by mixing one s orbital, three p orbitals, and two d

orbitals. The hybridization on sulfur is therefore sp^3d^2. (The six sp^3d^2 orbitals are used to form bonds to the six fluorine atoms.)

So far, we have discussed the valence bond theory only in terms of the formation of single bonds. Of course, many molecules also contain double or triple bonds. Almost every single bond is a **sigma (σ) bond.** Sigma bonds result from an end-on overlap of orbitals. **Pi (π) bonds** result from the side-to-side overlap of orbitals. A π bond usually occurs when p orbitals on adjacent atoms overlap—a situation that is found between atoms containing a double or triple bond. A double bond contains one σ bond and one π bond. A triple bond contains one σ bond and two π bonds. **Figure 9.19** in your text illustrates the formation of σ and π bonds.

Example 4 σ and π Bonds

How many sigma (σ) bonds and pi (π) bonds are in the following molecule, mimosine?

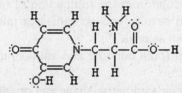

First Thoughts

To determine the number of σ bonds, add up the total number of single bonds in the molecule. To determine the number of π bonds, add up the total number of double bonds in the molecule.

Solution

σ bonds = 24
π bonds = 4

The use of hybridization is necessary so that we can accurately predict the bond angles of a molecule and have them agree with the experimentally measured bond angles. In addition, hybridization provides information about the relative lengths of bonds in molecules. **Table 9.4** in your text lists the relative sizes of some hybridized orbitals. **In general, sp^3 hybrid orbitals make bonds that are longer than those for sp^2 orbitals, and sp^2 hybrid orbitals make bonds that are longer than those for sp orbitals.**

Example 5   Bond Length

Three Lewis dot structures are shown below labeled (a), (b), and (c). Which has the longest C—C bond? Which has the shortest C—C bond? (The bond lengths in question are indicated by an arrow.)

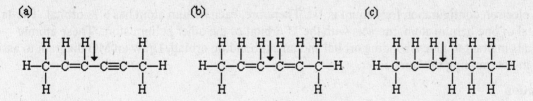

First Thoughts

Determine the hybridization of the two carbon atoms in the indicated bond for each molecule. Then envision the type of orbital overlap that you expect for each bond.

Solution

The longest C—C bond is found in compound (c). It results from the overlap of an sp^2 hybridized orbital with an sp^3 hybridized orbital. The shortest bond is found in compound (a). It results from the overlap of an sp^2 hybridized orbital with an sp hybridized orbital.

9.3   Molecular Orbital Theory

Here are the key points of the molecular orbital (MO) theory:

- The MO theory constructively and destructively adds atomic orbitals together to make molecular orbitals and achieve a lower energy state.
- A **bonding orbital** results from the addition of two overlapping atomic orbitals and indicates that there is some electron density between adjacent nuclei.
- An **antibonding orbital** results from the subtraction of two overlapping atomic orbitals and indicates a lack of electron density between adjacent nuclei (no bond exists). Antibonding orbitals are represented by an asterisk (*).
- Each new molecular orbital (both bonding and antibonding) can contain two electrons with opposite spins.
- Only atomic orbitals of similar symmetry (shape and orientation) and energy provide significant overlap in creating molecular orbitals.
- The strength of the bond between two atoms can be represented by calculating the bond order:

$$\text{Bond Order} = \frac{\text{Bonding Electrons - Antibonding Electrons}}{2}$$

  The larger the bond order, the greater the bond strength.
- MO diagrams graphically show the energies of the bonding and antibonding orbitals and help determine the bond order for a molecule. **Figure 9.23** in your text shows an MO diagram for H_2.
- The highest-energy occupied molecular orbital is known as **HOMO;** the lowest-energy unoccupied molecular orbital is known as **LUMO.**
- **Table 9.5** in your text shows the key features of constructing an MO diagram.

Example 6   **MO Theory**

Use the MO theory to determine whether He_2 exists. What is the bond order in this proposed molecule?

First Thoughts

The electron configuration for helium is $1s^2$. Therefore, each helium atom has a $1s$ orbital. The $1s$ orbital of one helium atom interacts with the $1s$ orbital of the other helium atom. These atomic orbitals interact to give a bonding orbital and an antibonding orbital. Draw an MO diagram to assist you in determining the bond order.

Solution

$$\text{He} \;\; \frac{\uparrow\downarrow}{1s} \qquad \frac{\quad}{\sigma_{1s}^*} \qquad \frac{\uparrow\downarrow}{1s} \;\; \text{He}$$

$$\frac{\quad}{\sigma_{1s}}$$

Place two electrons (one from each He atom) into the σ_{1s} orbital because it is lowest in energy.

$$\overline{\sigma_{1s}^*}$$

He $\underline{\uparrow}$ $\underline{\uparrow}$ He
$1s$ $1s$

$$\underline{\uparrow\downarrow}$$
$$\sigma_{1s}$$

Place the remaining two electrons into the higher-energy σ^*_{1s} orbital.

$$\underline{\uparrow\downarrow}$$
$$\sigma_{1s}^*$$

He $\underline{}$ $\underline{}$ He
$1s$ $1s$

$$\underline{\uparrow\downarrow}$$
$$\sigma_{1s}$$

Then, calculate the bond order using the equation given earlier:

$$\text{Bond Order} = \frac{(2-2)}{2} = 0$$

He_2 should not exist. There should not be any bonds between the two atoms. Conditions are more stable when the helium atoms exist freely than when they bond together.

When constructing MO diagrams, you will notice that some species contain one or more unpaired electrons (which is different than Lewis structures). When this occurs, the molecule is considered **paramagnetic**—that is, it is attracted to a magnetic field. The opposite of paramagnetism is **diamagnetism,** in which all of the electrons in the MO diagram of a species are paired.

Example 7   Paramagnetism versus Diamagnetism

Determine the bond orders for F_2 and F_2^-. State whether each one is paramagnetic or diamagnetic.

First Thoughts

Draw the MO diagrams for F_2 and F_2^-. Bonding for these two species will involve the $2s$ and $2p$ orbitals. Each fluorine atom in F_2 and F_2^- contains seven valence electrons; the F_2^- ion also contains one extra electron from the -1 charge.

Solution

F_2:

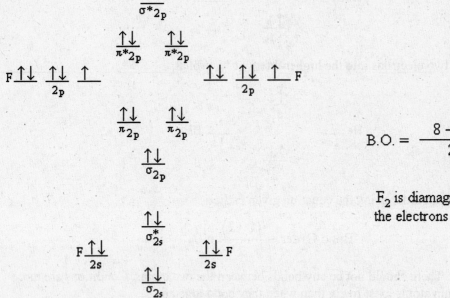

$$B.O. = \frac{8-6}{2} = \frac{2}{2} = 1$$

F_2 is diamagnetic since all of the electrons are paired.

The MO diagram for the F_2^- ion looks the same except that one extra electron is placed into the σ^*_{2p} orbital.

F_2^-:

$$B.O. = \frac{8-7}{2} = \frac{1}{2} = 0.5$$

F_2^- is paramagnetic since one of the electrons is unpaired.

Further Insight

F_2 has a stronger bond strength than F_2^- because it has a higher bond order. In addition, for F_2, HOMO is the π^*_{2p} orbital and LUMO is the σ^*_{2p} orbital. For F_2^-, HOMO is the σ^*_{2p} orbital and LUMO is the σ_{3s} orbital.

Example 8   Single Photon of Light

N_2 and O_2 each interact with a single photon of light. After the promotion of an electron, which molecule has the smaller bond order?

First Thoughts

When a molecule interacts with a single photon of light, one electron is promoted from a bonding orbital to an antibonding orbital of similar symmetry. Incorporate this change into your MO diagram before determining the bond order.

Solution

N_2:

$$B.O. = \frac{7-3}{2} = \frac{4}{2} = 2$$

O_2:

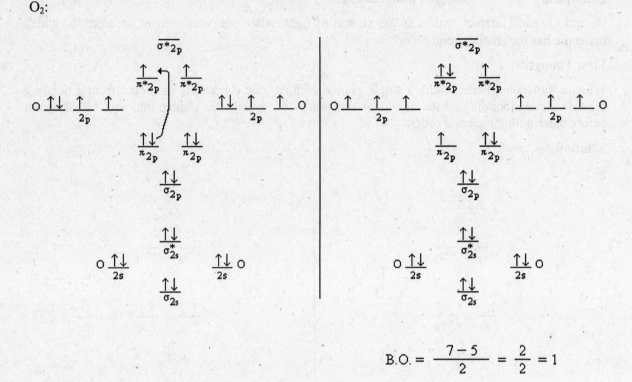

$$B.O. = \frac{7-5}{2} = \frac{2}{2} = 1$$

O_2 has the smaller bond order when interacting with a single photon of light.

9.4   Putting It All Together

Using a combination of all these bonding theories—Lewis dot structures, VSEPR theory, and valence bond theory/molecular orbital theory—helps to better explain the observed properties of many covalent compounds. Your text shows how to use this combination of theories when drawing the molecule benzene, C_6H_6. It also discusses and illustrates the concept of **delocalization,** in which the electrons in π orbitals are spread out over the atoms in a structure. Be sure to understand what this means and how to draw delocalization in a molecule.

Exercises

Section 9.1

1. True or false? Larger orbitals overlap more than smaller orbitals.

2. True or false? Orbitals with similar sizes overlap more than orbitals with mismatched sizes.

3. True or false? Orbitals with similar energies overlap more than orbitals with very different energies.

4. True or false? Orbitals that extend farther from the nucleus result in longer bonds.

5. True or false? Orbitals that exhibit more overlap result in less electron sharing and weaker covalent bonds.

6. What is the major weakness of the VSEPR model?

7. Arrange the following compounds in order from strongest to weakest bond strength (without looking at Table 9.2): NaH, H_2, LiH. Justify your answer.

8. Arrange the following compounds in order from strongest to weakest bond strength (without looking at Table 9.2 in your text): F_2, HBr, HF.

Section 9.2

9. What is hybridization?

10. True or false? A new set of hybridized orbitals has the properties of all of the orbitals from which they were made.

11. True or false? New hybridized orbitals that are created are degenerate and have the same shape, but are oriented in different directions.

12. True or false? The bond angles between the resulting hybridized orbitals are not the same as the bond angles determined by the VSEPR model.

13. Determine the shape and hybridization of the central atom in $BeCl_2$.

14. Determine the shape and hybridization of the central atom in NF_3.

15. Determine the shape and hybridization of the central atom in NO_3^-.

16. Determine the shape and hybridization of the central atom in XeF_4.

17. Determine the shape and hybridization of the central atom in SF_4.

Consider the following molecules or ions to answer Exercises 18–22:

     HCN Br_3^- NH_4^+ PH_3 ICl_3 SiF_4

18. Which of these molecules or ions contain sp orbitals about the central atom?

19. Which of these molecules or ions contain sp^2 orbitals about the central atom?

20. Which of these molecules or ions contain sp^3 orbitals about the central atom?

21. Which of these molecules or ions contain sp^3d orbitals about the central atom?

22. Which of these molecules or ions contain sp^3d^2 orbitals about the central atom?

23. How many sigma (σ) bonds and pi (π) bonds are in the molecule benzene?

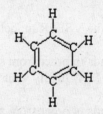

24. Cholesterol ($C_{27}H_{46}O$) has the following structure:

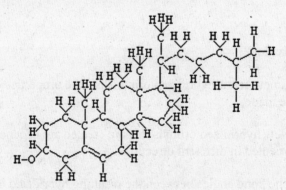

    State the number of σ bonds and π bonds in cholesterol.

Consider the following molecule for Exercises 25–29.

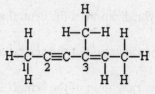

25. What is the hybridization of C_1?

26. What is the hybridization of C_2?

27. What is the hybridization of C_3?

28. How many σ bonds are present in this molecule?

29. How many π bonds are present in this molecule?

30. Order the following species with respect to nitrogen–nitrogen bond length (longest to shortest): N_2F_4, N_2F_2, N_2.

Section 9.3

31. What is the advantage of using the molecular orbital (MO) theory?

32. How does a bonding orbital form according to MO theory?

33. How does an antibonding orbital form according to MO theory?

34. True or false? The higher the bond order between two atoms in a molecule, the weaker the bond strength.

35. True or false? The electrons in an antibonding orbital are higher in energy than the electrons in a bonding orbital.

36. Draw a molecular orbital diagram for the electrons in the third principal energy level for Ar_2. Does this molecule exist? Explain your answer.

Consider the following species for Exercises 37–39.

$$H_2, H_2^+, H_2^-$$

37. Calculate the bond order for each species.

38. Identify which, if any, species are paramagnetic.

39. Which species has the strongest bond strength?

Consider the following species for Exercises 40 and 41.

$$He_2^{2+}, O_2^{2-}, B_2$$

40. Calculate the bond order for each species.

41. Identify which, if any, species are paramagnetic.

A diatomic homonuclear molecule has the following molecular orbital configuration:

$$(\sigma 1s)^2 (\sigma^* 1s)^2 (\sigma 2s)^2$$

42. What is the identity of this molecule?

43. Is this molecule paramagnetic or diamagnetic?

44. What is the bond order for this molecule?

45. Write the molecular orbital configuration for the species C_2.

Answers to Exercises

1. false

2. true

3. true

4. true

5. false

6. The VSEPR model does not adequately predict many properties of a molecule, such as bond length.

7. $H_2 > LiH > NaH$. The greater the amount of overlap, the stronger the bond. H_2 has the greatest amount of orbital overlap due to the similarity of the orbitals (1s–1s). LiH has more overlap than NaH because the two orbitals are more similar in energy (1s–2s versus 1s–3s).

8. $HF > HBr > F_2$

9. Hybridization is the process of mathematically combining two or more orbitals to provide new orbitals of equal energy.

10. true

11. true

12. false

13. linear, sp

14. bent, sp^3

15. trigonal planar, sp^2

16. square planar, sp^3d^2

17. see-saw, sp^3d

18. HCN

19. None of the compounds contain sp^2 hybridization.

20. NH_4^+, PH_3, SiF_4

21. Br_3^-, ICl_3

22. None of the compounds contain sp^3d^2 hybridization.

23. $\sigma = 12, \pi = 3$

24. $\sigma = 77, \pi = 1$

25. sp^3

26. sp

27. sp^2

28. 16

29. 3

30. N_2F_4, N_2F_2, N_2

31. The MO theory works best at describing the behavior of electrons located in the bonds.

32. A bonding orbital results from the addition of two overlapping atomic orbitals.

33. An antibonding orbital results from the subtraction of two overlapping atomic orbitals.

34. false

35. true

36. No, this molecule should not exist. The bond order is zero, so a bond does not exist between the two Ar atoms.

37. $H_2 = 1$, $H_2^+ = 0.5$, $H_2^- = 0.5$

38. H_2^+, H_2^-

39. H_2^-

40. $He_2^{2+} = 1$, $O_2^{2-} = 1$, $B_2 = 1$

41. None of the species are paramagnetic. They are all diamagnetic.

42. Li_2

43. diamagnetic

44. bond order = 1

45. $(\sigma 1s)^2 (\sigma * 1s)^2 (\sigma 2s)^2 (\sigma * 2s)^2 (\sigma 2p)^2 (\pi 2p)^2$

Chapter 10

The Behavior and Applications of Gases

The Bottom Line

When you finish this chapter, you will be able to:

- Identify properties of gases.
- Describe how real gases deviate from ideal behavior.
- Differentiate between different units of pressure.
- Understand and apply Dalton's law of partial pressures, Avogadro's law, Boyle's law, Charles's law, the combined gas law, the van der Waals equation, the root-mean-square (rms) speed of a gas and Graham's law of effusion.
- Explain the behavior of gases using the kinetic molecular theory.

10.1   The Nature of Gases

Under normal conditions, most gases behave as **ideal gases,** as outlined in **Table 10.3** in your text. Here are some other points about gases:

- Gases have low densities.
- Gases are compressible.
- Gases are greatly affected by changes in temperature and pressure.
- Gases behave more ideally at low pressures and high temperatures.

10.2   Production of Hydrogen and the Meaning of Pressure

Because gas particles contain mass and can accelerate, they exert force on the walls of their container. When large numbers of gas particles exert a certain amount of force to a given area, the gas has **pressure.** There are several units for pressure including pascals (Pa), atmospheres (atm), millimeters of mercury (mm Hg), torr, bars, and pounds per square inch (psi). **Table 10.4** in your text lists several useful pressure unit conversion factors. At 0°C (273 K) and 1.00 atm, a gas is at **standard temperature and pressure (STP).**

Example 1   **Pressure Conversions**

At 25,000 ft, the barometric pressure measured outside an airplane is 335 mm Hg. What is the pressure in atm, torr, kPa, and psi?

First Thoughts

Use Table 10.4 in your text to assist you in making your pressure conversions.

Solution

$$(335 \text{ mm Hg})\left(\frac{1 \text{ atm}}{760 \text{ mm Hg}}\right) = 0.441 \text{ atm}$$

$$(335 \text{ mm Hg})\left(\frac{760 \text{ torr}}{760 \text{ mm Hg}}\right) = 335 \text{ torr}$$

$$(335 \text{ mm Hg})\left(\frac{101{,}325 \text{ Pa}}{760 \text{ mm Hg}}\right)\left(\frac{1 \text{ kPa}}{1000 \text{ Pa}}\right) = 44.7 \text{ kPa}$$

$$(335 \text{ mm Hg})\left(\frac{14.7 \text{ psi}}{760 \text{ mm Hg}}\right) = 6.48 \text{ psi}$$

10.3   Mixtures of Gases: Dalton's Law and Food Packaging

Dalton's law of partial pressures states that for a mixture of gases in a container, the total pressure is equal to the sum of the pressures that each gas would exert if it were alone. Each gas in the container exerts a **partial pressure**.

Example 2   **Partial Pressure**

Hydrogen gas at a pressure of 469 torr is in a 5-L gas tank. Argon gas at 312 torr is in a different 5-L gas tank. The two gases are combined in a 5-L steel vessel at the same temperature. What is the pressure of the gas mixture?

Solution

Using Dalton's law of partial pressures, the total pressure inside the vessel is the sum of the partial pressures of hydrogen and argon.

$$P_{Total} = P_{H_2} + P_{Ar}$$
$$P_{Total} = 469 \text{ torr} + 312 \text{ torr}$$
$$P_{Total} = 781 \text{ torr}$$

10.4 The Gas Laws: Relating the Behavior of Gases to Key Properties

Several gas laws that describe the behavior of ideal gases. A summarized version of these laws can be found in **Table 10.5** in your text.

Avogadro's Law: The volume is directly proportional to the amount of a gas expressed in moles, at constant temperature and pressure ($V = kn$)

$$\frac{V_{initial}}{n_{initial}} = k = \frac{V_{final}}{n_{final}} \quad \text{OR} \quad \frac{V_{initial}}{n_{initial}} = \frac{V_{final}}{n_{final}}$$

where V = volume occupied by the gas
       n = amount of a gas expressed in moles
       k = a constant for a given temperature and pressure

Example 3   Avogadro's Law

If 0.281 mol of neon gas occupies a volume of 6.87 L at 25°C and 1.00 atm, what volume would 0.113 mol of neon occupy under the same conditions?

First Thoughts

Because the temperature and the pressure remain constant, only the volume and moles of gas change, which are directly related.

$$V_{initial} = 6.87 \text{ L} \qquad\qquad V_{final} = ?$$

$$n_{initial} = 0.281 \text{ mol} \qquad\qquad n_{final} = 0.113 \text{ mol}$$

Solution

Using Avogadro's law, we get

$$\frac{V_{initial}}{n_{initial}} = \frac{V_{final}}{n_{final}} \quad \text{OR} \quad \frac{6.87 \text{ L}}{0.281 \text{ mol}} = \frac{V_{final}}{0.113 \text{ mol}}$$

$$V_{final} = 2.76 \text{ L}$$

Further Insight

Our answer of 2.76 L is reasonable. Because the number of moles of gas gets smaller, the neon gas occupies a smaller volume.

Boyle's Law: The volume and pressure of a given amount of gas are inversely proportional at constant temperature ($PV = k'$)

$$P_{initial}V_{initial} = P_{final}V_{final}$$

where P = pressure of the gas
        V = volume occupied by the gas

Example 4   Boyle's Law

The pressure on a 1.50-L sample of helium gas is increased from 755 torr to 1.80 atm at constant temperature. What is the new volume of the helium gas sample?

First Thoughts

Because the temperature and the amount of helium (moles) remain constant, only the volume and pressure change, which are inversely proportional.

$$V_{initial} = 1.50 \text{ L} \qquad\qquad\qquad\qquad V_{final} = ?$$

$$P_{initial} = (755 \text{ torr})\left(\frac{1 \text{ atm}}{760 \text{ torr}}\right) = 0.993 \text{ atm} \quad P_{final} = 1.80 \text{ atm}$$

Solution

Using Boyle's law, we get

$$P_{initial}V_{initial} = P_{final}V_{final}$$

$$(0.993 \text{ atm})(1.50 \text{ L}) = (1.80 \text{ atm})V_{final}$$

$$V_{final} = 0.828 \text{ L}$$

Further Insight

Our answer of 0.828 L is reasonable. Because the pressure of helium is increasing, more collisions with the walls of its container must be occurring. For this to happen, the volume must decrease.

Charles's Law: The volume of a gas is directly proportional to the temperature at constant pressure and number of moles of gas ($V = k''T$)

$$\frac{V_{initial}}{T_{initial}} = \frac{V_{final}}{T_{final}}$$

where V = volume occupied by the gas
         T = temperature of the gas in kelvins

Temperature is a measure of the average kinetic energy of the gas particles in a system. The coldest possible temperature is known as **absolute zero,** which occurs at −273.15°C. At this temperature, the gas particles have no kinetic energy and, therefore, stop moving. When using Charles's law, make sure the temperature is measured in kelvins, because this scale measures the kinetic energy of the gas particles.

$$K = {}^{\circ}C + 273.15$$

Example 5   **Charles's Law**

If 0.440 L of a gas at 29°C is heated to 45°C at constant pressure, what is the new volume of the sample?

First Thoughts

Because the pressure and the moles of gas remain constant, only the volume and temperature change, which are directly related. In addition, the temperature must be in kelvins.

$V_{initial} = 0.440$ L $V_{final} = ?$

$T_{initial} = 29°C + 273.15 = 302$ K $T_{final} = 45°C + 273.15 = 318$ K

Solution

Using Charles's law, we get

$$\frac{V_{initial}}{T_{initial}} = \frac{V_{final}}{T_{final}} \quad \text{OR} \quad \frac{0.440\text{ L}}{302\text{ K}} = \frac{V_{final}}{318\text{ K}}$$

$$V_{final} = 0.463\text{ L}$$

Further Insight

Our answer of 0.463 L is reasonable. Because the temperature increases, the gas particles move faster and farther apart and, therefore, occupy more volume.

Combined Gas Equation: This equation combines Avogadro's law, Boyle's law, and Charles's law. We can use the combined gas equation to solve for the pressure, volume, amount, or temperature of a gas, as the gas changes conditions from one state to another. If one of the variables remains constant, that condition drops out of the equation, which simplifies our calculations.

$$\frac{P_{initial}V_{initial}}{n_{initial}T_{initial}} = \frac{P_{final}V_{final}}{n_{final}T_{final}}$$

Example 6 — Combined Gas Equation

If 1.25 moles of nitrogen gas occupies a volume of 28.0 L at STP, at what temperature (in °C) will the gas occupy a volume of 35.0 L at 1.78 atm of pressure?

First Thoughts

Only the moles of nitrogen gas remains constant. The temperature, volume, and pressure of the gas are changing, so we should use the combined gas equation.

$V_{initial} = 28.0$ L $V_{final} = 35.0$ L

$P_{initial} = 1.00$ atm (at STP) $V_{final} = 1.78$ atm

$T_{initial} = 273$ K (at STP) $T_{final} = ?$

Solution

Using the combined gas equation, we get

$$\frac{P_{initial}V_{initial}}{n_{initial}T_{initial}} = \frac{P_{final}V_{final}}{n_{final}T_{final}}$$

Because the moles of nitrogen are constant, $n_{initial}$ and n_{final} drop out of the equation:

$$\frac{P_{initial}V_{initial}}{T_{initial}} = \frac{P_{final}V_{final}}{T_{final}}$$

$$\frac{(1.00 \text{ atm})(28.0 \text{ L})}{(273 \text{ K})} = \frac{(1.78 \text{ atm})(35.0 \text{ L})}{T_{final}}$$

$$T_{final} = 607 \text{ K}$$

Finally, convert the final temperature to °C:

$$T_{final} = 607 - 273.15 = 334°C$$

$$T_{final} = 334°C$$

10.5 The Ideal Gas Equation

Avogadro's law, Boyle's law, and Charles's law can be combined into another equation, called the ideal gas law:

$$PV = nRT$$

Your text goes into more detail as to how the ideal gas equation is derived. R is known as the **ideal gas constant** and is equal to $0.08206 \text{ L} \cdot \text{atm/mol} \cdot \text{K}$.

Example 7   **Ideal Gas Law I**

What is the temperature (in K) if 50.0 g of oxygen gas (O_2) is in a 2.50-L vessel at a pressure of 15.1 atm?

First Thoughts

The ideal gas equation can be used to solve this problem but first we must convert the mass of oxygen gas to moles.

$$\left(50.0 \text{ g O}_2\right)\left(\frac{1 \text{ mol O}_2}{32.00 \text{ g O}_2}\right) = 1.56 \text{ mol O}_2$$

Solution

$T = ?$
$n = 1.56$ mol
$P = 15.1$ atm
$V = 2.50$ L
$R = 0.08206 \text{ L} \cdot \text{atm/mol} \cdot \text{K}$

Using the ideal gas equation, we get

$$PV = nRT$$
$$(15.1 \text{ atm})(2.50 \text{ L}) = (1.56 \text{ mol})(0.08206 \text{ L} \cdot \text{atm/mol} \cdot \text{K})(T)$$
$$T = 295 \text{ K}$$

Example 8   **Ideal Gas Law II**

How many SO_2 gas molecules are contained in a 10.0-L container at 30.°C and 2.00 atm of pressure?

First Thoughts

Use the ideal gas equation to solve for the moles of SO_2 present. Then use Avogadro's number to calculate the number of SO_2 molecules.

Solution

$n = ?$
$P = 2.00$ atm
$V = 10.0$ L
$R = 0.08206$ L $\cdot$ atm/mol $\cdot$ K
$T = 30.°C + 273.15 = 303$ K

$$PV = nRT$$

$$(2.00 \text{ atm})(10.0 \text{ L}) = (n)(0.08206 \text{ L} \cdot \text{atm/mol} \cdot \text{K})(303 \text{ K})$$

$$n = 0.804 \text{ mol SO}_2$$

Use Avogadro's number to convert to molecules of SO_2:

$$\left(0.804 \text{ mol SO}_2\right)\left(\frac{6.022 \times 10^{23} \text{ molecules SO}_2}{1 \text{ mol SO}_2}\right) = 4.84 \times 10^{23} \text{ molecules SO}_2$$

Real gases deviate from ideal gas behavior, especially at high pressures and low temperatures:

- Real gases occupy discrete volumes, so the available volume of the container is not as large as if it were empty or we assumed that the gas particles had no volume. (See **Figure 10.20** in your text for an illustration.)
- Real gases interact with each other, especially when those atoms or molecules exhibit stronger intermolecular forces. This causes a change in pressure because the particles do not collide with the walls of their container as frequently.

The ideal gas equation must be modified to include these deviations for real gases. The **van der Waals equation** is used for this purpose:

$$P = \frac{nRT}{V - nb} - a\frac{n^2}{V^2}$$

Table 10.6 in your text lists the van der Waals constants (a and b) for some common gases.

Example 9   Deviations from Ideal Behavior

What is the pressure of 0.830 moles of argon gas in a 1.00-L balloon at 25°C? Calculate the pressure using both the ideal gas law and the van der Waals equation. Is there a difference in your answers?

First Thoughts

We will need to use Table 10.6 to find the values of a and b in the van der Waals equation. It is also a good idea to list all of the information we know that will help us solve the problem, as we have done in previous examples.

$P = ?$
$V = 1.000$ L
$T = 25°C + 273.15 = 298$ K
$n = 0.830$ mol Ar
$R = 0.08206$ L $\cdot$ atm/mol $\cdot$ K
$a = 1.35$ atm $\cdot$ L^2/mol^2
$b = 0.0322$ L/mol

168   Chapter 10

Solution

Using the ideal gas law, we get

$$PV = nRT$$

$$(P)(1.000 \text{ L}) = (0.830 \text{ mol})(0.08206 \text{ L} \cdot \text{atm/mol} \cdot \text{K})(298 \text{ K})$$

$$P_{\text{ideal}} = 20.3 \text{ atm}$$

Using the van der Waals equation, we get

$$P = \frac{nRT}{V - nb} - a\frac{n^2}{V^2}$$

$$P = \frac{(0.830 \text{ mol})(0.08206 \text{ L} \cdot \text{atm/mol} \cdot \text{K})(298 \text{ K})}{(1.000 \text{ L} - [0.830 \text{ mol} \times 0.0322 \text{ L/mol}])} - (1.35 \text{ atm} \cdot \text{L}^2/\text{mol}^2)\left(\frac{(0.830 \text{ mol})^2}{(1.000 \text{ L})^2}\right)$$

$$P = 20.854 \text{ atm} - 0.930 \text{ atm}$$

$$P_{\text{real}} = 19.9 \text{ atm}$$

(when significant figures are taken into account)

Yes, there is a difference in pressures between argon behaving "ideally" versus under "normal" conditions.

Further Insight

As you can see, the calculations are a lot more involved for real gases. To make our calculations easier, we will assume that all gases behave ideally unless we are told otherwise.

10.6   Applications of the Ideal Gas Equation

There are many ways that the ideal gas equation can be used to solve real-world problems. For example, we can use the ideal gas equation and the density of a gas to determine the molar mass of a gas. Your text goes into more detail, but here is the short form:

$$\underline{M} = \frac{dRT}{P}$$

where $\underline{M}$ = molar mass of the gas
   d = density of the gas
   $R = 0.08206 \text{ L} \cdot \text{atm/mol} \cdot \text{K}$
   T = temperature of the gas in kelvins
   P = pressure of the gas

Example 10   Molar Mass of a Gas

At 139°C and 1.00 atm, the density of a gas is 2.55 g/L. What is the molar mass of this gas?

Solution

$$M = \frac{dRT}{P}$$

$$M = \frac{(2.55 \text{ g/L})(0.08206 \text{ L} \cdot \text{atm/mol} \cdot \text{K})\left(\left[139°C + 273.15\right]K\right)}{1.00 \text{ atm}}$$

$$\underline{M} = 86.2 \text{ g/mol}$$

The ideal gas equation can also be useful when solving problems involving chemical reactions.

Example 11   Volume of Gas Produced

Solid calcium carbonate produces solid calcium oxide and carbon dioxide gas when a lot of heat is applied. What volume of carbon dioxide, measured at 1.00 atm and 25°C, is produced if 20.0 g of calcium carbonate is heated?

First Thoughts

Before determining how much carbon dioxide can be produced, we must write a balanced chemical equation: $CaCO_3(s) \rightarrow CaO(s) + CO_2(g)$

Solution

Determine the number of moles of $CaCO_3$ first present and then use the mole ratio between CO_2 and $CaCO_3$ to calculate the moles of CO_2 produced.

$$\left(20.0 \text{ g CaCO}_3\right)\left(\frac{1 \text{ mol CaCO}_3}{100.09 \text{ g CaCO}_3}\right)\left(\frac{1 \text{ mol CO}_2}{1 \text{ mol CaCO}_3}\right) = 0.200 \text{ mol CO}_2 \text{ produced}$$

Next, use the ideal gas equation to determine the volume of CO_2 produced.

$V = ?$
$P = 1.00$ atm
$T = 25°C + 273.15 = 298$ K
$R = 0.08206$ L $\cdot$ atm/mol $\cdot$ K
$n_{CO_2} = 0.200$ mol

$$PV = nRT$$

$$\left(1.00 \text{ atm}\right)\left(V\right) = \left(0.200 \text{ mol}\right)\left(0.08206 \text{ L} \cdot \text{atm/mol} \cdot \text{K}\right)\left(298 \text{ K}\right)$$

$$V = 4.89 \text{ L}$$

10.7 Kinetic Molecular Theory

The **kinetic molecular theory (KMT)** is very important to our understanding of gases. It allows us to explain the behavior of gases and the gas laws by thinking about the nature of the gas particles themselves. This theory agrees with experiments done on gases in the laboratory. The KMT is based on statements that model gas behavior; these statements are listed in your text. Make sure you study and understand them. Most importantly, you must be able to use the KMT to explain the observations you make about gases in the real world.

Example 12   **Kinetic Molecular Theory**

Using the ideas of the KMT, explain why a party balloon's volume gets larger when you add helium at constant temperature and pressure. Include in your explanation how pressure and temperature play a role.

Solution

Because the helium atoms are in constant random motion at a certain temperature and pressure, they contain kinetic energy and collide with the walls of the party balloon with a certain amount of force. Adding helium gas to the balloon increases the number of moles of gas particles present and momentarily increases the number of collisions with the walls of the balloon. However, the balloon will expand until the pressure inside the balloon equals the pressure outside the balloon. So pressure and temperature essentially remain constant (from initial to final conditions). Because the balloon expands, the volume increases, which means there is more room for the helium particles to move around.

10.8 Effusion and Diffusion

At a given temperature, heavier gas particles move more slowly than their lighter-weight counterparts. The **root-mean-square speed** is related to the molar mass of the gas as follows:

$$u_{rms} = \sqrt{\frac{3RT}{M}}$$

where M = molar mass of the gas in kg/mol
  $R = 8.314 \ kg \cdot m^2/s^2 \cdot mol \cdot K$ (the ideal gas constant, just in different units)
  T = temperature of the gas in kelvins

Example 13   **rms Speed of Chlorine Gas**

What is the rms speed of chlorine gas at 35°C?

First Thoughts

Remember that chlorine gas is a diatomic molecule, Cl_2, and that the molar mass must be in units of kg/mol when using this equation.

Solution

$R = 8.314 \ kg \cdot m^2/s^2 \cdot mol \cdot K$
$T = 35°C + 273.15 = 308 \ K$
$$M = \left(\frac{70.9 \ g \ Cl_2}{1 \ mol \ Cl_2}\right)\left(\frac{1 \ kg \ Cl_2}{1000 \ g \ Cl_2}\right) = 0.0709 \ kg \ Cl_2 \ / \ mol$$

$$u_{rms} = \sqrt{\frac{3RT}{\underline{M}}} = \sqrt{\frac{3\left(8.314 \text{ kg} \cdot \text{m}^2/\text{s}^2 \cdot \text{mol} \cdot \text{K}\right)\left(308 \text{ K}\right)}{0.0709 \text{ kg/mol}}}$$

$$u_{rms} = 329 \text{ m/s}$$

The speeds of different gases can also be compared to one another. The relative rate with which gases pass through small openings into a very-low-pressure region is called **effusion.** This process can be quantified by using **Graham's law of effusion:**

$$\frac{re_1}{re_2} = \sqrt{\frac{\underline{M_2}}{\underline{M_1}}}$$

where re_1/re_2 is the ratio of effusion rates for two different gases and $\underline{M_1}$ and $\underline{M_2}$ represent their molar masses. A related phenomenon is called **diffusion,** which involves the mixing of one gas with another gas or with more of the original gas.

Example 14   Effusion of Gases

O_2 and Ne are placed in a long, thin tube that is completely closed on one end and capped on the other end with a porous barrier. Which gas would effuse through the pores in the barrier faster? How much faster would it effuse?

First Thoughts

Let's designate Ne as gas 1 and O_2 as gas 2.

Solution

$\underline{M_1} = 20.18$ g/mol (Ne)
$\underline{M_2} = 32.00$ g/mol (O_2)

$$\frac{re_1}{re_2} = \sqrt{\frac{\underline{M_2}}{\underline{M_1}}} = \sqrt{\frac{32.00 \text{ g/mol}}{20.18 \text{ g/mol}}}$$

$$\frac{re_1}{re_2} = 1.26$$

Because Ne has a lower molar mass, it will effuse 1.26 times faster through the porous barrier than the O_2 gas.

10.9   Industrialization: A Wonderful, Yet Cautionary, Tale

The gases that surround us have a huge impact on our lives and the environment that we live in. Become familiar with the importance of the Earth's ozone layer and the greenhouse effect, as thoroughly discussed in your text.

Exercises

Section 10.1

1. Which of the following statements concerning gases are true?
    a) Gases have high densities.
    b) Gases are not affected by changes in temperature and pressure.
    c) The atoms or molecules of ideal gases behave as individual units.
    d) Gases are not compressible.
    e) All of the above statements about gases are true.

Section 10.2

2. Convert: 475 torr = _____ atm

3. Convert: 2.31 atm = _____ Pa = _____ kPa

4. Convert: 233 mm Hg = _____ psi

5. Convert: 27.4 in Hg = _____ bars

6. Convert: 18.9 psi = _____ kPa

7. The barometric pressure on a very rainy day is measured to be 28.95 in Hg. What is this pressure in atm, torr, and kPa?

Section 10.3

8. Helium gas at a pressure of 1.50 atm is in a 1.00-L container. Chlorine gas at 3.00 atm is in a different 1.00-L container. The two gases are combined in a separate 1.00-L container at the same temperature. What is the pressure of the gas mixture?

9. Samples of 20.0 g of O_2 and 50.0 g of Kr are combined together in a mixture that exerts a pressure of 2.33 atm on its container. Calculate the partial pressure of O_2 and Kr in the mixture.

10. A mixture of gases contains H_2 and Ar in a 4:1 ratio. If the partial pressure of Ar in the mixture is 0.821 atm, what is the partial pressure of H_2? What is the total pressure of the mixture?

11. Consider a mixture of equal masses of neon and bromine gases. Which gas exerts the greater partial pressure? By what factor?

12. You have nitrogen gas on each side of a two-bulbed container connected by a valve as shown below. Initially the valve is closed.

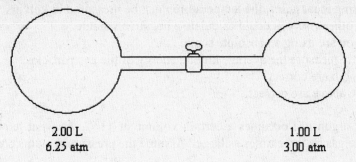

2.00 L
6.25 atm

1.00 L
3.00 atm

The left side of the container is at a pressure of 6.25 atm and has a volume of 2.00 L. The right side is at a pressure of 3.00 atm and has a volume of 1.00 L. What is the *total pressure* of the nitrogen gas after the stopcock between the two bulbs is opened? (Assume constant temperature, ideal gas behavior, and that the volume of the valve is too small to worry about.)

13. A 4.00-L sample of $N_2(g)$ was collected over water at a total pressure of 792 torr and 25°C. When the $N_2(g)$ was dried (the water vapor was removed), the gas had a volume of 3.88 L at 25°C and 792 torr. What is the vapor pressure of water at 25°C?

Section 10.4

14. Calculate the missing values below using Avogadro's law. Assume constant temperature and pressure.
    a) $V_i = 2.00$ L; $n_i = 0.500$ mol; $V_f = 6.59$ L; $n_f = ?$ mol
    b) $V_i = 335$ mL; $n_i = 1.23$ mol; $V_f = ?$ mL; $n_f = 0.440$ mol

15. Calculate the missing values below using Boyle's law. Assume moles of gas and temperature are constant.
    a) $P_i = 3.00$ atm; $V_i = 1.75$ L; $P_f = 2.11$ atm; $V_f = ?$ L
    b) $P_i = ?$ torr; $V_i = 0.690$ L; $P_f = 801$ torr; $V_f = 0.340$ L

16. Calculate the missing values below using Charles's law. Assume moles of gas and pressure are constant.
    a) $V_i = 1.33$ L; $T_i = 25$°C; $V_f = ?$ L; $T_f = 48$°C
    b) $V_i = 5.00$ L; $T_i = ?$ °C; $V_f = 2.00$ L; $T_f = 15$°C

17. Calculate the missing values below using the combined gas equation. Assume moles of gas are constant.
    a) $P_i = 1.00$ atm; $V_i = 2.00$ L; $T_i = 25$°C; $P_f = 2.00$ atm; $V_f = ?$ L; $T_f = 40.$°C
    b) $P_i = 225$ torr; $V_i = 1027$ mL; $T_i = 19$°C; $P_f = 749$ torr; $V_f = 513$ mL; $T_f = ?$ °C

18. The lever of a large syringe is pulled out so that the syringe fills with 45.0 mL of air at 1.00 atm of pressure. You place your finger over the other end of the syringe (the end that does not contain a needle) so that the air won't escape. You push down on the lever and compress the air to a volume of 10.0 mL. Should the pressure of the air increase or decrease? Why? What is the new pressure?

19. You transfer a gas at 25°C from a volume of 2.49 L and 0.950 atm to a vessel at 32°C that has a volume of 8.26 L. What is the new pressure of the gas?

20. When analyzing ideal gases, the temperature must be measured in kelvins
    a) Because otherwise you could calculate a negative volume.
    b) So that you are using an absolute scale.
    c) To directly measure the average kinetic energy of the gas particles.
    d) Both a and b are correct.
    e) All of the above are correct.

21. A sample of argon gas occupies a certain volume at 11°C. At what temperature would the volume of the gas be three times as large? Assume the pressure remains constant.

    a) 33.0°C   b) 852°C   c) 579°C   d) 109°C   e) 94.7°C

22. A flexible weather balloon contains helium gas at a volume of 855 L. Initially, the balloon is at sea level, where the temperature is 25°C and the barometric pressure is 0.947 atm. The balloon then rises to an altitude of 6000 ft, where the pressure is 0.796 atm and the temperature is 15°C. What is the *change* in the volume of the balloon as it ascends from sea level to 6000 ft?

23. A compressed gas cylinder, at 135 atm and 23°C, is in a room where a fire occurs. The fire raises the temperature of the gas to 475°C. What is the new pressure in the cylinder?

24. True or false? A sample of oxygen gas at 70.°C is twice as hot as a sample of oxygen gas at 35°C. Perform mathematical calculations to support your explanation.

25. You have a party balloon that is partially filled with 2.00 g of helium gas. You then add 3.00 g of hydrogen gas to the balloon. Assuming constant temperature and pressure, how many times bigger is the party balloon, comparing before and after the hydrogen gas has been added?

Section 10.5

26. What is the volume (in liters) of a 0.512-mol gas sample at 0.892 atm and 32°C?

27. a) A sample of oxygen gas (O_2) has a volume of 7.51 L at a temperature of 19°C and a pressure of 1.38 atm. Calculate the number of moles of O_2 molecules present in this gas sample.
    b) Determine the number of oxygen gas molecules present in this gas sample.

28. How many grams of CO_2 are in 21.0 L of $CO_2(g)$ at 0.953 atm and 9°C?

29. A 25.0-L steel vessel is filled with 100.0 g argon gas at 23°C. What is the pressure of the argon gas in the vessel?

30. The largest a party balloon can get before bursting is 8.00 L at 25°C and 1.00 atm. Suppose you fill the balloon with only oxygen gas. How many grams of oxygen can you add to the balloon before it pops? Assume you have not yet added any gas to the balloon at all (so essentially the balloon is flat).

31. Assume a hydrogen balloon is at 25°C and 1.00 atm and has a volume of 1.00 L. How many grams of argon gas must be added to the hydrogen balloon to achieve a volume of 3.00 L at constant temperature and pressure?

32. a) Solid carbon dioxide sublimes (goes directly from a solid to a gas) at a rate of 0.726 g/min. How many gaseous carbon dioxide molecules are present after 5 min?
    b) If the pressure builds to 3.75 atm at a temperature of 21.1°C, what is the final volume of the bag after 5 min? (Assume the volume of the solid carbon dioxide is negligible.)

33. What is the pressure of 200.0 g of chlorine gas (Cl_2) in a 1.000-L container at 25°C? Calculate the pressure using the ideal gas law and the van der Waals equation. Is there a difference between the two answers?

Section 10.6

34. What is the density of krypton gas at STP?

35. The density of a certain gas is 1.74 g/L. At 25°C and 1.00 atm, what is the molar mass of this gas?

36. Consider the following reaction:
$$CaC_2(s) + 2H_2O(l) \rightarrow Ca(OH)_2(aq) + C_2H_2(g).$$

Calculate the volume (in liters) of C_2H_2 produced at 19°C and 1.33 atm from 0.0888 mol H_2O and excess CaC_2.

37. Consider the following reaction: $2LiOH(s) + CO_2(g) \rightarrow Li_2CO_3(s) + H_2O(l)$.
What mass of lithium hydroxide is required to react with 3.01 L of carbon dioxide gas at 28°C and 1.22 atm?

38. An air bag is deployed by utilizing the following reaction:
$2NaN_3(s) \rightarrow 2Na(s) + 3N_2(g)$. What mass of NaN_3 must be used to inflate an air bag to 85.0 L at 1.00 atm and 25°C?

Section 10.7

39. Which of the following statements are true concerning *real* gases?
    a) A real gas behaves more like an ideal gas at high pressures and low temperatures.
    b) The individual gas particles have no volume.
    c) The individual gas particles are not attracted to one another.
    d) The particles collide with the walls of their container and exert pressure against it.
    e) The kinetic energy of the gas particles is directly proportional to the temperature of the gas in degrees Celsius.

Consider the following scenario when answering Exercises 40 and 41:

There are two balloons, both the *same size and temperature*, except that one balloon contains hydrogen and the other balloon contains carbon dioxide. The hydrogen balloon floats, and the carbon dioxide balloon sinks in the air.

40. How do the pressures inside the two balloons compare? How do you know this? Be sure to incorporate ideas of the kinetic molecular theory into your answer—specifically, pressure and temperature.

41. Which balloon contains the larger number of molecules? How do you know this?

Section 10.8

42. What is the rms speed of nitrogen gas at 50.°C?

43. What is the temperature (in kelvins) of a sample of oxygen gas that has a rms speed of 252 m/s?

44. Rank the following gases in order of increasing effusion rate: C_2H_6, CO_2, O_3, Ar .

45. How many times faster does helium effuse compared to krypton?

46. The effusion rate of pure methane (CH_4) is 47.8 mL/min. What is the effusion rate of H_2 under identical experimental conditions?

47. Gas A diffuses 1.11 times as fast as gas B. If gas B has a molar mass of 80.07 g/mol, what is the molar mass of gas A?

Section 10.9

48. What is the importance of the Earth's ozone layer?

49. What common component of refrigerants is known for depleting the ozone layer?

50. What causes the greenhouse effect?

Answers to Exercises

1. c

2. 0.625 atm

3. 2.34×10^5 Pa; 234 kPa

4. 4.51 psi

5. 0.928 bar

6. 130. kPa

7. 0.968 atm; 735 torr; 98.0 kPa

8. 4.50 atm

9. $P_{O_2} = 1.19$ atm ; $P_{Kr} = 1.14$ atm

10. $P_{H_2} = 3.28$ atm ; $P_{Total} = 4.10$ atm

11. Neon exerts a greater partial pressure because there are more neon atoms present to collide with the walls of the container (more moles); 7.92 times greater.

12. 5.17 atm

13. 24 torr

14. a) $n_f = 1.65$ mol
    b) $V_f = 120.$ mL

15. a) $V_f = 2.49$ L
    b) $P_i = 395$ torr

16. a) $V_f = 1.43$ L
    b) $T_i = 447°C$

17. a) $V_f = 1.05$ L
    b) $T_f = 213°C$

18. Because the temperature of the gas and the amount of gas in the syringe are constant, the pressure should increase; there is a smaller amount of space for the gas particles to move around in, so they will collide with the walls of the syringe more frequently. $P_f = 4.50$ atm.

19. 0.293 atm

20. e

21. c

22. $\Delta V = 128$ L

23. 341 atm

24. False. The average kinetic energy of the gas particles is directly proportional to the Kelvin temperature of the gas; therefore the temperature of the oxygen gas is measured on the Kelvin scale, not the Celsius scale. Converting the temperatures to Kelvin, we get

    K = 70. + 273 = 343 K   K = 35 + 273 = 308 K

    343 K is not twice the temperature (or kinetic energy) of 308 K.

25. 4.00 times bigger

26. 14.4 L

27. a) 0.433 mol O_2
    b) 2.60×10^{23} O_2 molecules

28. 38.1 g CO_2

29. 2.43 atm

30. 10.5 g O_2

31. 3.27 g Ar

32. a) 4.97×10^{22} CO_2 molecules
    b) 531 mL

33. $P_{ideal} = 69.0$ atm; $P_{real} = 30.0$atm; Yes, there is a significant difference in pressure between chlorine behaving "ideally" versus under "normal" conditions.

34. 3.74 g/L

35. 42.5 g/mol

36. 0.800 L

37. 7.12 g LiOH

38. 151 g NaN_3

39. d

40. The molecules are in constant random motion at a certain temperature, so they contain a certain amount of kinetic energy and collide with the walls of the balloons, thus exerting pressure. Because the balloon is neither expanding nor contracting, the pressures inside the balloons are equal to the pressure outside the balloons (around 1 atm). Therefore, the pressures in both balloons are essentially the same.

41. Both balloons contain the same number of molecules because they are both at the same volume, temperature, and pressure; therefore both have the same number of moles of gas (due to $PV = nRT$). Because the number of moles of gas is directly related to Avogadro's number (1 mole = 6.022×10^{23}), both balloons contain the same number of molecules.

42. 536 m/s

43. 81.5 K

44. O_3 < CO_2 < Ar < C_2H_6

45. 4.58 times faster

46. 135 mL/min

47. 65.0 g/mol

48. The Earth's ozone layer protects the Earth's surface from the Sun's harmful UV radiation.

49. chlorofluorocarbons (CFCs)

50. The greenhouse effect is caused by the accumulation of gases in the atmosphere that permit light to enter, but prevent some heat from exiting.

Chapter 11

Chemistry of Water and the Nature of Liquids

The Bottom Line

When you finish this chapter, you will be able to:

- Explain how the polarity of water is related to its structure.
- Appreciate the effects of intermolecular forces on the physical properties of water.
- Draw and interpret a phase diagram.
- Calculate solution concentrations based on moles and on mass.
- Determine the effects of temperature and pressure on the solubility of a gas in a liquid.
- Calculate colligative properties of solutions.

11.1   The Structure of Water

Water is a V-shaped molecule with an H—O—H bond angle of 104.5°. Water, a polar molecule, has a permanent dipole resulting from the shape of the molecule and the fact that the oxygen atom is more electronegative than the hydrogen atoms. In the liquid state, water molecules remain relatively close to one another, allowing the negative oxygen on one molecule to attract the positive hydrogen on another molecule. This interaction and subsequent arrangement of the molecules are examples of **intermolecular forces** at work. In water, the association occurs between 3 and 6 water molecules, averaging about 4.5 molecules. Intermolecular forces are much weaker than bonds (intramolecular forces).

11.2   A Close Look at Intermolecular Forces

Intermolecular forces result from the interaction of oppositely charged poles. The three types of intermolecular forces are collectively referred to as van der Waals forces:

- **London dispersion forces** (also known as **induced dipoles**) are the result of instantaneous dipoles, which appear in a molecule due to a momentary imbalance in electron distribution. Polarized electrons produce an induced dipole on one molecule that affects the electron distribution on a neighboring molecule (see **Figure 11.6** in the text). The larger the number of electrons in the system, the larger the polarizability, and hence the larger the induced dipole.

- **Permanent dipole–dipole forces** are the result of partial positive and negative charges of neighboring polar covalent molecules attracting one another. These forces are about 1% as strong as intramolecular polar covalent bonds.

- **Hydrogen bonds** result when hydrogen atoms on one molecule are attracted by highly electronegative atoms such as fluorine, oxygen, or nitrogen on another molecule. Hydrogen bonds are the strongest of the three intermolecular forces, and are about 10% as strong as an intramolecular covalent bond.

Example 1   **London Dispersion Forces**

The boiling point of neon is 27.3 K; the boiling point of xenon is 166.1 K. Why is neon's boiling point so low? Why is xenon's boiling point so much higher?

First Thoughts

Both Ne and Xe are noble gases that have a complete octet of valence electrons. These elements do not possess permanent dipoles. Your text points out that boiling points increase with increasing molar mass because having a larger number of electrons means that a molecule can induce more momentary dipoles.

Solution

Neon is small (molar mass = 20.2 g/mol), making it difficult for neon to establish a momentary dipole so that it could induce a dipole on a neighboring molecule. Xenon is much larger (molar mass = 131.3 g/mol) than neon, increasing the likelihood of momentary dipoles. It has greater polarizability than neon.

Example 2   Boiling Points and Intermolecular Forces

Arrange each of the following substances in order of increasing boiling point:

$$HF, H_2, CO, F_2$$

First Thoughts

The boiling point depends partially on the strength of intermolecular forces in the liquid. It also increases with increasing molar mass.

Solution

$H_2 < F_2 < CO < HF$. Hydrogen and chlorine are nonpolar; hence they have only weak London dispersion forces. Hydrogen (2.02 g/mol) is smaller than fluorine (38.0 g/mol). Carbon monoxide is polar; hence it has dipole–dipole forces. HF is hydrogen bonded; hydrogen bonding is the strongest of the intermolecular forces.

Example 3   Types of Intermolecular Forces

List the intermolecular forces that you would expect in the following substances:

$$SO_2, CH_4, C_2H_5OH$$

Solution

SO_2: dipole–dipole forces and London dispersion forces. It is a bent molecule with a permanent dipole.

CH_4: London dispersion forces. Methane is nonpolar.

C_2H_5OH: hydrogen bonding and London dispersion forces. Ethanol contains an oxygen atom that can hydrogen bond to the hydrogen on the OH group of a neighboring molecule.

11.3   Impact of Intermolecular Forces on the Physical Properties of Water: I

The following important definitions are drawn from the discussion in Section 11.3 of your text:

- **Evaporation:** the process of molecules leaving the liquid phase and entering the vapor phase.
- **Condensation:** the process of molecules leaving the vapor phase and entering the liquid or solid phase.
- **Sublimation:** the escape of molecules from the solid phase directly to the vapor phase.
- **Deposition:** molecules returning from the vapor phase directly to the solid phase.

- **Dynamic equilibrium:** the state of a system in which two opposing processes are occurring at the same rate.
- **Vapor pressure:** the pressure of a vapor over a liquid at equilibrium.
- **Boiling point:** the pressure of the liquid's vapor is equal to the surrounding pressure.
- **Normal boiling point:** the boiling point of a liquid if the surrounding pressure is 1 atmosphere.
- **Melting point:** the temperature at which a solid changes to a liquid.
- **Heat of fusion ($\Delta_{fus}H$):** the amount of heat needed to convert a solid to a liquid at its melting point and constant pressure.
- **Heat of vaporization ($\Delta_{vap}H$):** the amount of heat needed to convert a liquid to a vapor at its normal boiling point.

Example 4   Comparing Vapor Pressures

Arrange the following compounds in order of increasing vapor pressure:

$$\text{propane, } C_3H_8; \text{ ethanol, } C_2H_5OH; \text{ methane, } CH_4$$

First Thoughts

Compounds having the strongest intermolecular forces will have the lowest vapor pressure. If the compounds have London dispersion forces, then the strength of the dispersion force will increase with molar mass.

Solution

$C_2H_5OH < C_3H_8 < CH_4$. Ethanol is hydrogen bonded, so it will have the strongest intermolecular forces and hence the lowest vapor pressure. Methane and propane have only dispersion forces. Propane has a larger molar mass than methane, so propane has a stronger intermolecular force and lower vapor pressure.

   Heating curves show how much heat would be necessary to take a piece of a solid substance, liquefy it, bring the liquid to a boil, and convert all of the liquid to a vapor. A diagram of this process is shown in the text in **Figure 11.16**.

Example 5   Ice to Steam

How much heat is necessary to bring 10.0 g of ice at $-10.0°C$ to a boil at $100.0°C$? The specific heat of ice is 2.05 J/g °C, the heat of fusion of water is 334 J/g, the specific heat of water is 4.184 J/g °C and the heat of vaporization of water is 2.44 kJ/g.

First Thoughts

There are four steps to consider: (1) warming the ice; (2) melting the ice; (3) heating the water; and (4) boiling the water.

Solution

Let's do this in steps.

Step 1. Warming:

$$q_{warming} = (m_{ice})(SH_{ice})(\Delta t)$$
$$= (10.0 \text{ g})(2.05 \text{ J/g°C})(0.0°C - (-10.0°C))$$
$$= 205 \text{ J}$$

Step 2. Melting:

$$q_{\text{melting}} = (m_{\text{ice}})(\Delta_{\text{fus}}H)$$
$$= (10.0 \text{ g})(334 \text{ J/g})$$
$$= 3340 \text{ J}$$

Step 3. Heating:

$$q_{\text{heating}} = (m_{\text{water}})(SH_{\text{water}})(\Delta t)$$
$$= (10.0 \text{ g})(4.184 \text{ J/g}^\circ\text{C})(100.0^\circ\text{C} - 0.0^\circ\text{C})$$
$$= 4184 \text{ J}$$

Step 4. Boiling:

$$q_{\text{boiling}} = (m_{\text{water}})(\Delta_{\text{vap}}H)$$
$$= (10.0 \text{ g})(2.44 \times 10^3 \text{ J/g})$$
$$= 2.44 \times 10^4 \text{ J}$$

The total heat is

$$q_{\text{total}} = q_{\text{warming}} + q_{\text{melting}} + q_{\text{heating}} + q_{\text{boiling}}$$
$$= 205 \text{ J} + 3340 \text{ J} + 4184 \text{ J} + 2.44 \times 10^4 \text{ J}$$
$$= 3.21 \times 10^4 \text{ J}$$

11.4 Phase Diagrams

A **phase diagram** is a graph that shows the phase of a substance as a function of temperature and pressure (refer to **Figure 11.18** in the text). Lines on the graph known as **phase boundaries** separate the phases. Here are some important definitions that you should know:

- **Triple point:** the point on a phase diagram representing the temperature and pressure at which the three phases of a substance coexist in equilibrium.
- **Critical temperature:** the temperature above which the liquid state can no longer exist at any pressure.
- **Critical pressure:** the vapor pressure at the critical temperature.
- **Critical point:** the point on the vapor pressure curve that is defined by the critical temperature and the critical pressure.

Example 6   **Phase Diagrams**

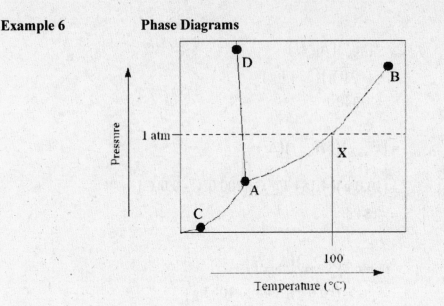

Using the phase diagram, identify the triple point and the critical point. In which phase is the substance at point X? Describe what will happen to the substance at point X, if it is cooled by 50°C at constant pressure.

Solution

Point A is the triple point, at which the three phases coexist. Point B is the critical point. Point X is in the vapor phase. If the substance is cooled at constant pressure, it will condense from the vapor phase becoming a liquid.

11.5 Impact of Intermolecular Forces on the Physical Properties of Water: II

This section details properties that are characteristic of liquids:

- **Viscosity:** the resistance to flow.
- **Surface tension:** a measure of the energy per area on the surface of a liquid.
- **Capillary action:** the upward rise of a liquid in a small-diameter tube caused by the adhesion of the molecules to the surface of the tube. The liquid forms a meniscus, which is the concave shape of the liquid in the tube.

11.6 Water: The Universal Solvent

Water is often referred to as the universal solvent because it can dissolve so many substances, thereby forming aqueous solutions. The solubility of substances in one another varies widely. As a general rule, "like substances will dissolve like substances"; in other words, substances with similar intermolecular forces will dissolve in one another. Because water is both polar and hydrogen bonded, we would expect polar or ionic substances to dissolve easily in water.

As discussed in your text, the dissolving of an ionic substance in water can be explained in terms of three steps: (1) solute separation, an endothermic process; (2) solvent separation, also an endothermic process; and (3) electrostatic interaction (solvation) between the solvent and solute ions, an exothermic process. The sum of these processes is exothermic, which favors the formation of a solution.

Example 7   **Predicting Solubility**

Which of the following substances are soluble in water: methanol, CH_3OH; acetic acid (vinegar), CH_3COOH; and octane, C_8H_{18}?

First Thoughts

Using the "like dissolves like" principle, look for hydrogen bonding or evidence of polarity to decide whether the substance will dissolve in water.

Solution

Both methanol and acetic acid are hydrogen bonded, so we would expect them to dissolve in water. Octane is a nonpolar hydrocarbon, so it will not dissolve in water.

11.7   Measures of Solution Concentration

Measures Based on Moles

There are three concentration units based on moles of solute:

Molarity (M): moles per liter of solution.

$$M = \frac{\text{mol solute}}{\text{L solution}}$$

Molality (m): moles per kilogram of solvent.

$$m = \frac{\text{mol solute}}{\text{kg solvent}}$$

Mole fraction (χ_i): the ratio of the number of moles of substance to the total number of moles of substances in solution.

$$\chi_i = \frac{\text{number of moles of i}}{\text{total number of moles}}$$

Example 8   **Mole-Based Units of Concentration**

Calculate the molarity, molality, and mole fraction of a solution of 53.0 g of acetic acid, CH_3COOH, made up to a volume of 500. mL with water. The density of the solution is 1.05 g/mL.

Solution

Each of the concentration units depends on the number of moles of solute, which we will calculate first.

$$\text{Moles acetic acid} = 53.0 \text{ g}\left(\frac{1 \text{ mol}}{60.04 \text{ g}}\right) = 0.833 \text{ mol}$$

$$\text{Molarity acetic acid} = \frac{0.883 \text{ mol}}{0.500 \text{ L}} = 1.77 \ M$$

To calculate the molality, we need the mass of the solvent:

$$\text{Mass of solution} = 500.\text{ mL}\left(\frac{1.05\text{ g}}{\text{mL soln}}\right) = 525\text{ g}$$

Mass of solvent (water) = mass solution - mass of acetic acid = 525 g - 53.0 g = 472 g

$$\text{Molality of acetic acid} = \frac{0.883\text{ mol}}{0.472\text{ kg}} = 1.87\ m$$

The mass of water has to be converted to moles for the mole fraction calculation:

$$\text{Moles water} = 472\text{ g}\left(\frac{1\text{ mol}}{18.0\text{ g}}\right) = 26.2\text{ mol}$$

$$\chi_{\text{acetic acid}} = \frac{0.883\text{ mol acetic acid}}{0.883\text{ mol acetic acid} + 26.2\text{ mol water}} = 0.0326$$

Measures Based on Mass

There are two mass-based concentration measures: weight percent and parts per million, per billion, and per trillion (see Chapter 4).

Weight percent (wt %): the mass fraction of a substance in a solution expressed as a percentage.

$$\text{wt \%} = \frac{\text{g substance}}{\text{g solution}} \times 100\%$$

Parts per million, parts per billion, and parts per trillion (ppm, ppb, ppt)

$$\text{ppm} = \frac{1\text{ g solute}}{10^6\text{g of solution}} \approx \frac{1\text{ mg solute}}{\text{L solution}}$$

$$\text{ppb} = \frac{1\text{ g solute}}{10^9\text{g of solution}} \approx \frac{1\ \mu\text{g solute}}{\text{L solution}}$$

$$\text{ppt} = \frac{1\text{ g solute}}{10^{12}\text{g of solution}} \approx \frac{1\text{ ng solute}}{\text{L solution}}$$

Example 9   Conversion Between Moles and Mass Concentration Units

The concentration of lead ion (Pb^{2+}) was determined to be 1.75×10^{-5} M. Express this concentration in ppm.

Solution

Using dimensional analysis:

$$\left(\frac{1.75\times10^{-5}\text{ mol }Pb^{2+}}{L}\right)\left(\frac{207.2\text{ g }Pb^{2+}}{\text{mol }Pb^{2+}}\right)\left(\frac{1000\text{ mg }Pb^{2+}}{1\text{ g }Pb^{2+}}\right) = 3.63\text{ ppm}$$

11.8 Effect of Temperature and Pressure on Solubility

Temperature effects: The solubility of a gas decreases with increasing temperature. The solubility of ionic solids *generally* increases with increasing temperature.

Pressure effects: The solubility of gases in water depends on the gas pressure above the liquid. Henry's law gives the relationship at constant temperature.

$$P_{gas} = k_{gas} C_{gas}$$

P_{gas} = pressure of the gas above the liquid
C_{gas} = concentration of the gas
k_{gas} = Henry's law constant

Example 10   Henry's Law

A liter of water dissolves 0.0404 g of oxygen at 25°C at a pressure of 760. torr. What would be the concentration of oxygen (in g/L) if the pressure were reduced to 183 torr at the same temperature?

Solution

Since we are dealing with the same gas we do not need to look up the Henry's law constant. Rather, we can solve this as a ratio:

$$\frac{P_1}{C_1} = \frac{P_2}{C_2}$$

$$\frac{0.0404 \text{ g/L}}{760. \text{ torr}} = \frac{C_2}{183 \text{ torr}}$$

$$C_2 = 9.73 \times 10^{-3} \text{ g/L}$$

11.9 Colligative Properties

Colligative properties depend on the number of nonvolatile solute particles in the solution. They include: vapor pressure lowering, boiling point elevation, freezing point depression and osmotic pressure.

Vapor Pressure Lowering

The vapor pressure of a solution is given by **Raoult's law:**

$$P_{solution} = \chi_{solvent} P^o_{solvent}$$

where $P^o_{solvent}$ is the vapor pressure of the pure solvent. The addition of a nonvolatile solute to a pure solvent will always lower the vapor pressure of the solvent.

Example 11   Vapor Pressure Lowering

What will be the vapor pressure of a solution made by dissolving 9.96 g of glucose, $C_6H_{12}O_6$, in 31.5 g of water at 25°C? How much was the vapor pressure of the pure water lowered? The vapor pressure of water at 25°C is 23.8 torr.

Solution

Begin by calculating the mole fraction of glucose and the mole fraction of water, and then apply Raoult's law.

$$\text{Moles glucose} = 9.96 \text{ g} \left(\frac{1 \text{ mol}}{180. \text{ g}}\right) = 0.0553 \text{ mol glucose}$$

$$\text{Moles water} = 31.5 \text{ g}\left(\frac{1 \text{ mol}}{18.0 \text{g}}\right) = 1.75 \text{ mol water}$$

$$\chi_{water} = \frac{1.75}{1.75+0.0553} = 0.969$$

$$P_{soln} = \chi_{water} P^o_{water} = (0.969)(23.8 \text{ torr}) = 23.1 \text{ torr}$$

Vapor pressure lowering = 23.8 torr - 23.1 torr = 0.7 torr

Boiling-Point Elevation

This colligative property depends on the molality (m) of the solution:

$$\Delta T_b = K_b m$$

ΔT_b = change in the boiling point
K_b = boiling-point elevation constant for the solvent
m = molality of the solution

Because colligative properties depend on the number of particles in solution, we must account for the ionization that occurs when ionic compounds are dissolved in water. For example, a 3.0 m $NiCl_2$ solution,

$$NiCl_2(s) \xrightarrow{H_2O} Ni^{2+}(aq) + 2Cl^-(aq)$$

would have a total molal concentration of 9.0 m (3.0 m Ni^{2+} + 6.0 m Cl^-). To account for the effect of the ions, we use the van't Hoff i factor:

$$\Delta T_b = i K_b m$$

i = sum of the coefficients of the ions (i = 1 for molecular compounds)

Example 12 Boiling-Point Elevation

A solution is prepared by dissolving 61.0 g of KCl in 250. g of water. Calculate the boiling point of the solution. K_b is 0.512°C/m.

Solution

To calculate ΔT_b, first calculate the molality and then determine the number of ions. KCl dissociates into K^+ and Cl^-, so $i = 2$.

$$\text{Moles of KCl} = 61.0 \text{ g}\left(\frac{1 \text{ mol}}{74.6 \text{ g}}\right) = 0.818 \text{ mol}$$

$$m_{KCl} = \frac{\text{mol KCl}}{\text{kg water}} = \frac{0.818 \text{ mol}}{0.250 \text{ kg}} = 3.27 \ m$$

$$\Delta T_b = i K_b m = (2)(0.512°C/m)(3.27 \ m) = 3.35°C$$

The boiling point of the solution is 100.00°C + 3.35°C = 103.35°C

Freezing-Point Depression

This colligative property is analogous to boiling-point elevation.

$$\Delta T_f = i\, K_f m$$

ΔT_f = change in the freezing point
i = van't Hoff factor
K_f = freezing-point depression constant for the solvent
m = molality of the solution

Example 13   Freezing-Point Depression

Compare the change in the freezing points of two solutions: (1) 255 g of ethylene glycol, EG, (antifreeze), $C_2H_6O_2$, in 1000. g of water; and (2) 255 g of salt, NaCl, in 1000. g of water. K_f (H_2O) = 1.86°C/m.

First Thoughts

First, calculate the molality of each solution. Second, determine the i factor: i = 1 for EG and i = 2 for NaCl because it ionizes to Na^+ and Cl^-.

Solution

Ethylene glycol (EG):

$$\text{Moles EG} = 255\text{ g}\left(\frac{1\text{ mol}}{62.08\text{ g}}\right) = 4.11\text{ mol}$$

$$m_{EG} = \frac{4.05\text{ mol}}{1.000\text{ kg}} = 4.05\ m$$

$$\Delta T_f = i\, K_f m = (1)(1.86°\text{C}/m)(4.11\ m) = 7.64°\text{C}$$

NaCl:

$$\text{Moles NaCl} = 255\text{ g}\left(\frac{1\text{ mol}}{58.44\text{ g}}\right) = 4.36\text{ mol}$$

$$m_{EG} = \frac{4.36\text{ mol}}{1.000\text{ kg}} = 4.36\ m$$

$$\Delta T_f = i\, K_f m = (2)(1.86°\text{C}/m)(4.36\ m) = 16.2°\text{C}$$

The freezing-point depression of NaCl is a little more than twice that of ethylene glycol.

Osmotic Pressure

This colligative property depends on the molarity of the solution, not the molality.

$$\Pi = i\, MRT$$

Π = osmotic pressure of the solution
M = molarity of the solution
R = gas constant = 0.08206 L · atm/K · mol
T = temperature in kelvin
i = van't Hoff factor

Example 14   Osmotic Pressure I

What is the osmotic pressure of a solution made of 5.0 g of glucose, $C_6H_{12}O_6$, dissolved in 100. mL of water at 20.°C?

First Thoughts

This colligative property depends on the molarity, which we will calculate first. Because glucose is a molecular compound, $i = 1$.

$$\text{Moles glucose} = 5.0\ \text{g}\left(\frac{1\ \text{mol}}{180.\ \text{g}}\right) = 0.028\ \text{mol}$$

$$M_{\text{glucose}} = \frac{0.028\ \text{mol}}{0.100\ \text{L}} = 0.28\ \text{mol/L}$$

$$\Pi = (1)(0.28\ \text{mol/L})\left(0.08206\ \frac{\text{L}\cdot\text{atm}}{\text{K}\cdot\text{mol}}\right)([20.+273]\ \text{K}) = 6.7\ \text{atm}$$

Example 15   Osmotic Pressure II

The osmotic pressure of a protein solution at 25.0°C was measured to be 1.44 torr. The solution contained 3.50 mg of protein made up to a solution volume of 5.00 mL. What is the molar mass of the protein? (Assume $i = 1$.)

First Thoughts

From the osmotic pressure and the temperature, we can calculate the molarity of the solution. Using the molarity, mass, and volume, we can calculate the molar mass.

Solution

$$M = \frac{\Pi}{RT} = \frac{(1.44\ \text{torr})\left(\frac{1\ \text{atm}}{760\ \text{torr}}\right)}{\left(0.08206\ \frac{\text{L}\cdot\text{atm}}{\text{K}\cdot\text{mol}}\right)(298\ \text{K})} = 7.75\times10^{-5}\ \text{mol/L}$$

$$\text{Moles} = \left(7.75\times10^{-5}\ \text{mol/L}\right)\left(5.00\times10^{-3}\ \text{L}\right) = 3.87\times10^{-7}\ \text{mol}$$

$$\text{Molar mass} = \frac{\text{grams}}{\text{moles}} = \frac{3.50\times10^{-3}\ \text{g}}{3.87\times10^{-7}\ \text{mol}} = 9.03\times10^{3}\ \text{g/mol}$$

Exercises

Section 11.1

1. What makes water a polar molecule?

2. Explain the difference between intramolecular forces and intermolecular forces.

Section 11.2

3. How do London dispersion forces arise?

4. Which type(s) of molecules would you expect to exhibit only dispersion forces?

5. Which type(s) of molecules would you expect to exhibit dipole–dipole forces?

6. In addition to a hydrogen atom, which other atoms are necessary for hydrogen bonds to form?

7. Which of the intermolecular forces is the strongest? Which is the weakest?

8. Based on intermolecular forces, arrange the following molecules in order by increasing boiling point: hexane, C_6H_{14}; butane, C_4H_{10}; pentane, C_5H_{12}.

9. Which would have the higher boiling point: ethanol (CH_3CH_2OH) or propanol ($CH_3CH_2CH_2OH$)?

10. Arrange the following molecules in order by increasing boiling point and explain your reasoning: NH_3, SF_2, CH_3CH_3.

11. Which types of intermolecular forces would you expect the following molecules to exhibit: CS_2, Kr, CH_3OH, PF_3?

12. Which would have the largest London dispersion forces: CCl_4 or $SiCl_4$?

Section 11.3

13. Explain the difference between the boiling point of a liquid and the normal boiling point of a liquid.

14. Explain the difference between sublimation and vaporization.

15. Why would the heat of vaporization of HF be so much larger than the heat of vaporization of methane, CH_4?

16. Liquid butane, C_4H_{10}, is found in lighters. It has a heat of vaporization of 21.3 kJ/mol. How much heat would be absorbed by the evaporation of 4.23 g of butane?

17. How much heat is required to heat 5.00 g of ice at $-5.00^{\circ}C$ to $50.0^{\circ}C$? The specific heat of ice is 2.05 J/g $^{\circ}C$, the specific heat of water is 4.184 J/g $^{\circ}C$, and the heat of fusion of water is 334 J/g.

18. Calculate the amount of heat necessary to convert 61.0 g of water at 25.0°C to steam at 100.0°C given that the specific heat of water is 4.184 J/g °C and the heat of vaporization of water is 2.44 kJ/g.

Section 11.4

Use the following phase diagram for Exercises 19–22:

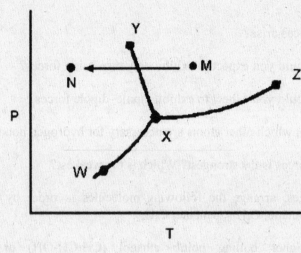

19. What does point Z represent for this substance?

20. What does point X represent for this substance?

21. Which boundary is represented by the line XY?

22. What happens to the substance as it moves from point M to point N?

Section 11.5

23. Define viscosity, and give some examples of viscous liquids.

24. Explain what causes the formation of a meniscus in a tube filled with a liquid.

25. Describe the meniscus formed by mercury in a glass tube. How does it differ from the meniscus for other liquids?

Section 11.6

26. Explain what is meant by the following statement: "Like substances will dissolve like substances."

27. Which would most likely dissolve in water: ethylene glycol, $HOCH_2CH_2OH$, or carbon tetrachloride, CCl_4?

28. Which would be more soluble in water: boric acid, H_3BO_3, or carbon tetrachloride, CCl_4?

29. Hexane, C_6H_{14}, is a nonpolar solvent. Which would be more soluble in hexane: benzene, C_6H_6, or methanol, CH_3OH?

Section 11.7

30. Calculate the molarity of a solution made by dissolving 5.23 g of Na_2CO_3 in water to make 250. mL of solution. What is the molarity of each of the ions?

31. Calculate the molality of a solution made by dissolving 4.85 g of $MgCl_2$ in 425 g of water. What is the molality of each of the ions?

32. How much water should be added to 3.50 g of glucose, $C_6H_{12}O_6$, to make a solution that is 1.25 m?

33. A solution is made by mixing 45.0 g of methanol, CH_3OH, with 55.0 g of ethanol, CH_3CH_2OH. What is the mole fraction of each alcohol?

34. How would you make 200. g of a 6.0% $CuSO_4$ solution?

35. Calculate the mass percentage of KF in a solution containing 11.7 g of KF in 443 g of water.

36. A solution contains 1.61 g of Cu^{2+} ion in 15.0 L of water. What is the concentration of Cu^{2+} in parts per million (ppm)?

37. Effluent water from a stream contained 0.00217 g of Ag^+ per liter. What is the concentration of Ag^+ in parts per million (ppm) and parts per billion (ppb)?

Section 11.8

38. The solubility of carbon dioxide in water is 0.161 g per 100 mL water at 20°C and 1.00 atm pressure. To carbonate beverages, a pressure of 5.75 atm is used. What is the solubility of CO_2 at that pressure?

39. The Henry's law constant for N_2 at 27°C is 6.0×10^{-4} M/atm. What would be the solubility of N_2 at a depth of 200 ft (about 6.1 atm)?

Section 11.9

40. Two solutions are made: one containing 50 g of glucose, $C_6H_{12}O_6$, in 1.0 kg of water, and the other containing 50 g of sucrose, $C_{12}H_{22}O_{11}$, in 1.0 kg of water. Are the vapor pressures above both solutions the same? Why or why not?

41. Calculate the vapor pressure of a solution made by dissolving 32.2 g of sucrose, $C_{12}H_{22}O_{11}$, in 86.5 g of water at 50°C. The vapor pressure of water at this temperature is 92.5 torr.

42. How much will the vapor pressure of benzene, C_6H_6, be lowered at 23°C, if 1.00 g of naphthalene, $C_{10}H_8$, is added to 24.0 g of benzene? Assume that the vapor pressure of naphthalene is negligible and the vapor pressure of benzene is 85.8 torr.

43. What is the boiling point of a solution made by adding 12.4 g of KCl to 96.0 g of water? K_b = 0.512°C/m.

44. Which aqueous solution has a higher boiling point: 0.12 m NaCl, 0.10 m CaCl$_2$, or 0.25 m C$_3$H$_8$O$_2$? Why?

45. An aqueous solution of a nonvolatile molecular compound boils at 101.7°C. What is the molality of the solution? $K_b = 0.512$°C/m.

46. What is the freezing point of an aqueous glycerol, C$_3$H$_8$O$_3$, solution made by dissolving 2.55 g of glycerol in 30.0 g of water? K_f(H$_2$O) = 1.86°C/m.

47. Which solution would depress the freezing point of water more: 1.0 m LiBr, 0.5 m MgBr$_2$, or 1.5 m glucose? Why?

48. Calculate the freezing-point depression of a solution made from dissolving 52.8 g of CoCl$_2$ in 225 g of water.

49. What is the osmotic pressure of an aqueous solution made by dissolving 1.25 g of urea, (NH$_2$)$_2$CO, in 100. mL of solution at 37.0°C?

50. A solution of 0.150 g of an enzyme in 210. mL of solution has an osmotic pressure of 0.955 torr at 25.0°C. What is the molar mass of this enzyme?

Answers to Exercises

1. Water is a polar molecule due to its V shape. The difference in the electronegativity of the O—H bond causes the formation of a permanent dipole.

2. Intramolecular forces are the bonds between atoms in a molecule or ionic compound. Intermolecular forces are much weaker forces between atoms or molecules.

3. London dispersion forces are time-dependent intermolecular forces. The asymmetrical arrangement of electrons in one molecule induces a temporary dipole on another molecule.

4. Nonpolar molecules (especially large, easily polarized molecules) are most likely to exhibit London dispersion forces.

5. polar molecules

6. In addition to hydrogen, the molecule must have a fluorine, oxygen, or nitrogen atom. All of these atoms are highly electronegative.

7. The strongest intermolecular forces are hydrogen bonds; the weakest are London dispersion forces.

8. Butane < pentane < hexane. All are nonpolar. The strength of the dispersion forces increases with molar mass. The stronger the intermolecular forces, the higher the boiling point.

9. Propanol. Both ethanol and propanol form hydrogen bonds, but the boiling point increases with molar mass.

10. CH_3CH_3 < SF_2 < NH_3. CH_3CH_3 is nonpolar and has only dispersion forces. SF_2 is polar and exhibits dipole–dipole forces, which are stronger than dispersion forces. NH_3 forms hydrogen bonds, the strongest of all the intermolecular forces.

11. CS_2: nonpolar, hence dispersion forces; Kr: dispersion forces; CH_3OH: hydrogen bonds; PF_3: polar, hence dipole–dipole forces

12. $SiCl_4$ is a larger molecule than CCl_4 and hence is more easily polarizable.

13. The boiling point of a liquid is the temperature at which the vapor pressure of the liquid equals the surrounding pressure. The normal boiling point is the temperature at which the vapor pressure of the liquid equals 1 atm.

14. Sublimation is the transformation of a substance from the solid phase directly to the vapor phase without becoming a liquid. Vaporization is the transformation of a liquid to the vapor phase.

15. HF is hydrogen bonded. It takes more energy to break the hydrogen bonds.

16. 1.55 kJ

17. 2.77×10^3 J

18. 168 kJ

19. the critical point

20. the triple point

21. the boundary between the solid and liquid phases

22. It freezes.

23. Viscosity is the resistance to flow. Examples are honey, molasses, and oil.

24. Capillary action makes the liquid "crawl" up the sides of the tube.

25. Liquids usually have a meniscus that "smiles." Mercury is so dense that its meniscus "frowns" (the edges turn downward).

26. Generally, substances that have similar intermolecular forces will dissolve in one another.

27. Ethylene glycol; it has hydrogen bonds like water.

28. Boric acid; it ionizes.

29. benzene, C_6H_6

30. 0.197 M Na_2CO_3; 0.394 M Na^+; 0.197 M CO_3^{2-}

31. 0.120 m $MgCl_2$; 0.120 m Mg^{2+}; 0.240 m Cl^-

32. 15.6 g

33. 0.541 CH_3OH, 0.459 CH_3CH_2OH

34. Dissolve 12.0 g of $CuSO_4$ in 188 g of water.

35. 2.57%

36. 107 ppm

37. 2.17 ppm, 2.17×10^3 ppb

38. 0.926 g/100 mL

39. 3.7×10^{-3} M

40. No, the sucrose solution has a smaller mole fraction than glucose.

41. 90.7 torr

42. 2.1 torr

43. 101.77°C

44. The $CaCl_2$ solution will have the highest boiling point. Because $i = 3$, it will have an effective molality of 0.30 m, thereby causing the highest boiling-point elevation.

45. 3.3 m

46. $-1.72°C$

47. The LiBr solution. Because $i = 2$, the effective molality of this solution is equal to 2.0 m. The molality of $MgBr_2$ is effectively 1.5 m, as is the molality of the glucose solution. The largest molality will cause the largest freezing-point depression.

48. 10.1°C

49. 5.30 atm

50. 1.39×10^4 g/mol

Chapter 12

Carbon

The Bottom Line

When you finish this chapter, you will be able to:

- Appreciate the importance of the millions of carbon-containing compounds in our world.
- Recognize and systematically name hydrocarbons.
- Write structures for structural and geometric isomers.
- Recognize and name compounds based on their functional groups.
- Explain and determine chirality in stereoisomers.

12.1   Elemental Carbon

Organic chemistry is the study of carbon-containing compounds. The name "organic" is attributed to the fact that all living organisms contain carbon. Given that more than 16 million synthetic and natural compounds are known, this is clearly a very important field of chemistry. Carbon has three **allotropes** (different forms of the same element):

- Diamond—a giant covalent network of carbon atoms (see **Figure 12.4** in the text). Each carbon atom is bonded to four other carbon atoms. This allotrope is extremely hard.
- Graphite—each carbon atom is bonded to three other carbon atoms, forming a hexagonal network in layers (**Figure 12.5**). This allotrope is soft and conducts electricity.
- Fullerenes—derivatives of C_{60}, in which 60 carbon atoms are bonded to one another in a "geodesic dome" structure (**Figure 12.6**).

12.2   Crude Oil: The Basic Resource

Crude oil or petroleum is not an allotrope of carbon. It is a complex mixture of hundreds of carbon containing compounds (see **Table 12.2** in the text). Petroleum is an important source of fuels and the starting materials for products such as plastics, fabrics, pharmaceuticals and countless other carbon containing materials.

12.3   Hydrocarbons

Hydrocarbons are compounds of hydrogen and carbon. They are classified as follows: alkanes, all carbon–carbon bonds are single bonds; alkenes, contain at least one carbon–carbon double bond; and alkynes, contain at least one carbon–carbon triple bond.

Alkanes

Alkanes are saturated hydrocarbons that are composed of all single-bonded carbons that are sp^3 hybridized. The general formula for these compounds is C_nH_{2n+2}. Alkanes can form straight chains (normal alkanes), branched chains, or rings. The first 10 normal alkanes are listed in **Table 12.4**.

Example 1   **Normal Alkanes**

Write the name and formula for the normal alkane with $n = 5$.

198

First Thoughts

Apply the formula C_nH_{2n+2} and then use Table 12.4 to identify the name.

Solution

C_5H_{12} is pentane. The condensed formula is $CH_3CH_2CH_2CH_2CH_3$. This is a five-carbon-long straight chain. Remember the bond angles between carbons are not $90°$ but rather $109.5°$, making a zigzag chain.

Branched Chains: Isomers

Isomers are compounds that have the same molecular formula but different structures. Let's look at pentane again. The straight chain is $H_3C—CH_2—CH_2—CH_2—CH_3$. An example of a structural isomer is

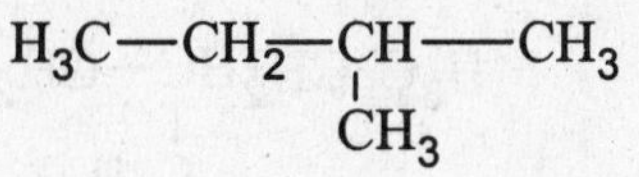

$$H_3C—CH_2—\underset{\underset{CH_3}{|}}{CH}—CH_3$$

Example 2   Isomers

Draw the remaining isomer(s) of pentane.

First Thoughts

Move the methyl group hanging down to the next carbon on the left. Moving it to left gives the same structure that is shown above. If you turn it around, you will see that the two isomers are the same. Try moving two carbons.

Solution

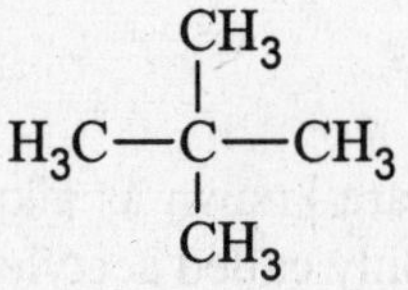

$$H_3C—\overset{\overset{CH_3}{|}}{\underset{\underset{CH_3}{|}}{C}}—CH_3$$

This is the only other isomer. There are a total of three: the straight chain and the two-branched chains.

Naming Alkanes

The rules for naming alkanes are given in **Section 12.3** of the text. One of the most important steps is to find the longest chain so that it can be numbered. If you review the names of the alkanes in **Table 12.4,** you will see that, after the first four, Greek prefixes are used.

Example 3   Naming Alkanes

Name the three isomers of pentane.

First Thoughts

Look for the longest chain to determine the molecule name. Substituted groups must have the lowest number. The longest straight-chain isomer is designated *n-* for normal.

Solution

$H_3C-CH_2CH_2CH_2CH_3$ $H_3C-CH_2-CH-CH_3$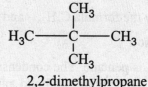
                                                          $|$
                                                         CH_3

n-pentane   2-methylbutane   2,2-dimethylpropane

Example 4   Drawing Alkanes

Draw structures for the following compounds:

a) 3-methylhexane

b) 3-ethyl-5-methyloctane

Solution

a) $H_3C-CH_2-CH-CH_2CH_2-CH_3$
                 $|$
                CH_3

b) $H_3C-CH_2HC-CH_2CH-CH_2-CH_2-CH_3$
              $|$ $|$
            C_2H_5 CH_3

Cyclic Alkanes

Examples of the alkanes that form a ring are shown in the text. The general formula for a cyclic alkane is C_nH_{2n}. Two hydrogens are lost when the carbons all bond to one another.

Alkenes

Hydrocarbons that contain at least one double bond between two carbons are known as alkenes. The names are analogous to alkanes with the suffix *-ene* replacing *-ane*. For example, ethane becomes ethene, $H_2C=CH_2$. The general formula for alkenes is C_nH_{2n}, identical to the general formula for a cyclic alkane.

Alkynes

Hydrocarbons that contain at least one triple bond between two carbons are known as alkynes. The general formula is C_nH_{2n-2}. The smallest alkyne is $HC\equiv CH$, ethyne, commonly called acetylene.

Geometric Isomers

Geometric isomers are compounds that are made up of the same types and numbers of atoms bonded together in the same sequence but with different spatial arrangements relative to the carbon–carbon double bond. If the substituted groups are on the same side of the double bond, the isomer is a *cis* isomer. If the substituted groups are on the opposite side of the double bond, the isomer is a ***trans*** isomer. (See Example 6).

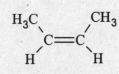

a *cis* isomer

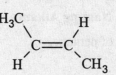

a *trans* isomer

Example 5   Cyclic Alkanes

What is the name of ⬠ ?

Solution

Each corner represents a carbon bonded to two hydrogens and to another carbon, so this is cyclopentane.

Example 6   **Naming Alkenes**

Name the following alkenes:

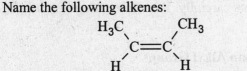
a)

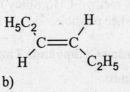
b)

First Thoughts

With alkenes, the double bond is given the lowest possible number. If similar groups are on the same side of the double bond, then you have a *cis* isomer. If they are opposite sides, it is a *trans* isomer.

Solution

a) *cis*-2-butene b) *trans*-3-hexene

Example 7   **Drawing Alkenes**

Draw the following alkenes:

a) 4-methyl-1-hexene   b) 4-methyl-*cis*-2-pentene

Solution

$H_2C=CH_2-CH_2-CH-CH_2-CH_3$
                    |
                    CH_3
a)

b)

$$\begin{array}{c} & & CH_3 \\ & & | \\ H_3C & HC-CH_3 \\ \quad \backslash & / \\ C=C \\ / & \backslash \\ H & H \end{array}$$

Example 8   **Naming Alkynes**

Name the following compounds:

$HC\equiv C-CH-CH_2-CH_3$
              |
              CH_3
a)

b) $H_3C-C\equiv CH$

Solution

The triple bond must have the lowest number.

a) 3-methyl-1-pentyne
b) 1-propyne

Aromatic Hydrocarbons

Aromatic hydrocarbons include one or more aromatic rings that contain one or more carbon–carbon double bonds. Many of these compounds are derivatives of benzene:

Benzene, C_6H_6, has six carbons in a ring with alternating double bonds. Each carbon is also bonded to one hydrogen. See the illustrations in the text.

Alkyl Groups

The text lists the alkyl groups for carbons 1–8 and the phenyl group. The names of alkyl groups are based on the stem alkane name. For example, the one-carbon alkane is methane, CH_4. The corresponding alkyl group is methyl, CH_3. Study the names carefully. Remember to assign the alkyl groups the lowest numbers when possible.

Example 9   **Naming Compounds That Contain Alkyl Groups**

Name the following compounds:

a)
$$H_3C-CH_2-\overset{\overset{\displaystyle CH_3}{|}}{\underset{\underset{\displaystyle CH_3}{|}}{C}}-\overset{\overset{}{}}{\underset{\underset{\displaystyle C_2H_5}{|}}{CH}}-CH_2-CH_3$$

b)
$$H_3C-\overset{}{\underset{\underset{\displaystyle CH_3}{|}}{CH}}-\overset{\overset{\displaystyle CH_3}{|}}{\underset{\underset{\displaystyle CH_3}{|}}{C}}-CH_3$$

Solution

a) 3-ethyl-4,4-dimethylhexane   b) 2,2,3-trimethylbutane

12.4   Separating Hydrocarbons by Fractional Distillation

The carbon-based compounds contained in crude oil can be separated by fractional distillation. **Figure 12.11** shows a schematic of a fractionating column and the fractions into which the oil is separated. The separation process utilizes the fact that the fractions have different boiling points. Because boiling point is also related to molecular size, the fractions that separate out will contain molecules that have approximately the same molecular size. **Table 12.5** lists the products of fractional distillation. Read the description of the fractional distillation operation in your text.

12.5   Processing Hydrocarbons

There are two major methods of processing:

Cracking: breaking long-chain hydrocarbons into smaller ones. There are four types:

- Thermal cracking uses heat alone.
- Catalytic cracking uses heat and a catalyst.
- Hydrocracking uses heat plus a catalyst in the presence of hydrogen gas.
- Steam cracking uses heat, steam, and a catalyst.

Reforming: straight-chain hydrocarbons are converted into a mixture containing more aromatic and branched-chain hydrocarbons.

Gasoline is a complex mixture of many hydrocarbons obtained by refining petroleum and blending fractions obtained by fractional distillation. The major components of gasoline are pentane, benzene, ethylbenzene, toluene, and xylenes.

12.6   Typical Reactions of Alkanes

Alkanes are not considered to be very reactive. However, in the presence of a free radical (see the discussion in the text) such as chlorine, a substitution reaction is possible.

$$CH_4 + Cl_2 \xrightarrow{hv} CH_3Cl + HCl$$
$$\text{chloromethane}$$

Copyright © Houghton Mifflin Company. All rights reserved.

The other type of reaction that alkanes can undergo is dehydrogenation, in which hydrogen is removed and a double bond forms to make the corresponding alkene.

$$H_3C{-}CH_3 \xrightarrow[500°C]{Pt} H_2C{=}CH_2 + H_2$$

12.7 The Functional Group Concept

Functional groups are atoms or groups of atoms with a characteristic set of chemical properties. **Table 12.6** in the text lists the important functional groups that distinguish types of organic compounds. You should learn these functional groups so that you will be able to identify the type of molecule.

Example 10 **Functional Groups**

Identify the type of molecule by looking at the functional group.

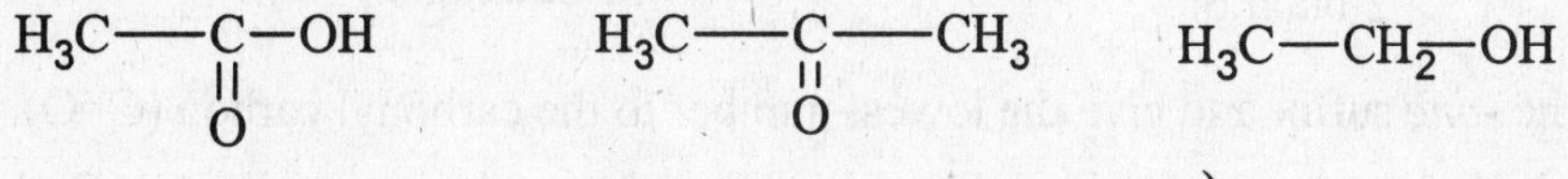

a) b) c)

Solution

By using Table 12.6, we find the compounds' identities:

a) carboxylic acid (COOH)
b) ketone (C=O)
c) alcohol (OH)

12.8 Ethene, the C=C Bond, and Polymers

The double bond in ethene (common name is ethylene) makes this molecule very reactive. Section 12.8 of the text discusses and illustrates the addition reactions with water and phosphoric acid that produce ethanol. The reaction with chlorine—that is, adding a chlorine atom to each carbon by breaking the double bond—produces 1,2-dichloroethane.

A commercially important use of ethene is the polymerization reaction to make polyethene (polyethylene). Polymers (commonly called "plastics") are made by reacting small molecules, referred to as monomers, in a such way that they attach end-to-end, forming very long polymeric chains. This method of making polymers is called **addition polymerization. Table 12.7** of the text lists some monomers and the polymers that are formed from them.

12.9 Alcohols

Alcohols all contain the —OH functional group. **Table 12.8** of the text lists some commonly used organic alcohols. Compounds may have more than one —OH group. If two groups are present, the compound is a dihydroxy alcohol; three groups makes a trihydroxy alcohol; and so on.

Primary alcohols have one carbon atom bonded to a carbon atom that bonds to a hydroxyl group. Secondary alcohols have two carbon atoms bonded to a carbon atom that bonds to a hydroxyl group. Tertiary alcohols have three carbon atoms bonded to an atom that bonds to a hydroxyl group. These are illustrated in this section of the text.

12.10 From Alcohols to Aldehydes, Ketones, and Carboxylic Acids

Oxidation of an alcohol leads to the formation of an aldehyde or a ketone. A primary alcohol forms an aldehyde:

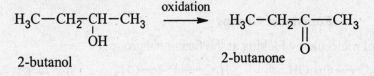

propanol   propanal

All aldehydes use the -al suffix.

A secondary alcohol forms a ketone:

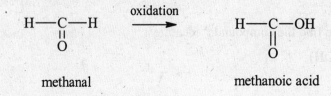

2-butanol   2-butanone

Ketones use the -one suffix and give the lowest number to the carbonyl carbon (C=O).

Tertiary alcohols cannot be oxidized to a compound containing a carbonyl. Both aldehydes and ketones can be further oxidized to a carboxylic acid:

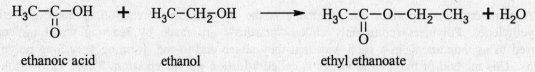

methanal   methanoic acid

Carboxylic acids use the suffix –oic in their names.

12.11 From Alcohols and Carboxylic Acids to Esters

An important reaction of alcohols and carboxylic acids is the condensation reaction, which yields an ester. This reaction involves the elimination of a water molecule. For example, consider the reaction of ethanoic acid with ethanol:

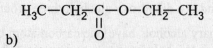

ethanoic acid   ethanol   ethyl ethanoate

Notice that the name of the ester has two parts. The -yl part comes from the alcohol using the alkyl name, and the -oate suffix is attached to the carboxylic acid stem.

Example 11   **Naming Esters**

Name the following esters:

a) H₃C—C—O—CH₃ (O)   b) H₃C—CH₂—C—O—CH₂—CH₃ (O)

First Thoughts

One carbon from the alcohol makes it methyl; two carbons from the acid makes it ethyl-anoate.

Solution

a) methyl ethanoate   b) ethyl propanoate

12.12 Condensation Polymers

Making polymers using a reaction that removes a small molecule as two larger molecules are connected is known as **condensation polymerization.** Polyesters are made from condensation reactions that join ester molecules to form the polymer. Instead of producing an ester, amines, which contain the —NH₂ functional group, are used to make polyamides commonly referred to as "nylons." Your text has illustrations of these polymers.

12.13 Polyethers

Another important group of polymers are those containing the —C—O—C—ether linkage. An example of a simple ether is CH₃—O—CH₂CH₃, methyl ethyl ether. Polyethers have repeating ether linkages within the long-chain polymer molecule. Epoxy resin is an example of a polyether.

12.14 Handedness in Molecules

Chiral compounds are non-superimposable mirror images. For example, a molecule that has a carbon atom with four different groups attached to it is chiral (see **Figure 12.16** in the text). The carbon atom is said to be "asymmetric." Molecules that do not have an asymmetric carbon atom are **achiral.**

The two mirror images of a chiral molecule are known as **stereoisomers.** The right-handed molecule and the left-handed molecule have similar physical properties. One important distinguishing characteristic of the stereoisomers is their ability to rotate polarized light (see **Figure 12.18** in the text). One of the stereoisomers will rotate polarized light to the left, while the other will rotate polarized light to the right.

Example 12　　　　**Chirality**

Which of the following molecules contains an asymmetric carbon atom?

a) CH₃CH₂COOH

b) b) CH₃CHOHCOOH

First Thoughts

An asymmetric carbon will have four different groups attached to it.

Solution

Draw the molecules to get a better view:

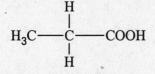 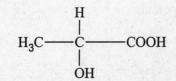

a) C has three different groups　　　　b) C has four different groups, making it asymmetric

12.15 Organic Chemistry and Modern Drug Discovery

Chapter 12 ends with a discussion of the impact that organic chemistry has made on pharmaceutical science. The example used is the anticancer drug Taxol. Taxol is derived from a natural product, the bark of the Pacific yew tree. The discussion centers on organic chemists' discovery that they could convert 10-deacetylbaccatin III, which can be extracted from the much more abundant needles of the tree, to Taxol using a modified total synthesis. This yielded the desired product in greater amounts and did not require the trees to be sacrificed.

Exercises

Section 12.1

1. What is an allotrope?

2. What structural properties account for the differences between diamond and graphite?

3. What type of bonding is found in diamond and graphite?

Section 12.2

4. What is crude oil?

5. Why is petroleum so important to the chemical industry?

Section 12.3

6. What are hydrocarbons?

7. What do "saturated" and "unsaturated" mean when referring to hydrocarbons?

8. How do alkanes, alkenes, and alkynes differ from one another?

9. List the names of the first five alkanes.

10. Write the condensed formulas for the first five alkanes.

11. What is a structural isomer?

12. Draw the structural isomers of butane.

13. Draw four isomers of heptane that have a five-member continuous chain and only methyl alkyl groups.

14. Name the following compounds:

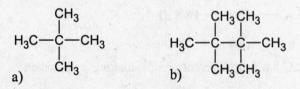

a)   b)

15. Name the following compounds:

a)   b)

16. Name the following compounds:
    a) $CH_3CH_2CH=CHCH_2CH_3$
    b) $CH_3CH=C(CH_3)_2$

17. Draw the structures of the compounds in Exercise 16.

18. Do the structures in Exercise 16 have geometric isomers? Why or why not?

19. Why do alkenes have geometric isomers but alkanes do not?

20. Which structure distinguishes aromatic hydrocarbons?

21. Which structural characteristic do all alkynes share?

22. Without drawing the structures, determine whether the following formulas might be an alkane, cycloalkane, alkene, or alkyne.
    a) C_6H_{12}
    b) C_4H_6
    c) C_5H_{12}
    d) C_3H_4

Section 12.4

23. Explain the process of separation by fractional distillation.

24. If crude oil were distilled at 200°C, which hydrocarbons would be separated out? See Figure 12.11 in the text.

Section 12.5

25. What are the processes of cracking and reforming of hydrocarbons?

26. Distinguish between the four processes of cracking hydrocarbons.

Section 12.6

27. Write the chemical equation for methane reacting with bromine to form bromomethane.

28. Which product would result from the dehydrogenation of propane?

Section 12.7

29. What are functional groups?

30. Using a chain of three carbons, write an example showing the functional group for the following types of compounds:
    a) alcohol
    b) aldehyde
    c) ketone
    d) carboxylic acid

31. How does a carboxylic acid differ from an ester?

32. Identify all of the functional groups in aspirin:

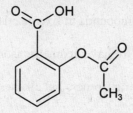

Section 12.8

33. How many carbon–carbon double bonds are found in these *straight-chain* hydrocarbons:
    a) C_4H_8 b) C_7H_8?

34. Draw the structure of 2-methyl-1-pentene.

35. Name the following structure:

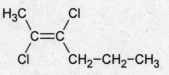

Section 12.9

36. What distinguishes a primary, secondary, and tertiary alcohol?

37. Name the following alcohols:
    a) $CH_3(CH_2)_4CH_2OH$ b) $CH_3CH_2CHOHCH_2CH_3$

38. Are the alcohols in Exercise 37 primary, secondary, or tertiary alcohols?

Section 12.10

39. What is the product of the oxidation of ethanol?

40. Draw the structure of pentanal.

41. Acetone's systematic name is 2-propanone. Why can't there be a 1-propanone?

42. Acetic acid's systematic name is ethanoic acid. Which aldehyde would be oxidized to make ethanoic acid?

43. Name the following acids:
    a) CH_3CH_2COOH b) $CH_3(CH_2)_4COOH$

Section 12.11

44. Which two organic compounds are reacted to form an ester?

45. Draw the structure of methyl pentanoate.

Section 12.12

46. Give two examples of commercially important condensation polymers.

Section 12.13

47. What is the name of $CH_3CH_2OCH_2CH_2CH_3$? Identify the functional group.

Section 12.14

48. What is the determining factor that makes a carbon atom in a compound chiral?

49. Which property allows a chemist to determine whether a molecule is left-handed or right-handed?

50. Draw the structure of $CH_3HCClOC_2H_5$ and identify the chiral carbon.

Answers to Exercises

1. An allotrope is a different form of the same element. For example, diamond and graphite are allotropes of carbon.

2. Diamond forms a covalent network in which each carbon bonds to four other carbons. Graphite forms a hexagonal network of layers in which each carbon bonds to three other carbons.

3. In diamond, the carbon is sp^3 hybridized into tetrahedrons of covalently bonded carbons. In graphite, the carbon is sp^2 hybridized into hexagons of covalently bonded carbons.

4. Crude oil or petroleum is a complex mixture of hundreds of carbon-containing compounds.

5. Petroleum is an important source of fuel energy and the starting material for the manufacture of plastics, fabrics, pharmaceuticals, and other carbon-containing products.

6. Hydrocarbons are compounds that contain only carbon and hydrogen.

7. Saturated hydrocarbons contain only single bonds. They are saturated because each carbon contains the maximum number of hydrogen atoms. Unsaturated hydrocarbons contain at least one double or triple bond between two carbons.

8. Alkanes have all carbon–carbon single bonds. Alkenes have at least one carbon–carbon double bond. Alkynes have at least one carbon–carbon triple bond.

9. methane, ethane, propane, butane, pentane

10. CH_4, CH_3CH_3, $CH_3CH_2CH_3$, $CH_3CH_2CH_2CH_3$, $CH_3CH_2CH_2CH_2CH_3$

11. Structural isomers have the same molecular formula but different bonding arrangements.

12.

$$H_3C{-}CH_2{-}CH_2{-}CH_3 \qquad\qquad H_3C{-}\underset{\underset{CH_3}{|}}{CH}{-}CH_3$$

n-butane   2-methylpropane

13.

$$H_3C{-}\underset{\underset{CH_3}{|}}{CH}{-}CH_2{-}\underset{\underset{CH_3}{|}}{CH}{-}CH_3 \qquad\qquad H_3C{-}\underset{\underset{CH_3}{|}}{HC}{-}\underset{\underset{CH_3}{|}}{CH}{-}CH_2{-}CH_3$$

2,4-dimethylpentane   2,3-dimethyl pentane

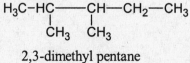

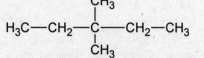

3,3-dimethylpentane   2,2-dimethylpentane

14. a) 2,2-dimethylpropane; b) 2,2,3,3-tetramethylbutane

15. a) cyclohexane; b) cycloheptane

16. a) 3-hexene; b) 2-methyl-2-butene

17.

a) $H_3C\!\!-\!\!CH_2\!-\!CH\!\!=\!\!CH\!\!-\!\!CH_2\!-\!CH_2\!-\!CH_3$

b)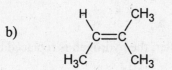

18. Structure (a) has a *cis* isomer if the hydrogens are on the same side of the double bond and a *trans* isomer if the hydrogens are on opposite sides of the double bond. In structure (b), it is not possible to spatially arrange two of the same groups on the same side or on the opposite side of the double bond.

19. Alkenes have geometric isomers due to the fact that the double bond prevents rotation and fixes the attached groups in position. The single bond of an alkane allows for free rotation around the bond.

20. Aromatic hydrocarbons are composed of ring compounds (usually six-membered rings) with double bonds on alternating carbons.

21. Alkynes have a carbon–carbon triple bond. The carbons are *sp* hybridized with 180° bond angles, making the molecule linear.

22. a) cycloalkane or alkene: both formulas are C_nH_{2n}; b) alkyne: C_nH_{2n-2}; c) alkane: C_nH_{2n+2}; d) alkyne: C_nH_{2n-2}

23. The petroleum mixture of hydrocarbons is heated in a boiler. The vaporized portion of the oil rises in the fractionating column, where it is separated by boiling point. The vapor cools and is collected as a liquid fraction. Because the boiling point is roughly related to the molecular size, each fraction contains molecules that are similar in terms of molecular size.

24. The fraction would contain molecules containing approximately 1 to 12 carbons.

25. Cracking breaks long-chain hydrocarbons into smaller ones. Reforming converts straight-chain hydrocarbons into a mixture that is richer in aromatic and branched-chain hydrocarbons.

26. Thermal cracking uses only heat. Catalytic cracking uses heat and a catalyst. Hydrocracking uses heat and a catalyst in the presence of hydrogen. Steam cracking uses heat, steam, and a catalyst.

27. $CH_4 + Br_2 \xrightarrow{hv} CH_3Br + HBr$

28. CH₃HC=CH₂ (propene)

29. A functional group is an atom or group of atoms with a characteristic set of chemical properties.

30.

a) H₃C—CH₂—CH₂—OH   b) H₃C—CH₂—CH   c) H₃C—C—CH₃
                                          ‖   ‖
                                          O   O

d) H₃C—CH₂—C—OH
              ‖
              O

31. The functional group for an acid is —COOH. For an ester, the hydrogen is replaced by a carbon —COOR (R = a carbon chain).

32.

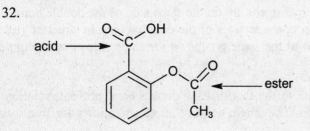

33. a) one; b) four

34.

```
                       CH₃
                        |
H₃C—CH₂—CH₂—C=CH₂
```

35. *trans*-2,3-dichloro-2-hexene

36. A primary alcohol has one carbon attached to the carbon bonded to the OH group. A secondary alcohol has two carbons attached to the carbon bonded to the OH group. A tertiary alcohol has three carbons attached to the carbon bonded to the OH group.

37. a) 1-hexanol; b) 3-pentanol

38. a) primary alcohol; b) secondary alcohol

39. ethanal (acetaldehyde)

40.

```
H₃C—CH₂—CH₂—CH₂—CH
                       ‖
                       O
```

41. A ketone has a carbon double-bonded to an oxygen and single-bonded to two other carbon atoms. If the double-bonded oxygen were on the first carbon, that carbon would be bonded to only one other carbon atom. The other bond would be to a hydrogen, making the compound an aldehyde (not a ketone).

42. ethanal (CH_3CH_2OCH)

43. a) propanoic acid; b) hexanoic acid

44. acid and alcohol

45.

$$H_3C—CH_2—CH_2—CH_2—\overset{\displaystyle \underset{\parallel}{}}{C}—OCH_3$$
$$\underset{O}{}$$

46. nylon and polyester

47. ethyl propyl ether; the function group for ethers is —C—O—C—

48. The carbon atom must be asymmetric; that is, it must have four different groups or atoms attached to it.

49. Chiral compounds rotate the plane of polarized light either to the right or to the left.

50. The carbon in the center is the chiral carbon:

$$\overset{\displaystyle H}{\underset{\displaystyle Cl}{H_3C—\overset{|}{\underset{|}{C}}—OC_2H_5}}$$

Chapter 13

Modern Materials

The Bottom Line

When you finish this chapter, you will be able to:

- Distinguish between various types of solids.
- Distinguish between the three main types of unit cells in a cubic lattice.
- Calculate the volume and density of each type of unit cell.
- Describe several properties of metals, and explain why metals exhibit these properties.
- Identify and describe different types of materials such as ceramics, alloys, plastics, and thin films.

13.1   The Structure of Crystals

Materials science is concerned with the chemistry and development of substances used to make the items we use everyday. Solids are extremely important in the field of material science, so it is crucial that we understand them on the atomic level. **Table 13.1** in your text lists the different types of solids and their properties. Here are some key points regarding solids:

- In a **crystalline solid,** the atoms, ions, or molecules are highly ordered in repeating units over long ranges. Sodium chloride is an example of a crystalline solid.
- **Amorphous solids** have fairly rigid and fixed locations of the atoms, ions, or molecules but lack the high degree of long-range order (while still having short-range order). Glass from a window is an example of an amorphous solid.
- **Ionic solids** are made up of ionic compounds that are held together by strong electrostatic forces of attraction between adjacent cations and anions. Ionic solids typically have high melting points and conduct electricity when put in solution.
- **Molecular solids** are made up of molecules held in a rigid structure. The intermolecular forces of attraction between adjacent molecules are not as strong as the electrostatic forces found in ionic solids. As a result, molecular solids have lower melting points and typically do not conduct electricity.

A structure known as a **crystal lattice** defines the shape of a crystalline solid. The smallest repeating unit of the lattice is the **unit cell.** There are three main types of cubic lattices:

- **Simple cubic unit cell (primitive):** The centers of the atoms, ions, or molecules are located only on the corners of the unit cell.
- **Body-centered cubic unit cell (bcc):** An additional atom, ion, or molecule is at the center of the simple cubic structure.
- **Face-centered cubic unit cell (fcc):** An atom, ion, or molecule is located on each of the faces of the cubic lattice.

Figure 13.4 in your text illustrates these unit cells. Try your best to picture these unit cells three-dimensionally in your head.

214

Metallic crystals can be further identified as follows:

- **Hexagonal closest packed structure (hcp):** The third layer of atoms in the crystal lattice lines up with the first layer. The unit cell is a hexagonal prism.
- **Cubic closest packed structure (ccp):** The third layer of atoms in the crystal lattice is staggered like the second layer so that it does not line up with the first layer.

Figures 13.6 and **13.7** illustrate each of these closest packed structures.

Example 1   Solids

Which of the following statements is false?

a) Solids can either be amorphous or crystalline.
b) Molecular solids generally have higher melting points than ionic solids.
c) The arrangement of particles within the crystal is a repeating unit called the unit cell.
d) The simple cubic, body-centered cubic, and face-centered cubic are examples of unit cells.
e) Metallic crystals can be classified as either hexagonal closest packed or cubic closest packed structures.

Solution

Statement (b) is false. Molecular solids generally have lower melting points than ionic solids.

The spheres (atoms) in a crystal are packed as closely together as possible, but some empty space will exist in a unit cell. To determine the amount of empty space, we need to know how much space the atoms in a unit cell occupy and how much space the entire cell occupies. The difference between the two values is the empty space. Here are the number of complete atoms (whole atoms) occupied by the various unit cells:

- **Bcc:** The unit cell contains two complete atoms (the central atom in the center of the cell and the eight 1/8ths that make up the corners). See **Figure 13.10** in your text to help you visualize this structure.
- **Primitive:** The unit cell contains one complete atom (the eight 1/8ths that make up the corners).
- **Fcc:** The unit cell contains four complete atoms (the eight 1/8ths that make up the corners and the six 1/2s that make up the sides).

Next, we can determine the volume of each type of unit cell. Your book goes into more detail, but here is a summary:

Bcc: $\text{Volume} = \left(\dfrac{4r}{\sqrt{3}}\right)^3$

Primitive: $\text{Volume} = \left(2r\right)^3 = 8r^3$

Fcc: $\text{Volume} = \left(r\sqrt{8}\right)^3$

Example 2   Percentage of Empty Space

What percentage of a bcc unit cell is occupied by empty space?

First Thoughts

To answer this problem, we first need to think about how much space a body-centered cubic unit cell occupies. Then, if we know how much space is actually taken up by atoms, we can determine the amount of empty space in the unit cell. In general,

$$\% \text{ occupied} = \frac{\text{volume occupied by atoms}}{\text{volume of unit cell}} \times 100$$

Solution

Looking at the volume, we know that a bcc unit cell occupies $\left(\frac{4r}{\sqrt{3}}\right)^3$. In addition, we know that the net number of atoms in a bcc unit cell is 2. Assuming each atom has a spherical shape, the volume of one sphere is $\frac{4}{3}\pi r^3$. Therefore, the volume of space occupied by the atoms is $2 \times \frac{4}{3}\pi r^3$. The percentage of occupied space is

$$\% \text{ occupied} = \frac{\text{volume occupied by atoms}}{\text{volume of unit cell}} \times 100$$

$$\% \text{ occupied} = \frac{2 \times \dfrac{4}{3}\pi r^3}{\left(\dfrac{4r}{\sqrt{3}}\right)^3} \times 100 = 68.0\%$$

The percentage of empty space is therefore $100.0\% - 68.0\% = 32.0\%$.

Once we know how to calculate the volume of a crystal, we can determine the density of the crystal using the following equation:

$$\text{Density } (d) = \frac{\text{mass}}{\text{volume}}$$

To find the mass of a unit cell, use the number of atoms that make up the unit cell and the mass of one of those atoms. See Example 3 for further explanation.

Example 3   Density of Cobalt

At a certain temperature, cobalt has an fcc arrangement. What is the density of cobalt (in g/cm^3) if the atomic radius is 125 pm?

First Thoughts

The first step is to convert the atomic radius to centimeters because the density needs to be in units of g/cm^3.

$$(125 \text{ pm})\left(\frac{1 \text{ m}}{10^{12} \text{ pm}}\right)\left(\frac{100 \text{ cm}}{1 \text{ m}}\right) = 1.25 \times 10^{-8} \text{ cm}$$

Solution

The volume of cobalt is

$$V = \left(r\sqrt{8}\right)^3 = \left((1.25 \times 10^{-8} \text{ cm})\sqrt{8}\right)^3 = 4.42 \times 10^{-23} \text{ cm}^3$$

Next, calculate the mass of cobalt in the unit cell. Remember, there are four atoms of cobalt in a unit cell.

$$(4 \text{ atoms})\left(\frac{1 \text{ mol Co}}{6.022 \times 10^{23} \text{ atoms Co}}\right)\left(\frac{58.93 \text{ g Co}}{1 \text{ mol Co}}\right) = 3.914 \times 10^{-22} \text{ g Co}$$

Thus, the density is

$$d = \frac{\text{mass}}{\text{volume}} = \frac{3.914 \times 10^{-22} \text{ g}}{4.42 \times 10^{-23} \text{ cm}^3} = 8.86 \text{ g/cm}^3$$

13.2 Metals

This section discusses some of the properties of metals and explains why metals behave as they do. Most metals

- Are ductile.
- Are malleable.
- Conduct electricity.
- Conduct heat.
- Are shiny.

One theory that tries to explain these properties is **band theory**. Band theory describes a metal as a lattice of metal cations spaced throughout a sea of delocalized electrons. Be sure to understand the difference between a **valence band** and a **conduction band** and how these bands play a role in accounting for the properties of metals.

Some metals, such as silicon, are classified as **semiconductors.** In semiconductors, the existence of a band gap suggests that a small amount of energy is needed to promote an electron from the valence band into the conduction band. In contrast to a semiconductor, an **insulator** has a large band gap, so very few electrons can be found in the conduction band at room temperature. In general, semiconductors conduct electricity and insulators do not. As the temperature of both increases, however, their conductivity increases.

In the real world, most of the metals we use are homogeneous mixtures of two or more metals called **alloys.** Typically, alloys are less ductile and malleable than pure metals. **Table 13.4** in your text lists some common alloys and their uses. Alloys can be classified as **substitutional alloys** or **interstitial alloys.** Examples of substitutional alloys include brass and sterling silver. The most common example of an interstitial alloy is steel.

13.3 Ceramics

Whereas metals are ductile and malleable, ceramic materials generally are not. **Ceramics** are nonmetallic substances that are made of inorganic compounds. Ceramics such as plaster and cement are often used to support a lot of weight. Other ceramics include mortar, glass, and superconductors.

Ceramic materials are:

- Amorphous or crystalline in their structure.
- Insulators with low electrical conductivity and high melting point.
- Resistant to corrosion.
- Not easily expandable when heat is applied.

Your text presents specific examples of ceramic materials and their chemistry.

13.4   Plastics

Plastics are polymers that can be molded into a shape and then hardened in that form. Plastics are good insulators and are lightweight. Some plastics are made through **crosslinking,** which involves linking the chains of adjacent polymer strands together to enhance the strength of the overall plastic. Other plastics, called **fibers,** are created by aligning the chains in one direction (parallel). This structure provides a high **tensile strength,** meaning the fibers stretch readily. **Tables 13.8** and **13.9** list several types of plastics.

Example 4   Identifying Materials

Indicate whether the following items are made of a metal, alloy, ceramic, or plastic.

a) glass window pane
b) Styrofoam coffee cup
c) silver dollar
d) brass trumpet

Solution

a) ceramic
b) plastic
c) metal
d) alloy

13.5   Thin Films and Surface Analysis

One strategy for using materials is to make the material very thin. A **thin film** is a film made from any material that is typically only 0.1 μm to 300 μm thick. Three common techniques have been developed to deposit films directly onto a surface (or else the thin film is too fragile and will break): physical deposition, sputtering, and chemical-vapor deposition. Your text goes into more detail about each one of these techniques. Be sure you understand them.

Example 5   Thin Film

How many atoms of calcium (atomic radius = 197 pm) are needed to coat a small chalkboard that is 4.0 ft × 3.0 ft? Assume that the calcium atoms are aligned in rows and columns as they cover the chalkboard.

First Thoughts

First, determine the area of the chalkboard; then, calculate the area occupied by one calcium atom (assuming it is square and not round). Because the units for both areas must be identical, let's convert the area of the chalkboard from ft^2 to pm^2.

Solution

Area of chalkboard:

$$(4.0 \text{ ft})\left(\frac{12 \text{ in}}{1 \text{ ft}}\right)\left(\frac{2.54 \text{ cm}}{1 \text{ in}}\right)\left(\frac{1 \text{ m}}{100 \text{ cm}}\right)\left(\frac{10^{12} \text{ pm}}{1 \text{ m}}\right) = 1.2 \times 10^{12} \text{ pm}$$

$$(3.0 \text{ ft})\left(\frac{12 \text{ in}}{1 \text{ ft}}\right)\left(\frac{2.54 \text{ cm}}{1 \text{ in}}\right)\left(\frac{1 \text{ m}}{100 \text{ cm}}\right)\left(\frac{10^{12} \text{ pm}}{1 \text{ m}}\right) = 9.1 \times 10^{11} \text{ pm}$$

$$\text{Area} = \text{length} \times \text{width} = (1.2 \times 10^{12} \text{ pm})(9.1 \times 10^{11} \text{ pm}) = 1.1 \times 10^{24} \text{ pm}^2$$

Area of the calcium atom:

Assuming the atom is a square, each side would be 394 pm (diameter = 2 × 197 pm).

$$\text{Area} = (394 \text{ pm})(394 \text{ pm}) = 1.55 \times 10^5 \text{ pm}^2$$

Number of atoms:

$$\text{Number of atoms} = \frac{\text{area of chalkboard}}{\text{area of one atom}} = \frac{1.1 \times 10^{24} \text{ pm}^2}{1.55 \times 10^5 \text{ pm}^2} = 7.2 \times 10^{18} \text{ Ca atoms}$$

13.6   On the Horizon: What Does the Future Hold?

Advances in the technology of modern materials occur every day. Some of the more recent developments include "green chemistry," biopolymers, and aerogels. Your text goes into detail about each of these areas.

Exercises

Section 13.1

1. Which of the following statements is false?
   a) The atoms in a crystalline solid are highly ordered in repeating units over long ranges.
   b) Molecular solids have stronger electrostatic forces than ionic solids.
   c) Amorphous solids lack the high degree of long-range order that crystalline solids have.
   d) Glass is an example of an amorphous solid.
   e) Sodium chloride is an example of a crystalline solid.

2. True or false? Ionic solids typically have low melting points but conduct electricity when put in solution.

3. True or false? Molecular solids are made up of molecules held in a rigid structure.

4. True or false? There are three main types of unit cells: the primitive cell, the body-centered cubic unit cell (bcc), and the face-centered cubic unit cell (fcc).

5. What is the difference between a hexagonal closest packed (hcp) structure and a cubic closest packed (ccp) structure?

6. Justify why the primitive unit cell contains only one complete atom.

7. Justify why the bcc unit cell contains two complete atoms.

8. Justify why the fcc unit cell contains four complete atoms.

9. What is the volume (in pm^3) of a bcc unit cell with an atomic radius of 129 pm?

10. What is the volume (in cm^3) of an fcc unit cell with an atomic radius of 185 pm?

11. What percentage of a primitive unit cell is occupied by empty space?

12. What is the density (in g/cm^3) of a metal X with a bcc arrangement if its atomic radius is 166 pm and its molar mass is 99.1 g/mol?

13. What is the density (in g/cm^3) of nickel, which has an fcc arrangement and an atomic radius of 136 pm?

14. Iridium (Ir) has an fcc arrangement and a density of 22.68 g/cm^3.
    a) What is the volume of the unit cell (in cm^3)?
    b) What is the atomic radius of iridium (in pm)?

15. Tungsten has a bcc unit cell and a density of 19.25 g/cm^3.
    a) What is the atomic radius of tungsten (in pm)?
    b) What is the edge length of the unit cell (in pm)?

Section 13.2

16. Which of the following is *not* a property of metals?
    a) shiny
    b) conduct electricity
    c) ductile
    d) malleable
    e) brittle

17. What is band theory?

18. What is the difference between a valence band and a conduction band?

19. Explain why metals are able to conduct heat.

20. Why are insulators poor conductors of electricity?

21. What is an alloy?

22. List three common examples of alloys.

23. What are some advantages of using alloys over pure metals?

24. What is an amalgam? Give an example of an amalgam.

Section 13.3

25. Which of the following statements are true about ceramics?
    a) Some types of ceramics can support a lot of weight.
    b) Ceramics are made from purely organic compounds.
    c) Ceramic materials have low melting points.
    d) Ceramic materials are very ductile and malleable.
    e) All of the above are true about ceramics.

26. Give three common examples of ceramic materials.

27. Which material (ceramic, alloy, metal) generally has the lowest coefficient of expansion (is the least likely to expand)?

28. Which material (ceramic, alloy, metal) generally has the lowest melting point?

29. Give an example of where a superconductor might be found.

Section 13.4

30. What are plastics?

31. List three common types of plastics and explain how they are used in our world today.

32. What type of plastic is used on the wings of some airplanes?

33. Indicate whether the following items are made of a metal, alloy, ceramic, or plastic.
    a) solid sodium
    b) PVC pipe
    c) cement

34. Indicate whether the following items are made of a metal, alloy, ceramic, plastic, or composite material.
    a) Kevlar
    b) lead crystal
    c) 2-L soda bottle
    d) sterling silver

Section 13.5

35. What is a thin film?

36. What are three common techniques designed to deposit thin films directly onto a surface?

37. Thin films of inorganic materials such as MgF_2 and SiO_2 are made by which technique?

38. Which technique is used to make thin films of metals, alloys, ceramics, and polymers?

39. Sputtering techniques are good for making thin films involving what elements?

40. How many atoms of titanium (atomic radius = 142 pm) are needed to cover an area that is 2.5 in × 4.8 in? Assume that the titanium atoms are aligned in rows and columns as they cover the area.

41. How many moles of gold (atomic radius = 144 pm) are needed to cover an area that is 2.9 m × 1.3 m? Assume that the gold atoms are aligned in rows and columns.

Section 13.6

42. What is "green chemistry"?

43. What are the six principles that guide chemists to a more "environmentally friendly" science?

44. What are biopolymers? What are some examples of biopolymers?

45. What is the importance of an aerogel?

Answers to Exercises

1. b

2. false

3. true

4. true

5. In hcp, the third layer of atoms in the crystal lattice lines up with the first layer. In ccp, the third layer does not line up with the first layer.

6. When you visualize a primitive unit cell, only 1/8th of an atom lies at each corner of the unit cell, but there are eight corners of a unit cell altogether. Therefore:

$$8 \text{ corners} \times \frac{1}{8} \text{ of atom per corner} = 1 \text{ atom}$$

7. When you visualize a bcc unit cell, there is one complete atom in the center of the cell in addition to the 1/8th piece in each corner.

$$1 \text{ atom} + \left[8 \text{ corners} \times \frac{1}{8} \text{ of atom per corner} \right] = 2 \text{ atoms}$$

8. When you visualize an fcc unit cell, only half of an atom lies at the face of each side of the unit cell and there are six sides to each cell. You also have to include the 1/8th piece in each corner.

$$\left[6 \text{ sides} \times \frac{1}{2} \text{ of atom per side} \right] + \left[8 \text{ corners} \times \frac{1}{8} \text{ of atom per corner} \right] = 4 \text{ atoms}$$

9. $2.64 \times 10^7 \text{ pm}^3$

10. $1.43 \times 10^{-22} \text{ cm}^3$

11. 47.6%

12. 5.84 g/cm^3

13. 6.85 g/cm^3

14. a) $5.629 \times 10^{-23} \text{ cm}^3$
    b) 135.5 pm

15. a) 137.1 pm
    b) 316.5 pm

16. e

17. Band theory describes a metal as a lattice of metal cations spaced throughout a sea of delocalized electrons.

18. A valence band contains valence electrons and is a band of closely spaced, filled bonding molecular orbitals. A conduction band contains the empty molecular orbitals. It is a band of closely spaced, unfilled antibonding orbitals.

19. When a metal is heated, some of the heat absorbed excites electrons into the conduction band. Because the electrons in that band are delocalized across the metal, the energy can be distributed from one end of the metal to the other end, enabling the metal to conduct heat.

20. The band gap in an insulator is large so there are hardly any electrons at room temperature in the conduction band.

21. a homogenous mixture of two or more metals

22. bronze, stainless steel, pewter

23. Alloys are stronger, less ductile, and less malleable than pure metals. In addition, only a limited number of pure metals exist in the real world.

24. An amalgam is an alloy containing the metal mercury. One example is a silver filling used by dentists.

25. a

26. cement, glass, plaster of Paris

27. ceramics

28. metals

29. MRI (magnetic resonance imager)

30. Plastics are polymers that can be molded into a shape and then hardened in that form.

31. polyethylene—plastic bags; Styrofoam—coffee cups; polypropylene—food packaging

32. polymeric fiber

33. a) metal
    b) plastic
    c) ceramic

34. a) composite material (also a type of plastic)
    b) ceramic (type of glass)
    c) plastic
    d) alloy

35. A thin film is a film made out of any material that is typically 0.1 μm to 300 μm thick.

36. physical deposition, sputtering, chemical-vapor deposition

37. physical deposition

38. chemical-vapor deposition

39. silicon, titanium, aluminum, gold, silver

40. 9.6×10^{16} Ti atoms

41. 7.5×10^{-5} mol Au

42. "Green chemistry" involves developing environmentally benign technologies.

43. The six principles include waste prevention, atom economy, reducing the use of hazardous chemicals, energy efficiency, use of catalytic reactions, and pollution prevention.

44. Biopolymers are natural polymers. They include DNA, RNA, enzymes, cellulose, chitin, and starch.

45. Aerogels are extremely effective insulators (especially for spacecraft).

Chapter 14

Thermodynamics

The Bottom Line

When you finish this chapter, you will be able to:

- Distinguish between a spontaneous process and a nonspontaneous process.
- Understand the meaning of entropy.
- Apply the second and third laws of thermodynamics to chemical systems.
- Determine the spontaneity of a chemical reaction by calculating the free energy.
- Appreciate the meaning of equilibrium as it relates to free energy and the rate of a chemical reaction.

14.1   Probability as a Predictor of Chemical Behavior

Probability is useful for predicting the direction of a spontaneous process. A microstate for a molecule in a system of molecules describes the possible arrangements available to the molecule. The total of the microstates in a system defines the macrostate of the system. For individual molecules, the microstate can include their translational motion, their vibrational–rotational state, and their electron configuration.

Example 1   **Probability**

Two boxes are connected by a sliding door. The box on the left contains three identical gas molecules. Determine the probability that only one atom will occupy one of the boxes.

Solution

Let's represent the bulbs by two boxes, putting the three molecules in one box:

ABC	

Now draw all possible arrangements (microstates) for the molecules. There will be $2^3 = 8$ of them.

ABC	
AB	C
AC	B
BC	A
A	BC
B	AC
C	AB
	ABC

Counting the number of microstates that have only one molecule in a bulb, we arrive at 6. The probability is 6/8 = 0.75 or 75%.

226

14.2   Why Do Chemical Reactions Happen: Entropy and the Second Law of Thermodynamics

The following are important concepts in this section:

- A **spontaneous process** occurs without continuous outside intervention.
- A **nonspontaneous process** does not occur without continuous outside intervention.
- In a chemical reaction, if the forward reaction is spontaneous, then the reverse reaction is nonspontaneous.
- **Thermodynamics** is the study of changes in energy in a reaction. It determines whether a reaction is possible.
- **Entropy (S)** measures how the energy and matter of a system are distributed throughout the system.
- **The second law of thermodynamics** states that a spontaneous process is accompanied by an increase in the entropy of the universe.

Mathematically, the second law is expressed as follows:

$$\Delta S_{universe} > 0$$

$$\Delta S_{universe} = \Delta S_{system} + \Delta S_{surroundings}$$

If $\Delta S_{universe} > 0$, the process is spontaneous.

If $\Delta S_{universe} < 0$, the process is nonspontaneous.

If $\Delta S_{universe} = 0$, the process is neither spontaneous nor nonspontaneous.

Table 14.2 in the text predicts spontaneity on the basis of the sign of $\Delta S_{surroundings}$, ΔS_{system}, and $\Delta S_{universe}$. Study this table carefully.

Example 2   Spontaneity in Common Processes

Which of the following processes are spontaneous?
  a) ice melting in a drink
  b) water decomposing into hydrogen and oxygen
  c) a boulder rolling uphill
  d) an iron tool rusting if left outside all winter

Solution

Experience tells us that processes a and d are spontaneous and processes b and c are not.

Example 3   Reaction Spontaneity

If $\Delta S_{system} = 325$ J/mol K and $\Delta S_{surroundings} = -205$ J/mol K for a reaction, will it be spontaneous?

Solution
A spontaneous process will have $\Delta S_{universe} > 0$, so

$$\Delta S_{universe} = \Delta S_{system} + \Delta S_{surroundings}$$
$$\Delta S_{universe} = 325 \text{ J/K mol} + (-205 \text{ J/K mol})$$
$$= 120 \text{ J/K mol}$$

The reaction is spontaneous.

14.3 Temperature and Spontaneous Processes

Exothermic reactions give off energy to the surroundings, increasing the kinetic energy of particles in the surroundings. If the motions of the atoms in the surroundings increase, $\Delta S_{surroundings}$ will be positive. Endothermic processes have the opposite effect on the surroundings, so $\Delta S_{surroundings}$ is negative.

The effect of temperature on entropy is given by the following equations:

$$\Delta S_{system} = \frac{\Delta H_{system}}{T}$$

$$\Delta S_{surroundings} = \frac{-\Delta H_{system}}{T}$$

Example 4   Entropy Change at a Phase Change

What is the entropy change of the system if we boil [$H_2O(l) \rightarrow H_2O(g)$] 212 g of water at 100.°C? Given: $\Delta_{vap}H = 44.0$ kJ/mol at 100.0°C.

First Thoughts

The entropy is calculated from the enthalpy and the temperature of the phase change. The temperature must be converted to kelvin: 100. + 273 = 373 K.

Solution

$$\Delta S_{system} = \frac{\Delta H_{system}}{T}$$

$$\Delta S_{system} = \frac{(44.0 \text{ kJ/mol})\left(\dfrac{1000 \text{ J}}{\text{kJ}}\right)}{373 \text{ K}} = 118 \text{ J/K mol}$$

$$= 212 \text{g } H_2O\left(\frac{1 \text{ mol } H_2O}{18.02 \text{ g } H_2O}\right)\left(\frac{118 \text{ J}}{\text{K mol}}\right) = 1.39 \times 10^3 \text{ J/K}$$

14.4 Calculating Entropy Changes in Chemical Reactions

The entropy change of a reaction is positive if the number of gaseous molecules increases. ΔS_{system} will be negative if the number of gaseous molecules decreases.

Example 5   Predict the Sign of the Entropy

Predict the sign of ΔS_{system} for the following reactions:

   a) $CS_2(l) \rightarrow CS_2(g)$
   b) $2Hg(l) + O_2(g) \rightarrow 2HgO(s)$
   c) $N_2(g) + 3H_2(g) \rightarrow 2NH_3(g)$

Solution

   a)   One mole of gaseous product, no gaseous reactants. The increase makes $\Delta S_{system} > 0$.
   b)   One mole of gaseous reactant, no gaseous products. The decrease makes $\Delta S_{system} < 0$.

c) Two moles of gaseous products, four moles of gaseous reactants. The decrease makes ΔS_{system} < 0.

The third law of thermodynamics states that the entropy of a pure perfect crystal at 0 K is zero. Under standard conditions of 298 K and 1 atm, standard molar entropies, S° values have been calculated for elements and compounds. **Table 14.4** in the text lists some of these values. Using standard molar entropies, the entropy change for a chemical reaction may be calculated using the following equation:

$$\Delta S^\circ = \sum nS^\circ{}_{products} - \sum nS^\circ{}_{reactants}$$

Example 6   **Calculating ΔS°**

Calculate ΔS° for the following reaction using data from the Appendix of your text and $S^\circ(Al_2O_3)$ = 52 J/K.

$$Fe_2O_3(s) + 2Al(s) \rightarrow 2Fe(s) + Al_2O_3(s)$$

First Thoughts

The standard molar entropies in the data table are given per mole. Consequently, you must multiply the value of S° of each reactant or product by the moles of that reactant or product.

Solution

$$\Delta S^\circ = \sum nS^\circ{}_{products} - \sum nS^\circ{}_{reactants}$$

$$= \left[2S^\circ{}_{Fe} + S^\circ{}_{Al_2O_3} \right] - \left[S^\circ{}_{Fe_2O_3} + 2S^\circ{}_{Al} \right]$$

$$= \left[2\ mol(27\ J/K\ mol) + 1\ mol(52\ J/K\ mol) \right] - \left[1\ mol(90\ J/K\ mol) + 2\ mol(28\ J/K\ mol) \right]$$

$$= -40.\ J/K$$

14.5 Free Energy

Free energy (G) provides a direct way of determining reaction spontaneity. It is experimentally used because it reflects $\Delta S_{universe}$. The textbook states two important relationships:

$$\Delta G = \Delta H_{system} - T\Delta S_{system}$$

$$\Delta S_{universe} = \frac{-\Delta G}{T}$$

The first equation provides us with a method of calculating the free energy if we know the enthalpy, the temperature, and the entropy of a reaction. It also implies that in some circumstances the temperature will determine whether a reaction is spontaneous (**see Table 14.5** in the text).

The second equation states that ΔG must be **negative** (less than 0) for the reaction to be spontaneous.

230 Chapter 14

Example 7   **Free Energy and Spontaneity**

Given the values of ΔH, ΔS, and T, determine whether each of the following sets of data represents a spontaneous or nonspontaneous reaction.

	ΔH (kJ)	ΔS (J/K)	T (K)
a)	+50	+400	300
b)	+50	+400	100
c)	−50	−400	300
d)	−50	−400	100

First Thoughts

To apply the Gibbs free energy equation, the units of ΔH and ΔS must be the same. Make them either joules or kilojoules. The temperature must be in kelvin.

Solution

$\Delta G = \Delta H - T\Delta S$; $\Delta G < 0$ indicates the reaction is spontaneous.

   a) $\Delta G = 50 \text{ kJ} - 0.400 \text{ kJ/K}(300 \text{ K}) = -70 \text{ kJ}$ spontaneous

   b) $\Delta G = 50 \text{ kJ} - 0.400 \text{ kJ/K}(100 \text{ K}) = +10 \text{ kJ}$ nonspontaneous

   c) $\Delta G = -50 \text{ kJ} - (-0.400 \text{ kJ/K})(300 \text{ K}) = +70 \text{ kJ}$ nonspontaneous

   d) $\Delta G = -50 \text{ kJ} - (-0.400 \text{ kJ/K})(100 \text{ K}) = -10 \text{ kJ}$ spontaneous

Another Way to Calculate ΔG

Because ΔG is a state function, depending only on the initial and final states, it may be calculated in the same manner as Hess's law was used for calculating ΔH. Example 8 illustrates this method.

Example 8   **Calculating ΔG by Combining Equations**

Given the following data:

   a) $N_2(g) + 2O_2(g) \rightarrow 2NO_2(g)$ $\Delta G = 104 \text{ kJ}$

   b) $2NO(g) + O_2(g) \rightarrow 2NO_2(g)$ $\Delta G = -70. \text{ kJ}$

Calculate ΔG for $N_2(g) + O_2(g) \rightarrow 2NO(g)$.

First Thoughts

Using the method learned from Hess's law, treat the reactions as if they were algebraic quantities. They may be added, reversed and then added, or multiplied by an integer.

Solution

   a) $N_2(g) + 2O_2(g) \rightarrow 2NO_2(g)$ $\Delta G = 104 \text{ kJ}$

   b) $2NO_2(g) \rightarrow 2NO(g) + O_2(g)$ $\Delta G = + 70. \text{ kJ}$

Adding a & b: $N_2(g) + O_2(g) \rightarrow 2NO(g)$ $\Delta G = 174 \text{ kJ}$

Free Energy of Formation

The spontaneity of a reaction can also be determined by using the standard free energy of formation ($\Delta_f G°$). The standard molar free energy of formation is the change in free energy of one mole of a substance in its standard state as it formed from elements in their standard states. Tables of $\Delta_f G°$ values are found in the textbook in **Table 14.4** and the Appendix. The free energy of a reaction is calculated as follows:

$$\Delta G° = \sum n\Delta_f G°_{products} - \sum n\Delta_f G°_{reactants}$$

Example 9   Calculating the Standard Free Energy from Table Values

Calculate the $\Delta_f G°$ for the following reaction:

$$C_3H_8(g) + 5O_2(g) \rightarrow 3CO_2(g) + 4H_2O(g)$$

First Thoughts

The values for each compound can be found in the tables. When looking up the value, be careful to use the correct one. For example, the value for liquid water is easily confused with the value for gaseous water. Elements in their standard state have a value of zero.

Solution

$$\Delta G° = \sum n\Delta_f G°_{products} - \sum n\Delta_f G°_{reactants}$$

$$= \left[3\Delta_f G°_{CO_2} + 4\Delta_f G°_{H_2O} \right] - \left[\Delta_f G°_{C_3H_8} + 5\Delta_f G°_{O_2} \right]$$

$$= \left[3 \text{ mol } (\text{-394 kJ/mol}) + 4 \text{ mol } (\text{-229 kJ/mol}) \right] - \left[1 \text{ mol } (\text{-24 kJ/mol}) + 5 \text{ mol } (0 \text{ kJ/mol}) \right]$$

$$= -2074 \text{ kJ}$$

14.6   When $\Delta G = 0$: A Taste of Equilibrium

If $\Delta G = 0$ for a reaction, a state of equilibrium has been reached. In other words, the rate of the forward reaction equals the rate of the reverse reaction, and neither the forward nor the reverse reaction is thermodynamically favored.

Because ΔG depends on the temperature of a reaction ($\Delta G = \Delta H - T\Delta S$), adjusting the temperature will affect the value of ΔG. The temperature at which $\Delta G = 0$ (equilibrium) can also be calculated as shown in Example 10.

Example 10   Equilibrium Temperature

For the reaction: $N_2(g) + 3H_2(g) \rightarrow 2NH_3(g)$

$\Delta H = -91.8$ kJ and $\Delta S = -197$ J/K. Calculate the temperature at which $\Delta G = 0$.

Solution

Set $\Delta G = 0$ and solve for T (remember to use the same units for ΔH and ΔS):

$$\Delta G = \Delta H - T\Delta S$$
$$0 = -91{,}800 \text{ J} - T(-197 \text{ J/K})$$
$$T = \frac{91800 \text{ J}}{197 \text{ J/K}}$$
$$= 466 \text{ K}$$
$$= 193^\circ\text{C}$$

Changes in Pressure Affect Spontaneity

The entropy of a reaction depends on pressure and, consequently, causes a change in the standard free energy. The following equation relates the free energy to the standard free energy and the pressure effects (pressure reaction quotient, Q_p):

$$\Delta G = \Delta G^\circ + RT \ln Q_p$$

R is the universal gas constant, 8.3145 J/mol K. Q_p is the expression of the ratio of the pressures of all of the gaseous products to the gaseous reactants raised to their stoichiometric coefficients.

Example 11   **Pressure Dependence of ΔG**

For the reaction:

$$N_2O_4(g) \rightarrow 2NO_2(g)$$

$\Delta G^\circ = 5.40$ kJ/mol. In a lab experiment, $P_{NO_2} = 0.122$ atm and $P_{N_2O_4} = 0.453$ atm at 298 K. Calculate ΔG and predict whether the reaction is spontaneous.

Solution

$$\Delta G = \Delta G^\circ + RT \ln Q_p$$
$$= \Delta G^\circ + RT \ln \frac{P_{NO_2}^2}{P_{N_2O_4}}$$
$$= 5.40 \times 10^3 \text{ J/mol} + (8.3145 \text{ J/mol K})(298 \text{ K})\left(\ln \frac{(0.122)^2}{0.453}\right)$$
$$= 5.40 \times 10^3 \text{ J/mol} - 8.46 \times 10^3 \text{ J/mol}$$
$$= -3.06 \times 10^3 \text{ J/mol}$$
$$= -3.06 \text{ kJ/mol}$$

ΔG is negative, so the reaction is spontaneous.

Changes in Concentrations Affect Spontaneity

ΔG can also be calculated by using the changes in concentrations of reactants and products:

$$\Delta G = \Delta G^\circ + RT \ln Q$$

In this equation, Q is the reaction quotient in terms of initial concentrations in units of moles per liter.

For the reaction:

$$N_2O_4(g) \rightarrow 2NO_2(g)$$

$$Q = \frac{[NO_2]_0^2}{[N_2O_4]_0}$$

where $[]_0$ is the initial molar concentration of the substance.

Coupled Reactions

Biochemical reactions often have a positive ΔG value, yet they are still important to living organisms. The idea of coupled reactions is that thermodynamically favorable reactions, such as the conversion of ATP to ADP, can be used to facilitate unfavorable reactions.

Proteins are an assembly of amino acids. Consider, as an example, the synthesis of a dipeptide (two amino acid chains) alanylglycine from the amino acids alanine and glycine:

alanine + glycine → alanylglycine ΔG = +29 kJ/mol

If this reaction is coupled with the conversion of ATP to ADP,

alanine + glycine → alanylglycine ΔG = +29 kJ/mol
ATP + H_2O → ADP + H_3PO_4 ΔG = −31 kJ/mol
_____

alanine + glycine + ATP + H_2O → alanylglycine + ADP + H_3PO_4

ΔG = −2 kJ/mol for the coupled reaction, making it thermodynamically possible.

Exercises

Section 14.1

1. What is a microstate and a macrostate?

2. Two chambers are connected by a trapdoor. The chamber on the left contains five identical gaseous atoms. How many microstates would there be if the trapdoor were opened?

3. What would be the probability that all the atoms in Exercise 2 would be in only one of the chambers?

4. What would the probability be that only one atom in Exercise 2 would be in a chamber?

Section 14.2

5. What is a spontaneous process? What is a nonspontaneous process?

6. Define entropy.

7. State the second law of thermodynamics.

8. Which of the following are spontaneous?
    a) A match bursts into flame.
    b) A sugar cube dissolves in hot coffee.
    c) A rusty nail turns shiny.
    d) Copper develops green coating.
    e) Rainwater evaporates from the pavement.

9. True or false? A spontaneous reaction is always a fast reaction.

10. True or false? The entropy of the universe will increase for a nonspontaneous process.

11. Which of the following reactions are spontaneous?
    a) $\Delta S_{system} = 125$ J/K and $\Delta S_{surroundings} = -225$ J/K
    b) $\Delta S_{system} = 400$ J/K and $\Delta S_{surroundings} = -200$ J/K
    c) $\Delta S_{system} = 100$ J/K and $\Delta S_{surroundings} = 100$ J/K

Section 14.3

12. Will an exothermic reaction increase or decrease $\Delta S_{surroundings}$?

13. Will an endothermic reaction increase or decrease ΔS_{system}?

14. Chloroform boils at 61.2°C and has a heat of vaporization of 29.6 kJ/mol. What is ΔS_{system} for the vaporization of 1.50 mol?

15. Methanol vaporizes at 298 K with a heat of vaporization of 37.4 kJ/mol. What is ΔS_{system} for the vaporization of 3.21 g?

Section 14.4

16. State the third law of thermodynamics. What is its significance?

17. How does an increase in gaseous products affect ΔS_{system}?

18. What is standard molar entropy?

19. What are the units of S°? Do elements have S° values?

20. Predict the sign of ΔS_{system} for the following reactions:
    a) $CO_2(s) \rightarrow CO_2(g)$
    b) $2Na(s) + Cl_2(g) \rightarrow 2NaCl(s)$
    c) $N_2(g) + 2O_2(g) \rightarrow 2NO_2(g)$

21. Predict the sign of ΔS_{system} for the following reactions:
    a) $2C(s) + O_2(g) \rightarrow 2CO(g)$
    b) $S_6(g) \rightarrow S_6(s)$
    c) $2CH_3OH(l) + 3O_2(g) \rightarrow 2CO_2(g) + 4H_2O(g)$

22. Calculate ΔS° for the following reactions. Use the Appendix of thermodynamic values in your text.
    a) $2Ca(s) + O_2(g) \rightarrow 2CaO(s)$
    b) $2C_2H_2(g) + 3O_2(g) \rightarrow 4CO(g) + 2H_2O(g)$

23. Calculate ΔS° for the following reactions. Use the Appendix of thermodynamic values in your text. S° for CS_2 = 151 J/mol K.
    a) $N_2(g) + 2O_2(g) \rightarrow 2NO_2(g)$
    b) $CS_2(l) + 3O_2(g) \rightarrow CO_2(g) + 2SO_2(g)$

24. Calculate ΔS° for the following reactions. Use the Appendix of thermodynamic values in your text.

    a) $CH_4(g) + 2O_2(g) \rightarrow CO_2(g) + 2H_2O(l)$
    b) $2MgO(s) \rightarrow 2Mg(s) + O_2(g)$

Section 14.5

25. Define free energy in terms of enthalpy and entropy. What are the units?

26. What does the sign of the free energy tell you about a chemical reaction?

27. If the free energy is negative for a reaction, will it occur rapidly?

28. How are the signs and magnitude of ΔG for the forward reaction and the reverse reaction related?

29. What is the free energy of formation? What are thermodynamic standard conditions, and how do they differ from STP for gases?

30. Calculate ΔG from the following data and determine whether the reaction is spontaneous or nonspontaneous.
   a) $\Delta H = 298$ kJ; $\Delta S = -313$ J/K; $T = 423$ K
   b) $\Delta H = -32.9$ kJ; $\Delta S = 275$ J/K; $T = 282$ K

31. Calculate ΔH from the following data:
   a) $\Delta G = -46.5$ kJ/mol; $\Delta S = 155$ J/mol K; $T = 37.0^{\circ}$C
   b) $\Delta G = 2.25$ kJ/mol; $\Delta S = 37.5$ J/mol K; $T = 25.0^{\circ}$C

32. Calculate ΔS from the following data:
   a) $\Delta G = 121$ kJ/mol; $\Delta H = -155$ kJ/mol K; $T = 337.0$ K
   b) $\Delta G = -52.8$ kJ/mol; $\Delta H = -95.2$ kJ/mol K; $T = 100.0^{\circ}$C

33. Calculate T from the following data:
   a) $\Delta G = 143$ kJ/mol; $\Delta H = -85.1$ kJ/mol K; $\Delta S = -195$ J/K
   b) $\Delta G = -193$ kJ/mol; $\Delta H = 302$ kJ/mol K; $\Delta S = 555$ J/K

34. Write the equation for the formation of one mole of the following compounds:
   a) $KCl(s)$
   b) $AsH_3(g)$

35. Given the following data:

$$S(s) + 3/2O_2(g) \rightarrow SO_3(g) \qquad \Delta G^{\circ} = -371 \text{ kJ}$$
$$2SO_2(g) + O_2(g) \rightarrow 2SO_3(g) \qquad \Delta G^{\circ} = -142 \text{ kJ}$$

   Calculate ΔG° for the following reaction:

$$S(s) + O_2(g) \rightarrow SO_2(g)$$

36. Given the following data:

$$C(s) + 2H_2(g) \rightarrow CH_4(g) \qquad \Delta G^{\circ} = -51 \text{ kJ}$$
$$2H_2(g) + O_2(g) \rightarrow 2H_2O(l) \qquad \Delta G^{\circ} = -474 \text{ kJ}$$
$$C(s) + O_2(g) \rightarrow CO_2(g) \qquad \Delta G^{\circ} = -394 \text{ kJ}$$

   Calculate ΔG° for the following reaction:
$$CH_4(g) + 2O_2(g) \rightarrow CO_2(g) + 2H_2O(l)$$

For Exercises 37–41, calculate ΔG° for the reaction using the Appendix of thermodynamic values in your textbook.

37. $2Mg(s) + O_2(g) \rightarrow 2MgO(s)$

38. $2H_2O_2(l) \rightarrow O_2(g) + 2H_2O(l)$ $\qquad$ ($\Delta_f G^{\circ}$ for $H_2O_2 = -118$ kJ/mol)

39. $2C_2H_6(g) + 7O_2 \rightarrow 4CO_2(g) + 6H_2O(l)$

40. $SO_2(g) + 2H_2S(g) \rightarrow 3S(s) + 2H_2O(g)$

41. $NH_4Cl(s) \rightarrow NH_3(g) + HCl(g)$

Section 14.6

42. In terms of rate, when has a reaction reached equilibrium?

43. What is the value of the free energy at equilibrium?

44. At equilibrium, what is the value of ΔS?

45. At equilibrium, which reaction is thermodynamically favored—the forward reaction or the reverse reaction?

46. A reaction has $\Delta H = 126$ kJ/mol and $\Delta S = 84$ J/mol K. At what temperature will the reaction become spontaneous?

47. Calculate ΔG for the following reaction at 298 K, if $\Delta G^\circ = 20.0$ kJ/mol:

$$PCl_5(g) \rightarrow PCl_3(g) + Cl_2(g)$$

The initial pressures are $P_{PCl_5} = 0.0029$ atm, $P_{PCl_3} = 0.25$ atm, and $P_{Cl_2} = 0.40$ atm.

48. Calculate ΔG for the following reaction at 298 K, if $\Delta G^\circ = 2.60$ kJ/mol:

$$H_2(g) + I_2(g) \rightarrow 2HI(g)$$

The initial pressures are $P_{H_2} = 0.010$ atm, $P_{I_2} = 0.015$ atm, and $P_{HI} = 0.20$ atm.

49. What is a coupled reaction?

50. What is the role of ATP in biochemical reactions?

Answers to Exercises

1. In terms of molecules, a microstate for a molecule describes one possible arrangement for that molecule. A macrostate is the collection of all possible microstates.

2. $2^5 = 32$

3. $2/32 = 6.25\%$

4. $10/32 = 31.25\%$

5. A spontaneous process is one that occurs without continuous outside intervention. A nonspontaneous process is one that does not occur without continuous outside intervention.

6. Entropy is a measure of how energy and matter are distributed throughout a system.

7. The second law of thermodynamics states that a spontaneous process is accompanied by an increase in the entropy of the universe.

8. b, d, e

9. False. Thermodynamic functions can predict the spontaneity of a process but do not give any information about how fast or slow a reaction is.

10. False. The entropy of the universe will increase for a spontaneous process.

11. b

12. $\Delta S_{surroundings}$ will increase if ΔH is negative.

13. ΔS_{system} will increase if ΔH is positive.

14. 133 J/K

15. 12.6 J/K

16. The entropy of a pure perfect crystal at 0 K is zero. This law allows for the calculation of standard molar entropies.

17. ΔS_{system} will increase.

18. The standard molar entropy is the entropy of one mole of a substance under standard conditions of 1 atm and 298 K.

19. The units of S^o are J/mol K. All substances, including elements in their standard states, have standard molar entropies.

20. a) positive
    b) negative
    c) negative

21. a) positive
    b) negative
    c) positive

22. a) –207 J/K
    b) 153 J/K

23. a) –122 J/K
    b) –56 J/K

24. a) –242 J/K
    b) 217 J/K

25. Free energy is the difference in the enthalpy and the entropy of a system at a given temperature. The units are J/mol or kJ/mol.

26. If ΔG is negative, the reaction is spontaneous. If ΔG is positive, the reaction is nonspontaneous.

27. The negative sign of ΔG indicates spontaneity. It says nothing about the rate of the reaction.

28. The signs of the forward and reverse reactions are opposites. The magnitude is the same.

29. The standard free energy of formation is the change in free energy for the formation of one mole of a substance from its constituent elements in their standard states. The standard state is 1 atm and 298 K. For gases, the standard conditions are 1 atm and 273 K.

30. a) 430. kJ; nonspontaneous
    b) –110. kJ; spontaneous

31. a) 1.55 kJ
    b) 13.4 kJ

32. a) –819 J/K
    b) –114 J/K

33. a) 1.17×10^3 K
    b) 892 K

34. a) $K(s) + \frac{1}{2}Cl_2(g) \rightarrow KCl(s)$
    b) $As(s) + 3/2H_2(g) \rightarrow AsH_3(g)$

35. –300 kJ

36. –817 kJ

37. –1138 kJ

38. –238 kJ

39. –2932 kJ

40. –90 kJ

41. 91 kJ

42. Equilibrium is reached when the rate of the forward reaction equals the rate of the reverse reaction.

43. $\Delta G = 0$

44. $\Delta S = -\Delta H / T$

45. Neither the forward reaction nor the reverse reaction is thermodynamically favored.

46. 1500 K

47. 28.8 kJ/mol

48. 16.4 kJ/mol

49. A coupled reaction is a reaction in which a thermodynamically favorable reaction is used to facilitate an unfavorable reaction.

50. The conversion of ATP to ADP is a reaction whose free energy can be used to facilitate unfavorable biochemical reactions.

Chapter 15

Chemical Kinetics

The Bottom Line

When you finish this chapter, you will be able to:

- Determine the rate of a reaction.
- Explain the meaning of the rate law of a reaction and of the reaction order.
- Use the integrated rate law equations to analyze kinetic data.
- Calculate the half-life of a reaction.
- Determine the rate law from experimental data.
- Use transition state theory to describe the energy of a reaction.
- Appreciate the effect of catalysts on reaction rate.

15.1 Reaction Rates

The average **reaction rate** of a chemical reaction is the change in the concentration of a reactant or product per unit time. The units are usually molarity per second (M/s). The **initial reaction rate** is the rate at the start of a reaction. The **instantaneous reaction rate** is the rate at a specific time.

Consider the following reaction:

$$2N_2O_5(g) \rightarrow 4NO_2(g) + O_2(g)$$

The rate of this reaction can be expressed as follows:

- Rate = rate of disappearance of N_2O_5 = $-\dfrac{\Delta\left[N_2O_5\right]}{\Delta t}$

- Rate = rate of appearance of NO_2 = $\dfrac{\Delta\left[NO_2\right]}{\Delta t}$

- Rate = rate of appearance of O_2 = $\dfrac{\Delta\left[O_2\right]}{\Delta t}$

By convention, the rate of appearance is positive and the rate of disappearance is negative. Furthermore, the rates can be related to one another as demonstrated in Example 1.

Example 1   Reaction Rates and Stoichiometry

Relate the rate of disappearance of N_2O_5 to the appearance of NO_2 and O_2.

First Thoughts

In the reaction, 2 moles of N_2O_5 disappear to form 4 moles of NO_2 and 1 mole of O_2. To relate the rates, divide the rate by the stoichiometric coefficient.

Solution

$$-\frac{1}{2}\frac{\Delta[N_2O_5]}{\Delta t} = \frac{1}{4}\frac{\Delta[NO_2]}{\Delta t} = \frac{\Delta[O_2]}{\Delta t}$$

The average rate, the initial rate, and the instantaneous rate are all calculated in the same manner. The only difference is the length of the time period. Example 2 shows a rate calculation.

Example 2   The Rate of Reaction

An experiment studying the decomposition of N_2O_5 begins by using 1.50 M N_2O_5. After 10.0 min, the concentration of N_2O_5 drops to 1.38 M. What is the average rate of reaction in M/s?

First Thoughts

The rate of reaction is the change in concentration divided by the change in time. The rate expression will be negative since N_2O_5 is disappearing.

Solution

$$\begin{aligned}
\text{Rate} &= -\frac{\Delta[N_2O_5]}{\Delta t}\\[6pt]
&= -\frac{(1.38\ M - 1.50\ M)}{(10.0\ \text{min} - 0.0\ \text{min})}\\[6pt]
&= -\frac{-0.12\ M}{10.0\ \text{min}}\\[6pt]
&= 0.012\ \frac{M}{\text{min}}\left(\frac{1\,\text{min}}{60\ \text{s}}\right)\\[6pt]
&= 2.0\times10^{-4}\ M/\text{s}
\end{aligned}$$

15.2   An Introduction to Rate Laws

The **rate law** is the relationship between the reactant concentration and the reaction rate. It has the general expression:

$$\text{Rate} = k[A]^m[B]^n$$

where k is the rate constant, **[A]** and **[B]** are the molar concentrations of A and B, and m and n are the orders of the individual compounds. The **overall reaction order** is the sum of individual orders—in this case, $m + n$. It is important to recognize that the order of a reaction is *experimentally* determined and cannot be determined by looking at the coefficients of the chemical reaction.

Example 3   The Rate Law

For the reaction

$$2NO(g) + Cl_2(g) \rightarrow 2NOCl(g)$$

the rate law was determined to be

$$\text{Rate} = [NO]^2[Cl_2]$$

What is the order of each reactant? What is the overall order?

Solution

The reaction is second order in [NO] and first order in [Cl_2]. The reaction overall is third order.

Collision Theory

Collision theory states that a reaction will occur when molecules collide, have the appropriate energy, and are properly oriented. This implies both that larger concentrations have faster reaction rates and that reactions at higher temperatures have faster reaction rates. The actual rates depend on the number of properly oriented collisions and are difficult to predict. The minimum energy required to make a product is called the **activation energy (E_a)**.

15.3   Changes in Time: The Integrated Rate Law

By using calculus, we can derive equations for first-, zero-, and second-order types of reactions.

First-Order Reaction

$$A \rightarrow Products$$

$$\ln\left(\frac{[A]_t}{[A]_0}\right) = -kt \quad \text{or} \quad [A]_t = [A]_0\, e^{-kt}$$

where $[A]_t$ = concentration of A at any time t, $[A]_0$ = concentration of A at time = 0, and k is the rate constant. As mentioned in your text, the plot of ln[A] versus t is linear for a first-order reaction.

The **half-life** of a first-order reaction is

$$t_{1/2} = \frac{0.693}{k}$$

This equation determines the time it takes for a reaction to proceed to 50% completion.

Zero-Order Reaction

In a zero-order reaction, the rate law does not depend on the concentration of any compound in the reaction. Consequently,

$$Rate = k[A]^0 = k$$

In other words, the rate is constant. For a zero-order reaction, a plot of $[A]_t$ versus t is linear and the half-life is

$$t_{1/2} = \frac{[A]_0}{2k}$$

Second-Order Reaction

There are two possibilities:

$$Rate = k[A]^2 \quad \text{and} \quad Rate = k[A][B]$$

The first rate law, which involves only one compound, can be integrated to

$$\frac{1}{[A]_t} = kt + \frac{1}{[A]_0}$$

244 Chapter 15

A plot of $1/[A]_t$ versus t will be linear and the half-life is

$$t_{1/2} = \frac{1}{k[A]_0}$$

The second-order reaction with two compounds is much more mathematically complicated. Your text discusses methods for experimentally manipulating conditions to simplify the calculations.

Example 4   **Integrated First-Order Rate Law**

If heated, cyclopropane will decompose to form propene. For this first-order reaction, $k = 5.40 \times 10^{-2}$ h^{-1} at 325 K. If the initial concentration of cyclopropane is 0.055 M, how many hours will elapse before the concentration of cyclopropane drops to 0.013 M?

Solution

$$\ln \frac{[\text{cyclopropane}]_t}{[\text{cyclopropane}]_0} = -kt$$

$$\ln \frac{0.013\ M}{0.055\ M} = -5.40 \times 10^{-2} h^{-1} (t)$$

$$= 26.7\ h$$

Example 5   **First-Order Half-life**

What is the half-life for the cyclopropane reaction if $k = 5.40 \times 10^{-2}$ h^{-1} at 325 K?

Solution

$$t_{1/2} = \frac{0.693}{k}$$

$$= \frac{0.693}{5.40 \times 10^{-2} h^{-1}}$$

$$= 12.8\ h$$

15.4   Methods of Determining Rate Laws

It is possible to arrive at the rate law for a reaction by carefully examining the experiments used to determine the initial rates. Example 6 illustrates this technique.

Example 6   **Determining the Rate Law from Initial Rates**

Consider the following reaction:

$$NO(g) + Cl_2(g) \rightarrow NOCl_2(g) \text{ at 400 K}$$

Determine the rate law and the value of the rate constant from the following experimental data:

Experiment	[NO]	[Cl$_2$]	Initial Rate (*M*/s)
1	0.250	0.250	1.40×10^{-4}
2	0.500	0.250	5.60×10^{-4}
3	0.250	0.500	2.80×10^{-4}

First Thoughts

Carefully examine the data. To determine the order of [NO], choose the experiments that hold [Cl$_2$] constant. Then use the experiments that hold [NO] constant to determine the order of [Cl$_2$]. Once you have identified the rate law, you can calculate k from any set of data.

Solution

The rate law will have the following form: Rate = k[NO][Cl$_2$]

Using experiments 1 and 2,

$$\frac{Rate_2}{Rate_1} = \left(\frac{[NO]_2}{[NO]_1} \right)^m$$

$$\frac{5.60 \times 10^{-4} \ M/s}{1.40 \times 10^{-4} \ M/s} = \left(\frac{0.500 \ M}{0.250 \ M} \right)^m$$

$$4 = 2^m$$

$$m = 2$$

The order of [NO] is 2. Now using experiments 1 and 3, we will determine the order of [Cl$_2$]:

$$\frac{Rate_3}{Rate_1} = \left(\frac{[Cl_2]_3}{[Cl_2]_1} \right)^n$$

$$\frac{2.80 \times 10^{-4} \ M/s}{1.40 \times 10^{-4} \ M/s} = \left(\frac{0.500 \ M}{0.250 \ M} \right)^n$$

$$2 = 2^n$$

$$n = 1$$

Then the rate law is **Rate = k[NO]^2[Cl$_2$]**, indicating that this is a third-order reaction.

To determine k, you can use any set of data:

$$Rate = k[NO]^2 [Cl_2]$$

$$2.80 \times 10^{-4} \ M/s = k(0.250 \ M)^2 (0.500 \ M)$$

$$k = 8.96 \times 10^{-3} \ M^{-2} s^{-1}$$

You can also use the **method of graphical analysis** to determine the rate law and the value of the rate constant. **Figure 15.18** in your textbook show graphs of zero-, first-, and second-order reactions. The following table summarizes what must be plotted and how to arrive at k.

Order	Plot	Slope equals
zero	[A] versus t	$-k$
first	ln[A] versus t	$-k$
second	1/[A] versus t	$+k$

This is a time-consuming method of determining the rate law. We can make the work easier by employing a graphical analysis program and using a linear regression line to fit the data. Each of the plots described in the preceding table is linear if the data fit the correct order.

15.5 Looking Back at Rate Laws

This section summarizes the discussion of reaction rates, rate laws, and determination of rate laws. **Table 15.3** in the textbook presents all of this information in a concise manner. It is a good summary to study.

15.6 Reaction Mechanisms

This section of the textbook discusses the importance of reaction mechanisms. A reaction mechanism can predict the order in which bonds are broken and formed, and the changes in relative positions of the atoms during the course of a reaction. You should read this section carefully. Here are some of the key concepts:

- A **reaction mechanism** is a series of steps by which a chemical reaction occurs.
- Each of the single steps in a chemical reaction is called an **elementary step.**
- The slow step in a reaction is known as the **rate-determining step.**
- The **molecularity** is the number of species that collide to produce the reaction indicated by an elementary step. Reactions can be **unimolecular** (involving one molecule), bimolecular (involving two molecules), or **termolecular** (involving three molecules).
- An **intermediate** is a compound that is formed and consumed during the course of a reaction.

Transition State Theory

Molecules require some minimum amount of energy to react. **Figure 15.17** in the text shows a reaction profile, which plots the potential energy versus the reaction progress. The reactants must have enough energy to cross the barrier known as the **activation energy. Transition state theory** describes how bonds in the reacting molecules are reorganized to form the bonds in the products. Atoms at the transition state are collectively known as the **activated complex.**

The **Arrhenius equation** relates the reaction rate to the energy of activation, the frequency of collisions with proper orientation, and the temperature:

$$k = Ae^{-E_a/RT}$$

where k is the rate constant, A is the frequency factor, and E_a is the activation energy.

The activation energy can be calculated from the rate of a reaction at two different temperatures:

$$\ln\left(\frac{k_2}{k_1}\right) = \frac{E_a}{R}\left(\frac{1}{T_1} - \frac{1}{T_2}\right)$$

where k_1 and k_2 are the rate constants measured at temperatures T_1 and T_2, respectively. The activation energy can also be calculated graphically by plotting $\ln k$ versus $1/T$. The resulting plot will be linear with a slope equal to $-E_a/R$.

Example 7   **Activation Energy**

For the reaction

$$N_2O_4(g) \rightarrow 2NO_2(g)$$

the rate constant $k = 4.50 \times 10^3$ s^{-1} at 275 K and 1.12×10^4 s^{-1} at 285 K. What is the energy of activation, E_a, in kJ/mol?

Solution

$$\ln\left(\frac{k_2}{k_1}\right) = \frac{E_a}{R}\left(\frac{1}{T_1} - \frac{1}{T_2}\right)$$

$$\ln\left(\frac{1.12 \times 10^4 s^{-1}}{4.50 \times 10^3 s^{-1}}\right) = \frac{E_a}{R}\left(\frac{1}{275} - \frac{1}{285}\right)$$

$$0.912 = \frac{E_a}{8.314 \text{ J/mol K}}\left(1.28 \times 10^{-4} \text{ K}^{-1}\right)$$

$$E_a = 59.4 \times 10^3 \text{ J/mol} = 59.4 \text{ kJ/mol}$$

15.7   Applications of Catalysts

A **catalyst** is a substance that changes the mechanism of the reaction to a new reaction pathway with a lower activation energy (see **Figure 15.21** and **Figure 15.22** in the textbook). Because its activation energy is lower, the new reaction pathway is faster. In this way, the catalyst effectively increases the rate of reaction. In addition, catalysts are not used up in a reaction, so they can recovered and reused.

A **homogenous catalyst** is present in the same phase as the reacting substance. A **heterogeneous catalyst** is present in a different phase than the reacting substance.

Exercises

Section 15.1

1. What is the average reaction rate? How does it differ from the initial rate?

2. For the reaction $2H_2(g) + O_2(g) \rightarrow 2H_2O(g)$, write a rate expression for the disappearance of the reactants.

3. For the reaction $4NH_3(g) + 5O_2(g) \rightarrow 4NO(g) + 6H_2O(g)$, write a rate expression for the appearance of the reactants.

4. In the reaction $N_2 + 3H_2 \rightarrow 2NH_3$, the molecular hydrogen reacts at a rate of 0.062 M/s. At what rate is ammonia formed?

5. The reaction forming ammonia (see Exercise 4) was started with 0.80 mol of nitrogen in a 2.00-L flask. After 120 s, there was 0.60 mol left. What is the reaction rate?

6. In the water reaction (see Exercise 2), the rate was followed by measuring the appearance of water. After 300. s, the concentration of water was 0.56 M. At 450. s, the concentration rose to 0.88 M. What is the reaction rate?

Section 15.2

7. What is meant by the rate law?

8. Define the overall reaction order.

9. What is the relationship between the coefficients of a balanced equation and the order of individual reactant concentrations?

10. If the rate law is Rate = $k[A][B]$, what is the order of the reaction?

11. If the rate law is Rate = $k[N_2][O_2]^2$, what is the rate with respect to each compound? What is the overall order?

12. In the two rate laws in Exercises 10 and 11, what is the significance of k?

13. According to collision theory, which of the following is necessary for a reaction to occur?
    a) Molecules must collide.
    b) Molecules must have the appropriate amount of energy.
    c) Molecules must be properly oriented during the collision.
    d) All of the above are necessary.
    e) None of the above is necessary.

14. True or false? At higher temperatures, a reaction will have a larger rate.

15. True or false? At smaller concentrations, reactions will have larger rates.

Section 15.3

16. The decomposition of H_2O_2 is a first-order reaction. If an experiment is started with 1.15 M hydrogen peroxide, what will be the concentration of peroxide after 5.65 min? The rate constant is 3.21×10^{-3} s^{-1}.

17. What is the half-life for the reaction in Exercise 16?

18. For the first-order decomposition of cyclobutane at 500 K, the rate constant is 2.45×10^{-3} s^{-1}. How long will it take for 15% of the sample to decompose?

19. The first-order decomposition of thionylchloride, SO_2Cl_2, has a half-life of 357 min. If the initial concentration of thionylchloride is 1.24×10^{-2} M, how long will it take for the concentration to decrease to 3.10×10^{-3} M?

20. For the reaction 2A→ products, Rate = $k[A]^2$ at 300 K and $k = 1.88 \times 10^{-4}$ $M^{-1}s^{-1}$. If the initial concentration of A is 0.226 M, how many minutes will it take for the concentration to decrease to 0.171 M?

21. What is the half-life of the reaction in Exercise 20?

22. The rate constant for the second-order reaction $2NOBr(g) \rightarrow 2NO(g) + Br_2(g)$ is 0.80 $M^{-1}s^{-1}$ at 285 K. Beginning with a NOBr concentration of 0.086 M, what would the concentration be after 22 s?

23. Radioactive tritium, ^3H, has a half-life of 12.5 yr. What is the first-order decay rate constant?

24. Tellurium-123 has a radioactive rate constant of 1.72×10^{-21} s^{-1}. The decay is first order. What is the half-life in years?

Section 15.4

25. Using the data in the following table, write the rate law and the rate constant for the following reaction:

$$A + B \rightarrow products$$

[A] (M)	[B] (M)	Rate (M/s)
1.50	1.50	3.20
1.50	2.50	3.20
3.00	1.50	6.40

26. For the reaction $2NO(g) + O_2(g) \rightarrow 2NO_2(g)$ at 320 K, use the data from the following table to write the rate law and the rate constant.

Experiment	[NO] (M)	[O_2] (M)	Rate (M/s)
1	0.0126	0.0125	0.0141
2	0.0252	0.0250	0.1128
3	0.0252	0.0125	0.0564

27. For the reaction $2NO(g) + Br_2(g) \rightarrow 2NOBr(g)$ at 275 K, use the following data to write the rate law and the rate constant.

Experiment	[NO] (M)	[Br_2] (M)	Rate (M/s)
1	0.10	0.20	24
2	0.25	0.20	150
3	0.10	0.50	60
4	0.35	0.50	735

Section 15.5

28. You are analyzing kinetic data from a reaction and suspect that it is first order. What would you plot to confirm this, and how will you determine the rate constant?

29. Data will be collected for determining the order of a reaction. You suspect that the reaction is zero order. What would you plot to confirm this, and what would the slope of the line tell you?

30. You have collected time and [A] data for the reaction $2A \rightarrow$ products. If it were a second-order reaction, what would you plot to confirm this? How would you calculate the rate constant?

31. Use the method of graphical analysis to determine the order of the reaction A → B. What is value of the rate constant?

[A] (M)	t (s)
0.200	0
0.100	40
0.050	80
0.025	160
0.0125	320

32. Show graphically that the reaction $2ClO_2(aq) + 2I^-(aq) \rightarrow 2ClO_2^-(aq) + I_2(s)$ is first order, and determine the rate constant from the plot.

t (s)	[ClO₂] (M)
0.00	4.77×10^{-4}
1.00	4.31×10^{-4}
2.00	3.91×10^{-4}
3.00	3.53×10^{-4}

Section 15.6

33. What is meant by a reaction mechanism?

34. What is an elementary step?

35. What is the molecularity of a reaction?

36. Explain the difference between unimolecular and bimolecular reactions.

37. Define activation energy.

38. What is the frequency factor in the Arrhenius equation?

39. True or false? At the transition state, all collisions will be successful and will form products.

40. For a reaction, the frequency factor was found to be 7.8×10^{12} s^{-1} and the activation energy was 99 kJ/mol at 85°C. What is the rate constant at this temperature?

41. A first-order reaction had $k = 2.25 \times 10^{-2}$ s^{-1} at 310. K. At 395 K, $k = 6.95 \times 10^{-2}$ s^{-1}. What is the activation energy?

42. Rate constants for some reactions are said to double for a 10-degree rise in temperature. If this were so, what would the activation energy be for a reaction carried out first at 298 K and then at 308 K?

43. A first-order reaction had $k = 3.60 \times 10^{-3}$ s^{-1} at 313 K and E_a = 121 kJ/mol. At what temperature will $k = 7.69 \times 10^{-2}$ s^{-1}?

44. Calculate the first-order rate constant at 50.°C, given that the reaction has E_a = 103 kJ/mol and $k = 2.4 \times 10^{-2}$ s^{-1} at 25°C.

45. How would you determine the activation energy graphically?

46. How would you determine the frequency factor graphically?

Section 15.7

47. What is a catalyst?

48. True or false? Catalysts lower a reaction's activation energy.

49. What is meant by a homogeneous catalyst?

50. Define heterogeneous catalyst.

Answers to Exercises

1. The average reaction rate is the change in concentration of a reactant or product per unit time. The initial rate is the rate at the start of the reaction.

2. $\text{Rate} = -\dfrac{1}{2}\dfrac{\Delta[H_2]}{\Delta t} = -\dfrac{\Delta[O_2]}{\Delta t}$

3. $\text{Rate} = \dfrac{1}{4}\dfrac{\Delta[NO]}{\Delta t} = \dfrac{1}{6}\dfrac{\Delta[H_2O]}{\Delta t}$

4. 0.041 M/s

5. 8.3×10^{-4} M/s

6. 2.1×10^{-3} M/s

7. The rate law is the relationship between the reactant concentration and the reaction rate.

8. The overall reaction order is the sum of the exponents to which all reactant concentrations appearing in the rate law are raised.

9. Any relationship between the coefficient of a reactant and the order of the concentration of that reactant is coincidental. The order must be derived from experimental data.

10. second order

11. first order in $[N_2]$; second order in $[O_2]$; overall, third order

12. k is the rate constant for the reaction

13. d

14. True

15. False. At higher concentrations, there will be more collisions and hence larger rates.

16. 0.387 M

17. 216 s

18. 66 s

19. 714 min

20. 126 min

21. 2.35×10^{4} s

22. 0.034 M

23. $0.0554 \ yr^{-1}$

24. $1.28 \times 10^{13} \ yr$

25. Rate = $k[A]$, $k = 2.13 \ M^{-1}s^{-1}$

26. Rate = $k[NO]^2[O_2]$, $k = 7.11 \times 10^3 \ M^{-2}s^{-1}$

27. Rate = $k[NO]^2[Br_2]$, $k = 1.2 \times 10^4 \ M^{-2}s^{-1}$

28. You would plot $\ln[A]$ versus t. If the data are linear, the reaction would be first order with slope = $-k$.

29. A plot of $[A]$ versus t should be linear for a zero-order reaction. The slope = $-k$.

30. A plot of $1/[A]$ versus t yielding a straight line would confirm that the reaction is second order in one reactant. The slope = $+k$.

31. Plots of $[A]$ versus t and $\ln[A]$ versus t are not linear. As shown below, the reaction is second order with $k = 0.24 \ M^{-1}s^{-1}$.

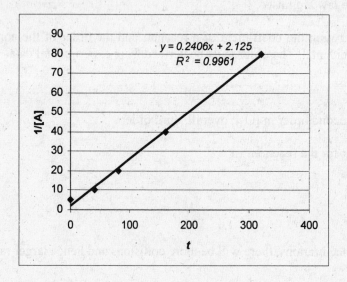

32. A plot of $\ln[ClO_2^-]$ versus t is linear with $k = 0.10$ s^{-1}

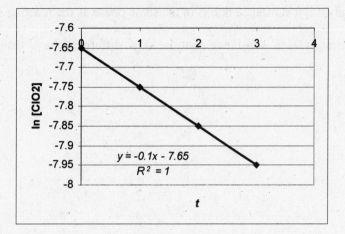

33. A reaction mechanism is a series of plausible steps illustrating how a reaction occurs.

34. A single step in a reaction mechanism is an elementary step.

35. The molecularity is the number of species that collide to produce the reaction in an elementary step.

36. A unimolecular reaction involves only one molecule in collisions. A bimolecular reaction involves two molecules in collisions.

37. The activation energy is the energy that the reactant species must attain to react.

38. The frequency factor indicates the number of properly oriented molecules undergoing collisions.

39. False. If molecules are not oriented properly, some collisions will not react to form products.

40. 0.028 s^{-1}

41. 13.5 kJ/mol

42. 52.9 kJ/mol

43. 335 K

44. 0.60 s^{-1}

45. Plot $\ln k$ versus $1/T$. $E_a = -$slope $\times R$.

46. The y-intercept of a plot of $\ln k$ versus $1/T$ is equal to $\ln A$.

47. A catalyst is a substance that increases the rate of a reaction by providing a reaction pathway that has a lower activation energy.

48. False. Catalysts do not lower the activation energy of a mechanism, but rather provide a reaction pathway that has a lower activation energy.

49. A homogeneous catalyst is a substance that is in the same phase as the reacting substance.

50. A heterogeneous catalyst is a substance that is in a different phase than the reacting substance.

Chapter 16

Chemical Equilibrium

The Bottom Line

When you finish this chapter, you will be able to:

- Define equilibrium in terms of free energy and the rates of forward and reverse reactions.
- Appreciate the meaning of dynamic equilibrium.
- Understand the implications of the magnitude of the equilibrium constant as it relates to the amount of products or reactants.
- Calculate equilibrium concentrations for a variety of equilibrium conditions.
- Calculate equilibrium constants from standard free energies.
- Apply Le Châtelier's principle to equilibrium systems.

16.1 The Concept of Chemical Equilibrium

In a chemical reaction, substances react to form a mixture of reactants and products and establish an equilibrium. The equilibrium is dynamic, in that the rate of the forward reaction, producing products from the reactants, equals the rate of the reverse reaction, producing reactants from the products. In this system, the free energy is at a minimum. There are two key ways to define equilibrium:

1) The free energy change for the forward and reverse reactions is zero: $\Delta G = 0$, and the free energy of the system is zero.
2) The rates of the forward and reverse reactions are equal.

The **mass-action** or **equilibrium expression** relates the molar concentrations of the reactants and the products to the equilibrium constant K. For a hypothetical reaction

$$a\text{A} + b\text{B} \cap c\text{C} + d\text{D}$$

the mass-action expression is

$$K = \frac{[\text{C}]^c [\text{D}]^d}{[\text{A}]^a [\text{B}]^b}$$

Remember that [A] means the molar concentration of substance A. Notice that the coefficients of the balanced equation become the exponents of the substances in the equilibrium expression.

Example 1   **The Mass-Action Expression**

Nitric oxide reacts with bromine to form nitrosyl bromide:

$$2\text{NO}(g) + \text{Br}_2(g) \cap 2\text{NOBr}(g)$$

Write the mass-action expression for the formation of nitrosyl bromide.

Solution

$$K = \frac{[\text{NOBr}]^2}{[\text{NO}]^2 [\text{Br}_2]}$$

16.2   Why Is Chemical Equilibrium a Useful Concept?

This section discusses the importance of the concept of chemical equilibrium. **Table 16.1** lists some of the chemical equilibrium processes that have a major impact on the chemical industry, the environment, and our own personal health.

16.3   The Meaning of the Equilibrium Constant

The magnitude of the equilibrium constant is a measure of the extent of a reaction.

- Processes that have very large values of K have mostly products present at equilibrium.
- Processes that have very small values of K have mostly reactants present at equilibrium.
- Processes with K values not too far from 1 have significant amounts of both reactants and products.

Example 2   **The Magnitude of the Equilibrium Constant**

Based on the magnitude of the equilibrium constant, state whether the following reactions are product favored or reactant favored.

- a) $2NO(g) + O_2(g) \cap 2NO_2(g)$ $K = 4.0 \times 10^{13}$ at 298 K
- b) $NiS(s) \cap Ni^+(aq) + S^{2-}(aq)$ $K = 3.0 \times 10^{-19}$ at 298 K

Solution

a)   The large value of K indicates that the reaction is product favored.
b)   The small value of K indicates that the solid does not dissolve very much in water and hence the reaction is reactant favored.

Homogeneous equilibrium reactions involve substances present in the same phase. **Heterogeneous equilibrium** reactions involve substances present in more than one phase. In the heterogeneous equilibrium expression, the concentrations of pure solids or liquids remain constant and are not incorporated into the equilibrium constant, K.

Example 3   **Heterogeneous Equilibrium Expressions**

Write the equilibrium expressions for the following reactions:

- a) $NiS(s) \cap Ni^+(aq) + S^{2-}(aq)$
- b) $2H_2O(l) \cap 2H_2(g) + O_2(g)$

Solution

The concentrations of pure solids or liquids do not appear in the expression.

a) $K = \left[Ni^{2+} \right]\left[S^{2-} \right]$

b) $K = \left[H_2 \right]^2 \left[O_2 \right]$

16.4   Working with Equilibrium Constants

The equilibrium chemical equations and thus the equilibrium expressions can be manipulated to change experimental conditions. Three methods are commonly used to manipulate the expressions. Examples 4–6 demonstrate each method.

Example 4   Multiplying an Equilibrium Reaction by a Number

The equilibrium constant of a reaction that has been multiplied by a number is equal to the equilibrium constant of the original reaction raised to a power equal to that number.

The equilibrium expression for $N_2O_4(g) \cap 2NO_2(g)$ is

$$K_1 = \frac{[NO_2]^2}{[N_2O_4]}$$

How is the equilibrium constant for $2N_2O_4(g) \cap 4NO_2(g)$ related to K_1?

Solution

The equilibrium expression for $2N_2O_4(g) \cap 4\,O_2(g)$ is

$$K_2 = \frac{[NO_2]^4}{[N_2O_4]^2} = \left(\frac{[NO_2]^2}{[N_2O_4]} \right)^2 = (K_1)^2$$

Example 5   Reversing an Equilibrium Reaction

The equilibrium constant for a reverse reaction is equal to the inverse of the equilibrium constant of the forward reaction.

The equilibrium expression for $N_2O_4(g) \cap 2NO_2(g)$ is

$$K_1 = \frac{[NO_2]^2}{[N_2O_4]}$$

How is the equilibrium constant for $2NO_2(g) \cap N_2O_4(g)$ related to K_1?

Solution

The equilibrium constant for $2NO_2(g) \cap N_2O_4(g)$ is

$$K_2 = \frac{[N_2O_4]}{[NO_2]^2} = \frac{1}{\dfrac{[NO_2]^2}{[N_2O_4]}} = \frac{1}{K_1}$$

Example 6   Finding the Equilibrium Constant for the Sum of Reactions

The equilibrium constant for a net reaction composed of two or more steps is the product of the equilibrium constants for the individual steps.

The reaction $H_2CO_3(aq) \cap 2H^+(aq) + CO_3^{2-}(aq)$ is actually composed of two equilibria:

1) $H_2CO_3(aq) \cap H^+(aq) + HCO_3^-(aq)$

2) $HCO_3^-(aq) \cap H^+(aq) + CO_3^{2-}(aq)$

Write the equilibrium constant for the overall reaction and relate it to the equilibrium constants for reactions 1 and 2.

Solution

The equilibrium expressions for reactions 1 and 2 are

$$K_1 = \frac{\left[H^+\right]\left[HCO_3^-\right]}{\left[H_2CO_3\right]}$$

$$K_2 = \frac{\left[H^+\right]\left[CO_3^{2-}\right]}{\left[HCO_3^-\right]}$$

The equilibrium expression for the overall reaction is

$$K = \frac{\left[H^+\right]\left[HCO_3^-\right]}{\left[H_2CO_3\right]} \times \frac{\left[H^+\right]\left[CO_3^{2-}\right]}{\left[HCO_3^-\right]}$$

$$K = \frac{\left[H^+\right]^2\left[CO_3^{2-}\right]}{\left[H_2CO_3\right]}$$

$$K = K_1 \times K_2$$

Calculating Equilibrium Concentrations from the Equilibrium Constant and K_p, the Equilibrium Constant Using Partial Pressures

The equilibrium expression involving the concentrations of the products divided by the reactants and the equilibrium constant allow us to calculate the concentration of a reactant or product if we know K and the concentrations of *some* of the reactants and products.

Example 7   **Calculating the Equilibrium Concentration from K**

Calculate the equilibrium concentration of NO_2, if the equilibrium concentration of N_2O_4 is 0.491 M and $K = 4.60 \times 10^{-3}$ at 298 K.

$$N_2O_4(g) \cap 2NO_2(g)$$

Solution

Write the equilibrium expression for the reaction. Substitute the values in for K and $[N_2O_4]$, and then solve for $[NO_2]$.

$$K = \frac{\left[NO_2\right]^2}{\left[N_2O_4\right]}$$

$$4.60 \times 10^{-3} = \frac{\left[NO_2\right]^2}{\left[0.491\right]}$$

$$\left[NO_2\right] = \sqrt{4.60 \times 10^{-3} \times 0.491}$$

$$= 0.0475 \ M$$

As discussed in the textbook, K is considered to be unitless. The concentrations are in mol/L or molarity (M).

Although chemists often work with concentration units, it is often easier to use partial pressures when working with gases. The equilibrium expression given in terms of partial pressures is analogous to the equilibrium expression given in terms of concentrations. To distinguish between the equilibrium constants, K_p is used in the equilibrium expression when it is stated in terms of partial pressures.

Example 8 $\qquad$ K_p

Write the equilibrium expression in terms of partial pressures for

$$N_2(g) + 3H_2(g) \rightleftharpoons 2NH_3$$

Solution

Instead of concentrations, partial pressures are used in this example. Once again, the equilibrium expression places products over reactants, each raised to the coefficient of the balanced equation.

$$K_p = \frac{P_{NH_3}^2}{P_{N_2} P_{H_2}^3}$$

The relationship between K and K_p is given by

$$K_p = K(RT)^{\Delta n} \text{ or } K = K_p (RT)^{-\Delta n}$$

where $R = 0.08206$ L atm/K mol, T = temperature in kelvins, and $\Delta n = n_{products} - n_{reactants}$ (moles of gaseous products minus moles of gaseous reactants).

Example 9 $\qquad$ Converting K to K_p

Given $K = 4.60 \times 10^{-3}$ at 298 K, calculate K_p for the following reaction:

$$N_2O_4(g) \rightleftharpoons 2NO_2(g)$$

Solution

$\Delta n = 2 - 1 = 1$, so using the conversion

$$K_p = K(RT)^{\Delta n}$$

$$= 4.60 \times 10^{-3} \left(\left(0.08206 \, \frac{L \, atm}{K \, mol} \right) (298 \, K) \right)^1$$

$$= 0.112$$

16.5   Solving Equilibrium Problems: A Different Way of Thinking

A. Predicting the Direction of Equilibrium

Using the **reaction quotient, Q,** it is possible to predict the direction of equilibrium. Q is the mass-action expression using *initial concentrations*, $[]_0$. For example, for the reaction

$$H_2(g) + I_2(g) \rightleftharpoons 2HI(g)$$

$$Q = \frac{[HI]_0^2}{[H_2]_0 [I_2]_0}$$

Q is then compared to K to determine the direction of equilibrium.

- If $Q = K$, the system is at equilibrium.
- If $Q > K$, the system will shift to the left to attain equilibrium.
- If $Q < K$, the system will shift to the right to attain equilibrium.

Example 10   *Q*: Predicting Equilibrium

A reaction was started by mixing 1.60 mol of hydrogen, 0.90 mol of iodine gas, and 1.20 mol of HI in a 2.0-L flask. If $K = 7.1 \times 10^2$ at 298 K for $H_2(g) + I_2(g) \rightleftharpoons 2HI(g)$, is the system at equilibrium? If not, in which direction will the system shift?

First Thoughts

First, we must calculate the initial concentrations of H_2, I_2 and HI. Then we evaluate Q and compare it to K to determine in which direction the system will shift.

Solution

$$[H_2]_0 = \frac{1.60 \text{ mol}}{2.0 \text{ L}} = 0.80 \ M$$

$$[I_2]_0 = \frac{0.90 \text{ mol}}{2.0 \text{ L}} = 0.45 \ M$$

$$[HI]_0 = \frac{1.20 \text{ mol}}{2.0 \text{ L}} = 0.60 \ M$$

$$Q = \frac{[HI]_0^2}{[H_2]_0 [I_2]_0}$$

$$Q = \frac{(0.60)^2}{(0.80)(0.45)} = 1.0$$

Since $Q < K$, the system is not at equilibrium and will shift to the right.

B. Equilibrium Calculations

Often in equilibrium problems, we want to find the amounts of substances present at equilibrium. Values of the equilibrium constants will fall into three general categories: small K values, large K values, and intermediate K values. We will deal with each technique separately in an example.

Example 11   Small Value of K

Vitamin C (ascorbic acid) is a weak acid with $K = 8.0 \times 10^{-5}$. The acid dissociates into hydrogen ion and ascorbate ion as follows:

$$HC_6H_7O_6(aq) \rightleftharpoons H^+(aq) + C_6H_7O_6^-(aq)$$

If we start with a solution of 0.50 M $HC_6H_7O_6$, what will be the equilibrium concentrations of the H^+ and $C_6H_7O_6^-$ ions?

First Thoughts

With a small K value, the reaction will stay far to the left and only a small amount will dissociate. To solve this problem, first construct an ICEA table. Second, solve for the concentrations of the ions. Finally, test the assumption against the "5% rule" (divide x, the amount that reacts, by M_0, the original concentration and multiply by 100%):

$$\frac{x}{M_0} \times 100\% \leq 5\%$$

Solution

Constructing the ICEA table,

$$HC_6H_7O_6(aq) \cap H^+(aq) + C_6H_7O_6^-(aq)$$

I	0.50 M	0 M	0 M
C	$-x$	$+x$	$+x$
E	$0.50 - x$	$+x$	$+x$
A	0.50	x	x

In the assumption row, we are assuming that the amount that dissociates, x, is so small compared to the original concentration that it can be neglected.

$$K = \frac{\left[H^+\right]\left[C_6H_7O_6^-\right]}{HC_6H_7O_6}$$

$$8.0 \times 10^{-5} = \frac{(x)(x)}{0.50}$$

$$x^2 = 4.0 \times 10^{-5}$$

$$x = 6.3 \times 10^{-3}$$

$$x = \left[H^+\right] = \left[C_6H_7O_6^-\right] = 6.3 \times 10^{-3} \ M$$

Now we must test this assumption against the 5% rule.

$$\frac{6.3 \times 10^{-3} \ M}{0.50 \ M} \times 100\% = 1.3\% \leq 5\%$$

The assumption passes the test.

Example 12   Small Value of K: Slightly Soluble Solids

Fluorite is the mineral name for calcium fluoride, CaF_2. Determine the concentrations of the calcium ion and the fluoride ion in a saturated solution of calcium fluoride.

$$CaF_2(s) \cap Ca^{2+}(aq) + 2F^-(aq) \qquad K = 3.4 \times 10^{-11}$$

First Thoughts

We can use the same approach that we used for the ascorbic acid problem, with one exception: CaF_2 is a solid, so it will not appear in the equilibrium expression.

Solution

Let x be the amount of CaF_2 that dissolves. It is also the solubility of CaF_2.

$$CaF_2(s) \rightleftharpoons Ca^{2+}(aq) + 2\,F^-(aq)$$

		Ca^{2+}	F^-
I	—	0	0
C	—	$+x$	$+2x$
E	—	x	$2x$

$$K = [Ca^{2+}][F^-]^2$$

$$3.4\times10^{-11} = (x)(2x)^2$$

$$3.4\times10^{-11} = 4x^3$$

$$\sqrt[3]{8.5\times10^{-12}} = x$$

$$2.0\times10^{-4}\ M = x = \left[Ca^{2+}\right]$$

$$[F^-] = 2x = 4.0\times10^{-4}\ M$$

Example 13   Large Value of K

For the hydrogen iodide equilibrium,

$$H_2(g) + I_2(g) \rightleftharpoons 2HI(g)$$

$K = 7.1 \times 10^2$ at 298 K. A reaction was started with 1.40 M hydrogen and 0.22 M iodine. Calculate the equilibrium concentrations of H_2, I_2, and HI.

First Thoughts

In this reaction, K is large. The reaction goes to completion, *except* for a small amount that has not reacted. From the stoichiometry, we see that I_2 is the limiting reactant and H_2 is in excess.

	H_2	$+$	I_2	$\rightleftharpoons$	2HI
I	1.40		0.22		0
C	$-0.22 + x$		$-0.22 + x$		$0.44 - 2x$
E	$1.18 + x$		x		$0.44 - 2x$
A	1.18		x		0.44

This ICEA table is different than the one we constructed in Example 12. Because K is large and the reaction essentially goes to completion, x represents the small amount that does not react to form product. Once again, we will assume that x is small relative to 1.18 and 0.44 and, therefore, can be neglected. We will have to test these assumptions against the 5% rule.

The approximation technique is a useful shortcut for solving these types of problems. You can always use the quadratic formula to solve quadratic equations. Given a calculator, this is not a difficult chore. (Example 14 solves the problem using the quadratic formula.)

Solution

$$K = \frac{[\text{HI}]^2}{[\text{H}_2][\text{I}_2]}$$

$$7.1 \times 10^2 = \frac{(0.44)^2}{(1.18)(x)}$$

$$x = 2.3 \times 10^{-4} \text{ M} = [\text{I}_2]$$

$$[\text{H}_2] = 1.18 \text{ M}$$

$$[\text{HI}] = 0.44 \text{ M}$$

Testing the assumptions:

$$\frac{x}{M_0} \times 100\% \leq 5\%$$

$$\frac{2.3 \times 10^{-4} \text{ M}}{1.18 \text{ M}} \times 100\% = 0.019 \leq 5\%$$

$$\frac{(2)(2.3 \times 10^{-4} \text{ M})}{0.44 \text{ M}} \times 100 = 0.10\% \leq 5\%$$

The assumptions are valid.

Example 14   **Intermediate Value of K**

The dissociation of 0.50 M hydrofluoric acid in water has an intermediate $K = 7.1 \times 10^{-4}$ at 298 K. Calculate the equilibrium concentrations of all species.

$$\text{HF}(aq) \cap \text{H}^+(aq) + \text{F}^-(aq)$$

First Thoughts

We will construct an ICE table for this problem. Because K has an intermediate value, we will not make any assumptions about neglecting x. Instead, we will solve the equation by using the quadratic formula.

Solution

	$\text{HF}(aq) \cap \text{H}^+(aq) + \text{F}^-(aq)$		
I	0.50	0	0
C	$-x$	x	x
E	$0.50 - x$	x	x

$$K = \frac{[\text{H}^+][\text{F}^-]}{[\text{HF}]} = 7.1 \times 10^{-4}$$

$$\frac{(x)(x)}{(0.50-x)} = 7.1 \times 10^{-4}$$

$$0 = x^2 + 7.1 \times 10^{-4} x - 3.6 \times 10^{-4}$$

Using the quadratic formula:

$$x = \frac{-b \pm \sqrt{b^2 - 4ac}}{2a}$$

$$= \frac{-7.1 \times 10^{-4} \pm \sqrt{\left(7.1 \times 10^{-4}\right)^2 - 4(1)\left(-3.6 \times 10^{-4}\right)}}{2(1)}$$

$$= \frac{-7.1 \times 10^{-4} \pm 0.038}{2}$$

$$= 1.9 \times 10^{-2} \ M \ \text{ or } \ -3.9 \times 10^{-2} \ M$$

The second solution is physically impossible, because the concentration of the ions cannot be negative.

$$[H^+] = [F^-] = 1.9 \times 10^{-2} \ M$$

$$[HF] = (0.50 - 1.9 \times 10^{-2}) \ M = 0.48 \ M$$

16.6 Le Châtelier's Principle

Le Châtelier's principle states that if stress is applied to an equilibrium system, it will change in such a way as to partially undo the applied stress and restore the equilibrium. There are three possible effects to consider: changing the concentration of a reactant or product, changing the partial pressure of a gaseous reactant or product, and changing the temperature.

- **Removing or Adding Products or Reactants.** If more reactant is added or if some product is removed from an equilibrium mixture, the reaction will shift to the right, producing more products. If more product is added or if some reactant is removed from an equilibrium mixture, the reaction will shift to the left, producing more reactants.

- **Changing the Pressure.** If the pressure is increased by decreasing the volume of the reaction mixture, the reaction will shift in the direction containing the fewer number of moles of gas.

- **Changing the Temperature.** Increasing the temperature favors an endothermic reaction. Decreasing the temperature favors an exothermic reaction.

An additional summary of these effects appears in **Table 16.4** in the text.

A catalyst alters an equilibrium system by increasing the rate at which equilibrium is attained. It does not change the composition of the equilibrium mixture.

Example 15   **Le Châtelier's Principle**

Given the reaction:

$$2CO_2(g) \rightleftharpoons 2CO(g) + O_2(g) \qquad \Delta H^\circ = 560 \text{ kJ}$$

How would each of the following changes affect the direction of equilibrium?

a) O_2 is removed as it is produced.
b) CO is added.
c) The pressure on the system is decreased.
d) The system is cooled.

Solution

a) The shift is to the right to make more products.
b) The shift is to the left to make more reactant.
c) The shift is to the right (the side with the highest amount of moles of gas).
d) The shift is to the left (the exothermic direction).

16.7   Free Energy and Equilibrium

At equilibrium, the standard free energy is

$$\Delta G^\circ = -RT \ln K_{eq}$$

where ΔG° is the standard free energy change, $R = 8.3145$ J/K mol, and T is the temperature in kelvins. This equation gives a convenient way of calculating the equilibrium constant.

Example 16   Free Energy and the Equilibrium Constant

Determine the equilibrium constant at 298 K for the reaction

$$AgCl(s) \cap Ag^+(aq) + Cl^-(aq)$$

using the standard energies of formation.

First Thoughts

First we must calculate ΔG° using the summation technique we learned in Chapter 14. That is, we take the free energies of the products and subtract the free energy of the reactant. Then we can use the above equation to calculate K. Recall that

$$\Delta G^\circ = \sum n \, \Delta_f G^\circ_{products} - \sum m \, \Delta_f G^\circ_{reactants}$$

$\Delta_f G^\circ$ values are found in the Appendix of the text.

Solution

$$AgCl(s) \cap Ag^+(aq) + Cl^-(aq)$$

$\Delta_f G^\circ$: -110 77 -131 kJ

$$\Delta G^\circ = \left[(77-131)-(-110)\right]$$
$$= 56 \text{ kJ}$$
$$\Delta G^\circ = -RT \ln K_{eq}$$
$$\ln K_{eq} = \frac{\Delta G^\circ}{-RT} = \frac{56\times10^3 \text{J}}{(8.3145 \text{ J/K mol})(298)} = -22.60$$
$$K_{eq} = e^{-22.60} = 1.5\times10^{-10}$$

Exercises

Section 16.1

1. What is the value of the free energy of an equilibrium system?

2. Why does the equilibrium constant depend on temperature? (*Hint:* How does temperature affect a rate constant?)

3. True or false? At equilibrium, the rate of the forward reaction is equal to the rate of the reverse reaction.

4. True or false? In a product-favored equilibrium reaction, the rate of forward reaction is greater than the rate of reverse reaction.

5. What is the meaning of the mass-action expression?

6. Write the equilibrium expressions for the following reactions:
    a) $3O_2(g) \cap 2O_3(g)$
    b) $2NO(g) + O_2(g) \cap 2NO_2(g)$

7. Write the heterogeneous equilibrium expressions for the following reactions:
    a) $2ZnS(s) + 3O_2(g) \cap 2ZnO(s) + 2SO_2(g)$
    b) $2NO_2(g) + 7H_2(g) \cap 2NH_3(g) + 4H_2O(l)$

8. Write the equilibrium expressions for the following reactions:
    a) $HCOOH(aq) \cap H^+(aq) + HCOO^-(aq)$
    b) $BaSO_4(s) \cap Ba^{2+}(aq) + SO_4^{2-}(aq)$

Section 16.2

9. What is the Haber process? Why is it important to agriculture?

Section 16.3

10. If the magnitude of the equilibrium constant is large, will the reaction be product favored or reactant favored?

11. If the magnitude of the equilibrium constant is small, will the reaction be product favored or reactant favored?

12. Consider the following reaction: $3O_2(g) \cap 2O_3(g)$. At 400 K, the equilibrium concentration of oxygen is 1.2×10^{-2} M and the ozone concentration is 3.3×10^{-2} M. What is the value of K?

Section 16.4

13. For the reaction $N_2O_4(g) \rightleftharpoons 2NO_2(g)$, $K = 4.7 \times 10^{-3}$ at 298 K. What is K for $2NO_2(g) \rightleftharpoons N_2O_4(g)$?

14. Given the information in Exercise 13, what is K for $4N_2O_4(g) \rightleftharpoons 8NO_2(g)$?

15. Given: $H_2CO_3(aq) \rightleftharpoons H^+(aq) + HCO_3^-(aq)$ $K = 4.3 \times 10^{-7}$

    $HCO_3^-(aq) \rightleftharpoons H^+(aq) + CO_3^{2-}(aq)$ $K = 4.8 \times 10^{-11}$

    What is K for $H_2CO_3(aq) \rightleftharpoons 2H^+(aq) + CO_3^-(aq)$?

16. A 2.00-mol sample of PCl_5 dissociates at 150°C in a 1.00-L vessel to form 0.32 mol of Cl_2 at equilibrium. What are concentrations of all species at equilibrium?

$$PCl_5(g) \rightleftharpoons PCl_3(g) + Cl_2(g)$$

17. In a 1.00-L flask, 4.00 mol of N_2O_4 is injected. It dissociates at 400 K to form 1.2 mol of NO_2 at equilibrium. What is the equilibrium concentration of N_2O_4?

$$N_2O_4(g) \rightleftharpoons 2NO_2(g)$$

18. For the PCl_5 equilibrium in Exercise 16, $K = 5.6$ at 400°C. If the equilibrium concentrations of PCl_3 and Cl_2 are both 0.82 M, what is the equilibrium concentration of PCl_5?

19. For the reaction $2NO(g) + O_2(g) \rightleftharpoons 2NO_2(g)$, $K = 1.4 \times 10^2$ at 320 K. At equilibrium, [NO] = 0.36 M and [O_2] = 0.14 M. What is [NO_2]?

20. For the equilibrium in Exercise 19, what is the value of K_p?

21. For the equilibrium in Exercise 16, $K_p = 1.24$ at 192°C. What is the value of K?

22. For the reaction $4Fe(s) + 3O_2(g) \rightleftharpoons 2Fe_2O_3(s)$, the partial pressure of oxygen is 0.025 atm. What is K_p for this equilibrium?

Section 16.5

23. What is the reaction quotient? How does it differ from the equilibrium constant?

24. For the reaction $2NO_2(g) \rightleftharpoons N_2O_4(g)$, $K = 170$ at 298 K. The initial concentrations are [NO_2] = 0.020 M and [N_2O_4] = 0.030 M. Is the system at equilibrium? If not, which way will it shift?

25. For the reaction $N_2(g) + O_2(g) \rightleftharpoons 2NO(g)$, $K = 4.4 \times 10^{-4}$ at 1900 K. The concentrations of N_2, O_2, and NO are 0.11 M, 0.11 M, and 4.0×10^{-3} M, respectively. Is the system at equilibrium? If not, in which way will the reaction proceed toward equilibrium?

26. For the reaction $H_2(g) + I_2(g) \rightleftharpoons 2HI(g)$, $K = 56$ at 700 K. If 1.0 mol of H_2 and 1.0 mol of I_2 are put in a 0.50-L flask and allowed to reach equilibrium, what would be the concentrations of all species?

27. For the reaction $2NO(g) \cap N_2(g) + O_2(g)$, $K = 482$ at 1200 K. If 1.15 mol of NO was placed in a 1.00-L flask until it reached equilibrium, what would be the concentrations of nitrogen and oxygen?

28. At 1500 K, the reaction $Br_2(g) \cap 2Br(g)$ has $K = 1.2 \times 10^{-4}$. If a reaction was started with $5.9 \times 10^{-2} M$ Br_2, what would be the equilibrium concentration of Br?

29. The equilibrium for butane $\cap$ isobutane has $K = 2.50$ at 298 K. If a reaction is started with 0.12 M isobutane, how much butane will there be at equilibrium?

30. Acetic acid is a weak acid with $K = 1.8 \times 10^{-5}$ for $CH_3COOH(aq) \cap H^+(aq) + CH_3COO^-(aq)$. What is the equilibrium concentration of H^+, if the initial concentration of CH_3COOH is 0.40 M?

31. A solution of 0.50 M HCN dissociates as follows: $HCN(aq) \cap H^+(aq) + CN^-(aq)$ with $K = 4.0 \times 10^{-10}$. What is the equilibrium concentration of CN^-?

32. Calcium carbonate is very slightly soluble in water, $CaCO_3(s) \cap Ca^{2+}(aq) + CO_3^{2-}(aq)$, $K = 3.4 \times 10^{-9}$. What is the equilibrium concentration of calcium ion?

33. Magnesium fluoride has $K = 5.2 \times 10^{-11}$ for $MgF_2(s) \cap Mg^{2+}(aq) + 2F^-(aq)$. What is the equilibrium concentration of fluoride ion?

34. Silver bromide is used by photographic film companies in making film emulsions. It has $K = 5.0 \times 10^{-13}$ for $AgBr(s) \cap Ag^+(aq) + Br^-(aq)$. What is the equilibrium concentration of the silver ion?

35. For the equilibrium $PCl_5(g) \cap PCl_3(g) + Cl_2(g)$, $K = 1.25$ at 300 K. Starting with 1.20 M PCl_5, what will be the equilibrium concentration of PCl_5?

36. Consider the reaction $C(graphite) + CO_2(g) \cap 2CO(g)$, for which $K = 0.021$ at 25°C. If a reaction is started with 0.10 M CO_2, what will be the equilibrium concentration of CO?

37. For the equilibrium $H_2(g) + Br_2(g) \cap 2HBr(g)$, $K = 46.5$ at 750 K. If 2.0 M H_2 is mixed with 1.0 M Br_2 in a reaction vessel, what will be the equilibrium concentration of HBr?

Section 16.6

38. State Le Châtelier's principle.

For Exercises 39–44, consider the Haber process: $2N_2(g) + 3H_2(g) \cap 2NH_3(g)$, $\Delta H° = -96$ kJ.

39. How would adding more nitrogen affect the equilibrium mixture?

40. In which direction would the equilibrium shift if the reaction were heated?

41. If the volume of the reaction vessel were decreased by a piston (thereby increasing the pressure), in which direction would the equilibrium shift?

42. If ammonia were removed as it formed, in which direction would the equilibrium shift?

43. How would adding a Pt catalyst affect the reaction? Would it shift the equilibrium?

44. To maximize the yield of nitric oxide, NO, how would you change the temperature and pressure of the following reaction: $4NH_3(g) + 5O_2(g) \cap 4NO(g) + 6H_2O(g)$, $\Delta H° < 0$?

45. Under what conditions of temperature and pressure would you run the equilibrium reaction $2H_2O(g) \cap 2H_2(g) + O_2(g)$, $\Delta H° = 484$ kJ, so as to maximize the decomposition of water?

Section 16.7

46. The standard free energy change for a certain reaction is 22.5 kJ at 37°C. What is the equilibrium constant?

47. What would be the standard free energy change for the hydrogen bromide equilibrium in Exercise 37?

48. What would be the standard free energy change for the equilibrium reaction in Exercise 36?

49. Use the standard molar free energies of formation found in the textbook's Appendix to calculate K at 298 K for $CaCO_3(s) \cap CaO(s) + CO_2(g)$.

50. Using the given standard molar free energies of formation, calculate K at 298 K for $CaF_2(s) \cap Ca^{2+}(aq) + 2F^-$ ($\Delta_f G°$ kJ/mol: $CaF_2 = -1162$, $Ca^{2+} = -553$, $F = -279$).

Answers to Exercises

1. $\Delta G = 0$

2. The equilibrium constant depends on the ratio of the rate of the forward reaction to rate of the reverse reaction. Because the rate of a reaction depends on the temperature. the equilibrium constant depends on the temperature as well.

3. True

4. False. At equilibrium. the rate of the forward reaction equals the rate of the reverse reaction.

5. The mass-action (equilibrium) expression is the ratio of the concentrations (or partial pressures) of the individual products raised to the power of their stoichiometric coefficients to the concentrations (or partial pressures) of the individual reactants raised to the power of their stoichiometric coefficients.

6. a) $K = \dfrac{[O_3]^2}{[O_2]^3}$ b) $K = \dfrac{[NO_2]^2}{[NO]^2[O_2]}$

7. a) $K = \dfrac{[SO_2]^2}{[O_2]^3}$ b) $K = \dfrac{[NH_3]^2}{[NO_2]^2[H_2]^7}$

8. a) $K = \dfrac{[H^+][HCOO^-]}{[HCOOH]}$ b) $K = [Ba^{2+}][SO_4^{2-}]$

9. The Haber process is the reaction used to produce ammonia. It relies on the following equilibrium: $N_2 + 3H_2 \rightleftharpoons 2NH_3$. Ammonia is used in agriculture as a fertilizer. It is a readily available source of nitrogen.

10. product-favored

11. reactant-favored

12. 6.3×10^2

13. 2.1×10^2

14. 4.9×10^{-10}

15. 2.1×10^{-17}

16. $[Cl_2] = [PCl_3] = 0.32\ M$, $[PCl_5] = 1.68\ M$

17. $3.4\ M$

18. 0.12 M

19. 1.6 M

20. 5.3

21. 3.2×10^{-2}

22. 6.4×10^{4}

23. The reaction quotient is the mass-action expression using *initial* concentrations. The equilibrium constant is calculated using *equilibrium* concentrations.

24. $Q = 75 < K$. The reaction will shift to the right to attain equilibrium.

25. $Q = 1.3 \times 10^{-3} > K$. The reaction will shift to the left.

26. [HI] = 0.78 M, [H_2] = [I_2] = 0.11 M

27. [N_2] = [O_2] = 0.56 M; this is correct.

28. 2.7×10^{-3} M

29. 0.034 M

30. 2.7×10^{-3} M

31. 1.4×10^{-5} M

32. 5.8×10^{-5} M

33. 4.7×10^{-4} M

34. 7.1×10^{-7} M

35. 0.45 M

36. 4.1×10^{-2} M

37. 1.86 M

38. If stress is applied to an equilibrium, it will change in such a way as to partially undo the applied stress and restore equilibrium.

39. The reaction shifts to the right to react the additional nitrogen.

40. The reaction shifts to the left, in the endothermic direction.

41. The reaction shifts to the right, the side with the fewest number of moles of gas.

42. The reaction shifts to the right to replenish the ammonia.

43. The equilibrium would be attained faster. No, adding a catalyst does not shift the equilibrium.

44. decrease the pressure and lower the temperature

45. decrease the pressure and increase the temperature

46. 1.6×10^{-4}

47. -23.9 kJ

48. 9.6 kJ

49. 1.1×10^{-23}

50. 1.2×10^{-9}

Chapter 17

Acids and Bases

The Bottom Line

When you finish this chapter, you will be able to:

- Differentiate between an acid, a base, a conjugate acid, and a conjugate base.
- Compare the strengths of acids and bases.
- Determine the pH of strong acids, strong bases, weak acids, and weak bases.
- Calculate the pH of salts and predict their acid–base behavior.

17.1 What Are Acids and Bases?

Many compounds are classified as either an acid or a base. Three models are used to describe acids and bases. Use the equation below to understand the differences between an acid and a base.

$$HBr(g) + H_2O(l) \rightleftharpoons Br^-(aq) + H_3O^+(aq)$$

Acids

- An **Arrhenius acid** is any species that produces hydrogen ions in solution. HBr is an Arrhenius acid because it produces H^+ ions in solution. The **hydronium ion, H_3O^+,** is the same as $H^+(aq)$ because $H^+(aq)$ is actually surrounded by water molecules in solution.
- A **Brønsted–Lowry acid** is defined as any species that donates a hydrogen ion (proton) to another species. HBr is considered a Brønsted–Lowry acid because it donates H^+ to the water forming H_3O^+.
- A **Lewis acid** accepts a previously nonbonded pair of electrons (a lone pair) to form a coordinate covalent bond. HBr is a Lewis acid because it accepts a lone pair of electrons from an oxygen in water. [Your text shows more specific examples with $Al(H_2O)_6^{3+}$ and BH_3.]

Bases

- An **Arrhenius base** is any species that produces hydroxide ions in solution. H_2O is an Arrhenius base in the reaction given earlier because when it dissociates into its ions, water produces OH^-.

$$H_2O \rightleftharpoons H^+ + OH^-$$

NaOH is also a good example of an Arrhenius base.
- A **Brønsted–Lowry base** is any species that can accept a hydrogen ion from an acid. In the earlier reaction involving HBr, H_2O is a Brønsted–Lowry base because it accepts an H^+ from HBr.
- A **Lewis base** donates a lone pair of electrons to form a coordinate covalent bond. Water is a Lewis base.

A unique feature of an acid and a base reacting with each other is the production of a conjugate acid and conjugate base. For the reaction with HBr, we already identified the acid and the base on the reactant side:

$$HBr(g) + H_2O(l) \rightleftharpoons Br^-(aq) + H_3O^+(aq)$$

acid   base

The bromide ion, Br^-, is called the **conjugate base** of HBr because this base results from HBr's donation of a proton to water. H_3O^+ is the **conjugate acid** of water because this acid results from water's acceptance of the proton from HBr.

$$HBr(g) + H_2O(l) \rightleftharpoons Br^-(aq) + H_3O^+(aq)$$

acid   base   conj. base conj. acid

Example 1   Conjugate Acid–Base Pairs

Write the products for the equations below and identify the conjugate acid–base pairs.

a) $HCN(aq) + H_2O(l) \rightleftharpoons$ _____ + _____

b) $NH_3(aq) + H_2O(l) \rightleftharpoons$ _____ + _____

First Thoughts

Water can act as either an acid or a base because it can either donate or accept a proton. The best approach is to identify whether the other reactant is an acid or a base first and then examine the water.

Solution

a)   HCN can donate a proton, so it is an acid, making H_2O the base.

$$HCN(aq) + H_2O(l) \rightleftharpoons CN^-(aq) + H_3O^+(aq)$$

acid   base   conj. base conj. acid

   HCN and CN^- are a conjugate acid–base pair.
   H_3O^+ and H_2O are a conjugate acid–base pair.

b) NH_3 can donate a lone pair to H_2O, so it is a base, making H_2O the acid.

$$NH_3(aq) + H_2O(l) \rightleftharpoons NH_4^+(aq) + OH^-(aq)$$

base   acid   conj. acid conj. base

   H_2O and OH^- are a conjugate acid–base pair.
   NH_4^+ and NH_3 are a conjugate acid–base pair.

17.2   Acid Strength

Acids and bases have different strengths. They range along a continuum from very strong to very weak. Here are some key points about acids:

- The common strong acids are H_2SO_4, HCl, HNO_3, and $HClO_4$.
- A **strong acid** completely dissociates (ionizes) to produce H^+ ion and the conjugate base. The reverse reaction from products to reactants is very unlikely to occur. An example is the dissociation of HCl:

$$HCl(aq) + H_2O(l) \rightarrow Cl^-(aq) + H^+(aq)$$

$$(\text{or } H_3O^+)$$

- A weak acid only partially dissociates (ionizes) to produce H^+ ion and the conjugate base. The reverse reaction from products to reactants also occurs. An example is the dissociation of HCN:

$$HCN(aq) + H_2O(l) \rightleftharpoons CN^-(aq) + H^+(aq)$$

- The equilibrium constant for the dissociation of an acid is called K_a. The principles of equilibrium that you learned in Chapter 16 also apply to acid–base reactions. **Table 17.4** and the Appendix in your text list several K_a values for some weak acids.
- Strong acids have K_a values much greater than 1, whereas weak acids have K_a values less than 1. The higher the K_a value, the stronger the acid.
- Strong acids have very weak conjugate bases, and weak acids have somewhat stronger conjugate bases.

Example 2   **Relative Acid Strength**

Using K_a values listed in the Appendix of your text, place the following acids in order from weakest to strongest:

$$HF, HClO_4, HC_3H_5O_2, HCN$$

First Thoughts

Look up the K_a values for each of the acids listed. Note that $HClO_4$ is a strong acid, so it has a K_a value much larger than 1.

Acid	K_a
HF	7.2×10^{-4}
$HClO_4$	$\gg 1$
$HC_3H_5O_2$	1.3×10^{-5}
HCN	6.2×10^{-10}

Solution

The larger the K_a value, the more the acid dissociates and is considered a stronger acid.

$$HCN < HC_3H_5O_2 < HF < HClO_4$$

Example 3   **Strength of Acids and Their Conjugates**

Consider the following acids in water: $HCl, HOBr, HClO_2$

a) Write the mass-action expression for the dissociation of each acid.
b) Place the acids in order from weakest to strongest.
c) Place the conjugate bases of these acids in order from weakest to strongest.

First Thoughts

Write the equilibrium reactions for each acid in water and look up the K_a values for each weak acid.

$$HCl(aq) + H_2O(l) \rightleftharpoons Cl^-(aq) + H^+(aq) \qquad K_a \gg 1$$
$$HOBr(aq) + H_2O(l) \rightleftharpoons OBr^-(aq) + H^+(aq) \qquad K_a = 2 \times 10^{-9}$$
$$HClO_2(aq) + H_2O(l) \rightleftharpoons ClO_2^-(aq) + H^+(aq) \qquad K_a = 1.2 \times 10^{-2}$$

Solution

a) Write the K_a expression for each reaction.

$$K_a(HCl) = \frac{[Cl^-][H^+]}{[HCl]} \qquad K_a(HOBr) = \frac{[OBr^-][H^+]}{[HOBr]} \qquad K_a(HOCl_2) = \frac{[ClO_2^-][H^+]}{[HClO_2]}$$

b) The higher the K_a value, the stronger the acid.

$$HOBr < HClO_2 < HCl$$

c) The stronger the acid, the weaker its conjugate base.

$$Cl^- < ClO_2^- < OBr^-$$

17.3   The pH Scale

The acidity of a solution is measured using the equilibrium hydrogen ion concentration, which depends on both the strength and the initial concentration of the acid. We can also express the acidity by calculating the **pH** of the acid.

$$pH = -\log[H^+]$$

Example 4   **pH and [H$^+$]**

Calculate the desired term below:

a) pH for $[H^+] = 5.5 \times 10^{-3}$ M
b) pH for $[H^+] = 1.2 \times 10^{-10}$ M
c) $[H^+]$ for pH = 3.39

Solution

a)
$$pH = -\log[H^+]$$
$$pH = -\log(5.5 \times 10^{-3} \, M)$$
$$pH = 2.26$$

(Note that pH is given to two decimal places because [H$^+$] is given to two significant figures.)

b)
$$pH = -\log(1.2 \times 10^{-10} \, M)$$
$$pH = 9.92$$

c)

$$pH = -\log[H^+]$$
$$[H^+] = 10^{-pH}$$
$$[H^+] = 10^{-3.39}$$
$$[H^+] = 4.1 \times 10^{-4}\ M$$

In our discussion so far, all of the acids and bases have been dissolved in water. Pure water can undergo **autoprotolysis,** in which the proton transfer involves only the solvent itself. As a result, water is **amphiprotic,** meaning it can act as either an acid or a base.

$$H_2O(l) \rightleftharpoons H^+(aq) + OH^-(aq) \qquad K_w = 1.0 \times 10^{-14}\ \text{at } 24^\circ C$$

The concentrations of H^+ and OH^- in pure water are both $1.0 \times 10^{-7}\ M$ with a pH of 7.0 (known as **neutral pH**). The pH scale ranges between 0 and 14. Aqueous solutions with a pH less than 7.0 are **acidic;** those with a pH greater than 7.0 are **basic.** As solutions become more acidic or basic, their pH moves farther away from 7.0, as shown in **Figure 17.14** in your text.

17.4 Determining the pH of Acid Solutions

pH of Strong Acid Solutions

For a strong monoprotic acid such as HCl, the hydrogen ion concentration in aqueous solution approximately equals the initial concentration of the acid (as long as the concentration is greater than $10^{-6}\ M$). At lower concentrations, the autoprotolysis of water is significant and will affect the final pH. This effect is discussed in more detail in Chapter 18.

Example 5   pH of a Strong Acid

What is the pH of a 0.025 M solution of HNO_3?

First Thoughts

HNO_3 is a strong acid, so it completely dissociates into H^+ and NO_3^-. Therefore, H^+ equals the initial concentration of the solution.

Solution

$$pH = -\log[H^+]$$
$$pH = -\log(0.025\ M)$$
$$pH = 1.60$$

Further Insight

Our answer of pH = 1.60 is reasonable. Because the solution contains a strong acid at a high concentration, the pH should be low.

pH of Weak Acid Solutions

Determining the pH of a weak acid solution requires more steps than determining the pH of a strong acid solution because weak acids do not completely dissociate. Use the same principles that you used in Chapter 16 to solve an equilibrium problem.

Example 6   **pH of a Weak Acid**

Calculate the pH of a 0.018 M HOCl solution. K_a for HOCl = 3.5×10^{-8}.

Solution

1. Determine the equilibria and the resulting species that are in solution.

$$HOCl(aq) \rightleftharpoons H^+(aq) + OCl^-(aq) \qquad K_a = 3.5 \times 10^{-8}$$
$$H_2O(l) \rightleftharpoons H^+(aq) + OH^-(aq) \qquad K_w = 1.0 \times 10^{-14}$$

2. Determine the equilibria that are the most important contributors to $[H^+]$ in the solution.

    The K_a of HOCl is much larger than the K_w of water, so the dissociation of HOCl is the only important contributor to H^+.

3. Write the equilibrium expression for HOCl.

$$K_a = \frac{[H^+][OCl^-]}{[HOCl]}$$

4. Set up a table of the initial and equilibrium concentrations of each pertinent species.

	HOCl	$\rightleftharpoons$	H^+	+	OCl^-
Initial	0.018 M		0 M		0 M
Change	$-x$		$+x$		$+x$
Equilibrium	$0.018 - x$		$+x$		$+x$
Assumption	0.018		$+x$		$+x$

5. Solve for the estimated concentration of H^+.

$$3.5 \times 10^{-8} = \frac{(x)(x)}{0.018} = \frac{x^2}{0.018}$$
$$x = [H^+] = [OCl^-] = 2.5 \times 10^{-5} \ M$$

6. Check our assumption of negligible ionization.

$$\% \ dissociation = \frac{[OCl^-]}{[HOCl]} \times 100\% = \frac{2.5 \times 10^{-5}}{0.018} \times 100\% = 0.14\%$$

Less than 5% of the original amount of HOCl ionized, so our assumption is acceptable.

7. Solve for the pH of the solution.

$$pH = -\log(2.5 \times 10^{-5} \ M)$$
$$pH = 4.60$$

Further Insight

In this particular example, our assumption followed the 5% rule in step 6. This is not always the case, and using the quadratic formula is sometimes required to solve for x. **Exercise 17.11** in your text shows an example where the 5% rule does not apply.

In addition, some solutions contain a mixture of acids that differ greatly in strength. Use the standard approach by again writing the relevant equilibria in the solution along with the K_a values. Look for the dominant equilibrium to determine the pH of the solution.

17.5 Determining the pH of Basic Solutions

Calculating the pH of strong and weak bases is similar to calculating the pH of strong and weak acids. Here are some key points to remember about bases:

- Compounds containing hydroxides are strong bases because they completely dissociate in water to produce hydroxide ions (OH^-). NaOH and $Ba(OH)_2$ are examples of strong bases.
- The K_b value of a strong base is much larger than 1, and the K_b value of a weak base is less than 1 (analogous to the K_a values for strong and weak acids).

Example 7   **pH of a Strong Base**

Calculate the pH of a 0.12 M aqueous solution of NaOH.

First Thoughts

NaOH is a strong base, so it completely dissociates into Na^+ and OH^-. Therefore, OH^- equals the initial concentration of the solution.

$$NaOH \rightarrow Na^+ + OH^-$$

Solution

You can determine the pH in one of two ways. Choose whichever method works best for you.

Method 1: Calculate the pOH of the solution and then subtract that value from 14.00 (the top of the pH scale).

$$[OH^-] = 0.12\ M, \text{ so}$$
$$pOH = -\log[OH^-] = -\log(0.12\ M)$$
$$pOH = 0.92$$

$$pH = 14.00 - 0.92$$
$$pH = 13.08$$

Method 2: Calculate the $[H^+]$ of the solution using the relationship $K_w = 1.00 \times 10^{-14} = [H^+][OH^-]$ and then determine the pH.

$$K_w = 1.0 \times 10^{-14} = [H^+][OH^-]$$
$$1.0 \times 10^{-14} = [H^+][0.12\ M]$$
$$[H^+] = 8.3 \times 10^{-14}\ M$$

$$pH = -\log(8.3 \times 10^{-14}\ M)$$
$$pH = 13.08$$

Further Insight

Our answer of pH = 13.08 is reasonable. Because the solution contains a strong base at a high concentration, the pH should be high.

Example 8   **pH of a Weak Base**

Calculate the pH of an aqueous solution of hydrazine, H_2NNH_2, at the same initial concentration as NaOH in Example 7 (0.12 M). K_b for $H_2NNH_2 = 3.0 \times 10^{-6}$.

First Thoughts

Solving for the pH of a weak base is similar to solving for the pH of a weak acid, except that you solve for $[OH^-]$ instead.

Solution

The equilibria in solution are

$$H_2NNH_2(aq) \rightleftharpoons H_2NNH_3^+(aq) + OH^-(aq) \qquad\qquad K_b = 3.0 \times 10^{-6}$$

$$H_2O(l) \rightleftharpoons H^+(aq) + OH^-(aq) \qquad\qquad K_w = 1.0 \times 10^{-14}$$

Because the K_b for H_2NNH_2 is much larger than the K_b for H_2O, the dissociation of H_2NNH_2 is the only important contributor to OH^-.

$$K_b = \frac{[H_2NNH_3^+][OH^-]}{[H_2NNH_2]}$$

$$H_2NNH_2 \quad \rightleftharpoons \quad H_2NNH_3^+ \quad + \quad OH^-$$

	H_2NNH_2	$H_2NNH_3^+$	OH^-
Initial	0.12 M	0 M	0 M
Change	$-x$	$+x$	$+x$
Equilibrium	$0.12 - x$	$+x$	$+x$
Assumption	0.12	$+x$	$+x$

$$3.0 \times 10^{-6} = \frac{x^2}{0.12}$$

$$x = [H_2NNH_3^+] = [OH^-] = 6.0 \times 10^{-4} \ M$$

Let's check our assumption:

$$\% \text{ dissociation} = \frac{[H_2NNH_3^+]}{[H_2NNH_2]} \times 100\% = \frac{6.0 \times 10^{-4}}{0.12} \times 100\% = 0.50\%$$

Our assumption is valid. Next, let's calculate the pOH of the solution.

$$pOH = -\log[OH^-] = -\log(6.0 \times 10^{-4} \ M)$$

$$pOH = 3.22$$

$$pH = 14.00 - 3.22$$

$$pH = 10.78$$

Further Insight

Our answer of pH = 10.78 is reasonable. The pH is higher than 7, which is consistent with a base, but not as high as the pH for NaOH with the same concentration, verifying that this is a weak base.

17.6   Polyprotic Acids

Polyprotic acids contain more than one acidic hydrogen, such as H_3PO_4 or H_2SO_4. As a result, when these acids are in solution, they have more than one K_a value because more than one acidic hydrogen participates in the equilibrium reaction. **Table 17.9** in your text lists the values for some common polyprotic acids. When determining the pH of polyprotic acids, assess all of the equilibria that occur in solution.

Example 9   **pH of a Polyprotic Acid**

Calculate the pH of a 4.5 M H_2CO_3 solution and the equilibrium concentrations of the species H_2CO_3, HCO_3^-, and CO_3^{2-}. (The K_a values are listed in Table 17.9.)

First Thoughts

Establish the equilibria of the species in solution that contribute hydrogen ions.

$$H_2CO_3 \rightleftharpoons HCO_3^- + H^+ \qquad K_{a_1} = 4.3 \times 10^{-7}$$

$$HCO_3^- \rightleftharpoons CO_3^{2-} + H^+ \qquad K_{a_2} = 5.6 \times 10^{-11}$$

$$H_2O \rightleftharpoons H^+ + OH^- \qquad K_w = 1.0 \times 10^{-14}$$

The K_{a_1} value is considerably larger than the other values, so we will assume that it is the important equilibrium.

Solution

$$K_{a_1} = \frac{[HCO_3^-][H^+]}{[H_2CO_3]}$$

	H_2CO_3	$\rightleftharpoons$	HCO_3^-	+	H^+
Initial	4.5 M		0 M		0 M
Change	$-x$		$+x$		$+x$
Equilibrium	$4.5 - x$		$+x$		$+x$
Assumption	4.5		$+x$		$+x$

$$4.3 \times 10^{-7} = \frac{x^2}{4.5}$$

$$x = [HCO_3^-] = [H^+] = 1.4 \times 10^{-3} \ M$$

Because 1.4×10^{-3} is less than 5% of 4.5, the assumption is acceptable.

$$pH = -\log[H^+] = -\log(1.4 \times 10^{-3} \ M)$$

$$pH = 2.85$$

So far the equilibrium concentrations are

$$x = [HCO_3^-] = [H^+] = 1.4 \times 10^{-3} \ M$$

$$[H_2CO_3] = 4.5 - x = 4.5 \ M - 1.4 \times 10^{-3} \ M = 4.5 \ M$$

To calculate $[CO_3^{2-}]$, we use the expression for K_{a_2} and the values of $[H^+]$ and $[HCO_3^-]$ calculated above:

$$K_{a_2} = \frac{[CO_3^{2-}][H^+]}{[HCO_3^-]}$$

$$5.6 \times 10^{-11} = \frac{[CO_3^{2-}][1.4 \times 10^{-3}]}{[1.4 \times 10^{-3}]}$$

$$[CO_3^{2-}] = 5.6 \times 10^{-11} \ M$$

17.7 Assessing the Acid–Base Behavior of Salts in Aqueous Solutions

Salts are ionic compounds, which means that they are made up of at least one cation and at least one anion. When salts dissociate in water, their ions could behave as an acid or a base. When determining whether a salt solution is acidic or basic, keep this relationship in mind:

$$1.0 \times 10^{-14} = K_w = K_a \times K_b$$

It will help you determine the K_b for a weak base if you know only the K_a for its conjugate acid (and vice versa).

Example 10   **pH of a Salt**

Determine the pH of a 0.50 M NaF solution. K_a for HF = 7.2×10^{-4}.

First Thoughts

NaF will dissociate into Na^+ and F^- ions in water. The Na^+ cation exhibits no important acid–base properties. The F^- will react with the water to produce OH^- ions, leading to a basic solution.

Solution

$$F^-(aq) + H_2O(l) \rightleftharpoons HF(aq) + OH^-(aq)$$

$$K_b = \frac{[HF][OH^-]}{[F^-]}$$

To calculate the value of K_b, use K_w and K_a:

$$K_w = K_a \times K_b$$

$$1.0 \times 10^{-14} = (7.2 \times 10^{-4})(K_b)$$

$$K_b = 1.4 \times 10^{-11}$$

Now use the table to solve for $[OH^-]$.

	F^-	$\rightleftharpoons$	HF	+	OH^-
Initial	0.50 M		0 M		0 M
Change	$-x$		$+x$		$+x$
Equilibrium	$0.50 - x$		$+x$		$+x$
Assumption	0.50		$+x$		$+x$

$$1.4 \times 10^{-11} = \frac{x^2}{0.50}$$

$$x = [HF] = [OH^-] = 2.6 \times 10^{-6} \ M$$

The assumption is acceptable by the 5% rule, so

$$pOH = -\log[OH^-] = -\log(2.6 \times 10^{-6} \ M)$$
$$pOH = 5.59$$

$$pH = 14.00 - 5.59$$
$$pH = 8.41$$

Further Insight

Our answer of pH = 8.41 is reasonable for a weak base.

Example 11   Acid–Base Property of a Salt

Predict whether an aqueous solution of NH_4F will be acidic, basic, or neutral.
$$K_a(NH_4^+) = 5.6 \times 10^{-10}; \ K_a(HF) = 7.2 \times 10^{-4})$$

First Thoughts

NH_4F will dissociate into NH_4^+ and F^- ions in water. The K_b for F^- is calculated as follows:

$$K_w = K_a \times K_b$$
$$1.0 \times 10^{-14} = (7.2 \times 10^{-4})(K_b)$$
$$K_b = 1.4 \times 10^{-11}$$

Solution

The K_a for NH_4^+ is larger than the K_b for F^-, so NH_4^+ is a stronger acid than F^- is a base. This solution will be acidic.

Further Insight

Your text goes into more detail about how to calculate the pH of salts with both acidic and basic properties; it also explores the acid–base properties of amino acids. **Table 17.10** summarizes the effects of the cation and anion on the pH of a salt.

17.8   Anhydrides in Aqueous Solution

This section discusses the set of reactions that causes acid deposition from the atmosphere. You should become familiar with them. Two important types of compounds are seen in these reactions:

- **Basic anhydrides** are binary compounds formed between metals with very low electronegativities and oxygen. An example of a reaction of a basic anhydride with water follows:

$$K_2O(s) + H_2O(l) \rightleftharpoons 2KOH(aq)$$

- **Acid anhydrides** are binary compounds formed between nonmetals and oxygen. An example of a reaction of an acid anhydride with water follows: $CO_2(g) + H_2O(l) \rightleftharpoons H_2CO_3(aq)$

Exercises

Section 17.1

1. Identify the following compounds as an acid or a base.
   a) H_2SO_4
   b) NaOH
   c) HCN

2. Identify the following compounds as an acid or a base.
   a) $HC_2H_3O_2$
   b) NH_3
   c) C_5H_5N

3. Write the conjugate base for each of the acids listed below.
   a) HNO_2
   b) HOBr
   c) H_3PO_4

4. Write the conjugate acid for each of the bases listed below.
   a) NH_3
   b) $C_2H_5NH_2$
   c) $(C_2H_5)_3N$

5. Write the products for the equations below and identify the conjugate acid–base pairs.
   a) $HClO_2(aq) + H_2O(l) \rightleftharpoons$ _____ + _____
   b) $HF(aq) + H_2O(l) \rightleftharpoons$ _____ + _____

6. Write the products for the equations below and identify the conjugate acid–base pairs.
   a) $NH_3(aq) + H_2O(l) \rightleftharpoons$ _____ + _____
   b) $C_6H_5NH_2(aq) + H_2O(l) \rightleftharpoons$ _____ + _____

7. Write the reactants for the equations below and identify the conjugate acid–base pairs.
   a) _____ + _____ $\rightleftharpoons HC_3H_5O_2(aq) + OH^-(aq)$
   b) _____ + _____ $\rightleftharpoons H_2NNH_2(aq) + H_3O^+(aq)$

Section 17.2

8. What are three common strong acids?

9. What is the difference between a strong acid and a weak acid?

10. Which of the following statements is true?
    a) Weak acids have K_a values much larger than 1.
    b) The higher the K_a value for an acid, the weaker its acid strength.
    c) Strong acids have strong conjugate bases because they completely dissociate in solution.
    d) Weak acids in solution always produce hydroxide ions.
    e) Weak acids generally produce somewhat stronger conjugate bases.

11. Using the K_a values listed in the Appendix of your text, place the following acids in order from weakest to strongest: H_3BO_3, HNO_2, $HC_2H_3O_2$.

12. Using the K_a values listed in the Appendix of your text, place the following acids in order from weakest to strongest: HCl, HCO_2H, NH_4^+.

13. Using the K_b values listed in the Appendix of your text, place the following bases in order from weakest to strongest: $(C_2H_5)_2NH$, C_5H_5N, NH_3.

14. Consider the following acids in water: $HClO_2$, HNO_3, $HOCl$.
   a) Write the mass-action expression for the dissociation of each acid.
   b) Place the acids in order from weakest to strongest.
   c) Place the conjugate bases of the acids in order from weakest to strongest.

15. Consider the following bases in water: NH_3, $HONH_2$, C_5H_5N.
   a) Write the mass-action expression for the dissociation of each base.
   b) Place the bases in order from weakest to strongest.
   c) Place the conjugate acids of the bases in order from weakest to strongest.

Section 17.3

16. What is the pH of a solution where $[H^+] = 9.3 \times 10^{-4}$ M at equilibrium?

17. What is the pH of a solution where $[H^+] = 0.013$ M at equilibrium?

18. What is the pH of a solution where $[H^+] = 4.0 \times 10^{-13}$ M at equilibrium?

19. A solution has a pH of 4.51. What is the $[H^+]$ of the solution? Is this solution acidic, basic, or neutral?

20. A solution has a pH of 13.60. What is the $[H^+]$ of the solution? Is this solution acidic, basic, or neutral?

21. What is the pOH of a solution where $[OH^-] = 1.5 \times 10^{-3}$ M at equilibrium? Is this solution acidic, basic, or neutral?

22. What is the pOH of a solution where $[OH^-] = 3.1 \times 10^{-12}$ M at equilibrium? Is this solution acidic, basic, or neutral?

23. What is the pH of a solution where $[OH^-] = 6.2 \times 10^{-5}$ M at equilibrium? Is this solution acidic, basic, or neutral?

24. What is the pH of a solution where $[OH^-] = 2.9 \times 10^{-10}$ M at equilibrium? Is this solution acidic, basic, or neutral?

25. What is the pH of a solution where $[OH^-] = 1.0 \times 10^{-7}$ M at equilibrium? Is this solution acidic, basic, or neutral?

26. A solution has a pH of 11.74. What is the $[OH^-]$ of the solution? Is this solution acidic, basic, or neutral?

27. A solution has a pOH of 10.93. What is the $[H^+]$ of the solution? Is this solution acidic, basic, or neutral?

Section 17.4

28. What is the pH of a 0.030 M solution of HCl?

29. What is the pH of a 0.012 M solution of $HClO_4$?

30. What is the hydrogen ion concentration of an HNO_3 solution with a pH of 1.83?

31. What is the pH of a 2.0×10^{-9} M solution of HCl?

32. Calculate the pH of a 0.024 M $HC_2H_3O_2$ solution. K_a for $HC_2H_3O_2 = 1.8 \times 10^{-5}$.

33. Calculate the pH of a 2.50 M HF solution. K_a for HF $= 7.2 \times 10^{-4}$. Justify the pH that you calculate.

34. What is the pH of a 0.095 M HCN solution? K_a for HCN $= 6.2 \times 10^{-10}$.

35. A HOBr solution has a pH of 5.50. What was the original concentration of the solution before reaching equilibrium? K_a for HOBr $= 2.0 \times 10^{-9}$.

36. A 0.040 M acid solution has a pH of 4.43. What is the K_a for this acid? Identify the acid using the Appendix in your text.

Section 17.5

37. Calculate the pOH and pH of a 0.25 M solution of KOH.

38. Calculate the pH of a 0.091 M LiOH solution.

39. Determine the pH of a 0.035 M $Ba(OH)_2$ solution.

40. Calculate the pH of a 0.022 M ammonia solution. K_b for ammonia $= 1.8 \times 10^{-5}$.

41. Determine the pH of a 4.8×10^{-4} M $HONH_2$ solution. K_b for $HONH_2 = 1.1 \times 10^{-8}$.

42. A C_5H_5N solution has a pH of 8.75. What was the original concentration of the solution before reaching equilibrium? K_b for $C_5H_5N = 1.7 \times 10^{-9}$.

Section 17.6

43. Calculate the pH of a 5.3 M H_3PO_4 solution. The K_a values are listed in Table 17.9.

44. Determine the pH of a 4.0 M ascorbic acid ($H_2C_6H_6O_6$) solution and the equilibrium concentrations of the species $H_2C_6H_6O_6$, $HC_6H_6O_6^-$, and $C_6H_6O_6^{2-}$. The K_a values are listed in Table 17.9.

45. Calculate the pH of a 2.0 M H_2SO_4 solution. What is the equilibrium concentration of SO_4^{2-}? K_{a_2} is listed in Table 17.9.

Section 17.7

46. Determine the pH of a 0.35 M $KC_2H_3O_2$ solution. K_a for $HC_2H_3O_2 = 1.8 \times 10^{-5}$.

47. Calculate the pH of a 0.28 M NH_4NO_3 solution. K_b for $NH_3 = 1.8 \times 10^{-5}$.

48. Determine the pH of a 0.49 M $NaCN$ solution. Is the salt solution acidic, basic, or neutral? K_a for $HCN = 6.2 \times 10^{-10}$.

49. Predict whether the following aqueous solutions will be acidic, basic, or neutral. Use the Appendix in your text to look up the K_a and K_b values.
    a) KNO_2
    b) $NaCl$
    c) NH_4ClO_2

Section 17.8

50. What is the difference between a basic anhydride and an acid anhydride? Give an example of each.

Answers to Exercises

1. a) acid, b) base, c) acid

2. a) acid, b) base, b) base

3. a) NO_2^-, b) OBr^-, c) $H_2PO_4^-$

4. a) NH_4^+, b) $C_2H_5NH_3^+$, c) $(C_2H_5)_3NH^+$

5. a) $HClO_2(aq) + H_2O(l) \rightleftharpoons ClO_2^-(aq) + H_3O^+(aq)$

    $HClO_2/ClO_2^-$; H_2O/H_3O^+ are conjugate acid–base pairs.

    b) $HF(aq) + H_2O(l) \rightleftharpoons F^-(aq) + H_3O^+(aq)$

    HF/F^-; H_2O/H_3O^+ are conjugate acid–base pairs.

6. a) $NH_3(aq) + H_2O(l) \rightleftharpoons NH_4^+(aq) + OH^-(aq)$

    NH_3/NH_4^+; H_2O/OH^- are conjugate acid–base pairs.

    b) $C_6H_5NH_2(aq) + H_2O(l) \rightleftharpoons C_6H_5NH_3^+(aq) + OH^-(aq)$

    $C_6H_5NH_2/C_6H_5NH_3^+$; H_2O/OH^- are conjugate acid–base pairs.

7. a) $C_3H_5O_2^-(aq) + H_2O(l) \rightleftharpoons HC_3H_5O_2(aq) + OH^-(aq)$

    $C_3H_5O_2^-/HC_3H_5O_2$; H_2O/OH^- are conjugate acid–base pairs.

    b) $H_2NNH_3^+(aq) + H_2O(l) \rightleftharpoons H_2NNH_2(aq) + H_3O^+(aq)$

    $H_2NNH_3^+/H_2NNH_2$; H_2O/H_3O^+ are conjugate acid–base pairs.

8. H_2SO_4, HCl, HNO_3

9. A strong acid completely dissociates to produce H^+ ion and the conjugate base. A weak acid only partially dissociates so the reverse reaction from products to reactants also occurs.

10. e

11. $H_3BO_3 < HC_2H_3O_2 < HNO_2$

12. $NH_4^+ < HCO_2H < HCl$

13. $C_5H_5N < NH_3 < (C_2H_5)_2NH$

14. a)

$$K_a(HClO_2) = \frac{[ClO_2^-][H^+]}{[HClO_2]} \qquad K_a(HNO_3) = \frac{[NO_3^-][H^+]}{[HNO_3]} \qquad K_a(HOCl) = \frac{[OCl^-][H^+]}{[HOCl]}$$

b) $HOCl < HClO_2 < HNO_3$

c) $NO_3^- < ClO_2^- < OCl^-$

15. a)

$$K_b(NH_3) = \frac{[NH_4^+][OH^-]}{[NH_3]} \qquad K_b(HNO_2) = \frac{[HONH_3^+][OH^-]}{[HONH_2]}$$

$$K_b(C_5H_5N) = \frac{[C_5H_5NH^+][OH^-]}{[C_5H_5N]}$$

b) $C_5H_5N < HONH_2 < NH_3$

c) $NH_4^+ < HONH_3^+ < C_5H_5NH^+$

16. 3.03

17. 1.89

18. 12.40

19. $3.1 \times 10^{-5}\,M$; acidic

20. $2.5 \times 10^{-14}\,M$; basic

21. 2.82; basic

22. 11.51; acidic

23. 9.79; basic

24. 4.46; acidic

25. 7.00; neutral

26. $5.5 \times 10^{-3}\,M$; basic

27. $8.5 \times 10^{-4}\,M$; acidic

28. 1.52

29. 1.92

30. $0.015\,M\,HNO_3$

31. 7.00

32. 3.18

33. 1.37; The pH is less than 7.0, which verifies that HF is an acid. Even though HF is a weak acid, it has a high concentration so the pH is low (more like a strong acid).

34. 5.11

35. 0.0050 M HOBr

36. 3.5×10^{-8}; HOCl

37. pOH = 0.60; pH = 13.40

38. 12.96

39. 12.85

40. 10.80

41. 8.36

42. 0.019 M C_5H_5N

43. 0.70

44. pH = 1.75; $[H_2C_6H_6O_6]$ = 4.0 M; $[HC_6H_6O_6^-]$ = 1.78×10^{-2} M, $[C_6H_6O_6^{2-}]$ = 1.6×10^{-12} M

45. pH = –0.30; $[SO_4^{2-}]$ = 1.2×10^{-2} M

46. 9.14

47. 4.90

48. 11.45; basic

49. a) basic, b) neutral, c) acidic

50. A basic anhydride is a binary compound formed between a metal and oxygen. An acid anhydride forms between a nonmetal and oxygen. Basic anhydride = Li_2O; acid anhydride = SO_2.

Chapter 18

Applications of Aqueous Equilibria

The Bottom Line

When you finish this chapter, you will be able to:

- Identify a buffer solution.
- Predict whether a buffer solution will be acidic, basic, or neutral.
- Calculate the pH of a buffer solution and explain how to make a buffer solution.
- Determine the buffer capacity of a solution and identify when the buffer capacity has been exceeded.
- Understand titrations and titration curves, including the titration midpoint and equivalence point.
- Choose indicators for titration experiments.
- Understand solubility equilibria, including K_{sp} and molar solubility, and determine whether a precipitate will form.
- Illustrate the stepwise formation of a complex ion.

18.1   Buffers and the Common-Ion Effect

In our continued study of acid–base equilibrium, buffers are extremely important. A **buffer** is a chemical system that is able to resist changes in pH. Here are some key points about buffers:

- A buffer is the combination of a weak acid and its conjugate base or the combination of a weak base and its conjugate acid.
- Buffers do not exist if a strong acid or strong base is paired with its conjugate.
- Buffers can accommodate the addition of strong acid and base, and can also withstand dilution, without large changes in the solution pH.
- When a buffer is present in a system, either the acid or base equilibrium expression may be used. However, we often work with the expression describing the reaction of the stronger conjugate.

Example 1   **pH of a Buffer Solution**

Predict whether a buffer solution containing 0.400 M HCN and 0.200 M NaCN would be acidic, basic, or neutral. Verify your prediction by calculating the pH of this buffer.
K_a for HCN = 6.2×10^{-10}.

First Thoughts

The key reactants in the buffer solution are HCN, NaCN, and H_2O. Because the K_w for H_2O is so small, we will consider only the equilibria of HCN and NaCN.

$$HCN(aq) + H_2O(l) \rightleftharpoons H^+(aq) + CN^-(aq) \qquad K_a = 6.2 \times 10^{-10}$$

$$CN^-(aq) + H_2O(l) \rightleftharpoons HCN(aq) + OH^-(aq) \qquad K_b = 1.6 \times 10^{-5}$$

(Recall from Section 17.7 that $K_b = K_w/K_a$.)

Solution

Given that the K_b for CN^- (NaCN) is larger than the K_a for HCN, we predict that the buffer solution will be basic.

Let's calculate the pH of the buffer solution using the same methods we used in Chapters 16 and 17.

The mass-action expression is

$$K_b = \frac{[HCN][OH^-]}{[CN^-]}$$

The table is:

	CN^-	$\rightleftharpoons$	HCN	+	OH^-
Initial	0.200 M		0.400 M		0 M
Change	$-x$		$+x$		$+x$
Equilibrium	$0.200 - x$		$0.400 + x$		$+x$
Assumption	0.200		0.400		$+x$

$$1.6\times10^{-5} = \frac{(0.400)(x)}{0.200}$$

$$x = [OH^-] = 8.0\times10^{-6}\ M$$

The approximations are valid by the 5% rule, so

$$pOH = -\log[OH^-] = -\log(8.0\times10^{-6}\ M)$$
$$pOH = 5.10$$
$$pH = 14.00 - 5.10$$
$$pH = 8.90$$

The pH verifies that the buffer solution is basic.

Further Insight

Besides using the approximation methods for equilibrium reactions (5% rule), you can also recognize that for buffer solutions, a high enough concentration of each conjugate exists initially so that the acid–base equilibria are effectively suppressed. As a result, the equilibrium concentrations of both species are roughly equal to the initial concentrations of these components.

A very useful way to determine the pH of a buffer solution and prepare a buffer solution involves using the **Henderson-Hasselbalch equation**:

$$pH = pK_a + \log\left(\frac{[Base]}{[Acid]}\right)$$

For example, we can use this equation as an alternative method for calculating the pH in Example 1.

$$pH = pK_a + \log\left(\frac{[CN^-]}{[HCN]}\right)$$

$$pH = -\log\left(6.2\times10^{-10}\right) + \log\left(\frac{[0.200\ M]}{[0.400\ M]}\right)$$

$$pH = 8.90$$

Example 2   Preparing a Buffer Solution

How many milliliters of 0.200 M NaOH must be added to 50.0 mL of a 0.150 M solution of $HC_2H_3O_2$ to prepare a buffer that has a pH of 5.40? K_a for $HC_2H_3O_2 = 1.8 \times 10^{-5}$.

First Thoughts

We have a solution of $HC_2H_3O_2$. A strong base (NaOH), which converts $HC_2H_3O_2$ to $C_2H_3O_2^-$, is added until the ratio of the weak acid to its conjugate base is the same as that in a pH 5.40 buffer solution. Let's determine what this ratio is by using the Henderson-Hasselbalch equation:

$$pH = pK_a + \log\left(\frac{[Base]}{[Acid]}\right)$$

$$5.40 = -\log\left(1.8\times10^{-5}\right) + \log\left(\frac{[C_2H_3O_2^-]}{[HC_2H_3O_2]}\right)$$

$$\frac{[C_2H_3O_2^-]}{[HC_2H_3O_2]} = 4.52$$

Solution

The moles of $HC_2H_3O_2$ initially present in the solution are

$$\left(0.0500\ L\ HC_2H_3O_2\right)\left(\frac{0.150\ mol\ HC_2H_3O_2}{1\ L\ HC_2H_3O_2}\right) = 0.0075\ mol\ HC_2H_3O_2$$

Because the volume of the buffer solution is the same for $HC_2H_3O_2$ and $C_2H_3O_2^-$, we can work with moles rather than molarity when looking at the ratio of base to acid:

$$\frac{mol\ C_2H_3O_2^-}{mol\ HC_2H_3O_2} = 4.52$$

$$mol\ C_2H_3O_2^- = \left(4.52\right)\left(mol\ HC_2H_3O_2\right)$$

Next, determine the moles of $HC_2H_3O_2$ and $C_2H_3O_2^-$ so that the sum equals 0.0075 mol.

$$\text{mol } C_2H_3O_2^- + \text{mol } HC_2H_3O_2 = 0.0075 \text{ mol}$$

$$\left[(4.52)\left(\text{mol } HC_2H_3O_2 \right) \right] + \text{mol } HC_2H_3O_2 = 0.0075 \text{ mol}$$

$$5.52 \text{ mol } HC_2H_3O_2 = 0.0075 \text{ mol}$$

$$\text{mol } HC_2H_3O_2 = 1.36 \times 10^{-3} \text{ mol}$$

$$\text{mol } C_2H_3O_2^- = (4.52)\left(1.36 \times 10^{-3} \text{ mol} \right)$$

$$\text{mol } C_2H_3O_2^- = 6.14 \times 10^{-3} \text{ mol}$$

Finally, we calculate how many milliliters of 0.200 M NaOH must be added to produce 6.14×10^{-3} mol $C_2H_3O_2^-$ and have a pH of 5.40.

$$\left(6.14 \times 10^{-3} \text{ mol NaOH} \right)\left(\frac{1 \text{ L}}{0.200 \text{ mol NaOH}} \right)\left(\frac{1000 \text{ mL}}{1 \text{ L}} \right) = 30.7 \text{ mL NaOH}$$

As stated previously, the Henderson-Hasselbalch equation is convenient when you want to determine the pH of a buffer solution. For you to use this equation, however, the system must contain both a weak acid and its conjugate base (or vice versa); otherwise, you will get incorrect results!

Another important task for chemists is to determine the buffer capacity of a system. The **buffer capacity** of a system measures the number of moles of strong acid or base that can be added while keeping the pH relatively constant (at the point at which all of the conjugate acid or base is reacted). For a monoprotic acid or base, the buffer capacity is greatest when [HA] = [A⁻].

Example 3   Buffer Capacity

Suppose we have 100.0 mL of a solution containing 0.200 M HCN and 0.300 M NaCN.

a) What will be the initial pH of the buffer solution?
b) What is the buffer capacity against a strong acid like HCl? In other words, how many moles of HCl can be added to the solution before the buffer capacity is exceeded?
c) If you add 75.0 mL of 1.00 M HCl to the buffer solution, will you exceed the buffer capacity? What will be the pH after that addition?

First Thoughts

The dominant equilibrium in solution is

$$CN^-(aq) + H_2O(l) \rightleftharpoons HCN(aq) + OH^-(aq) \qquad K_b = 1.6 \times 10^{-5}$$

Solution

a) Solving for the pH of the initial buffer solution is similar to Example 1.

	CN⁻	$\rightleftharpoons$	HCN	+	OH⁻
Initial	0.300 M		0.200 M		0 M
Change	$-x$		$+x$		$+x$
Equilibrium	$0.300 - x$		$0.200 + x$		$+x$
Assumption	0.300		0.200		$+x$

$$1.6 \times 10^{-5} = \frac{(0.200)(x)}{0.300}$$

$$x = [OH^-] = 2.4 \times 10^{-5} \ M$$

The approximations are valid by the 5% rule, so

$$pOH = -\log[OH^-] = -\log(2.4 \times 10^{-5} \ M)$$
$$pOH = 4.62$$
$$pH = 14.00 - 4.62$$
$$pH = 9.38$$

(The buffer solution is basic.)

b) First calculate the moles of each component initially in solution.

$$\text{moles HCN}_{initial} = (0.1000 \text{ L HCN})\left(\frac{0.200 \text{ mol HCN}}{1 \text{ L HCN}}\right) = 0.0200 \text{ mol HCN}_{initial}$$

$$\text{moles CN}^-_{initial} = (0.1000 \text{ L CN}^-)\left(\frac{0.300 \text{ mol CN}^-}{1 \text{ L CN}^-}\right) = 0.0300 \text{ mol CN}^-_{initial}$$

Because we are adding HCl, the H^+ ions in HCl will react with the CN^- ions in the buffer solution. The CN^- could react with up to 0.0300 mol HCl without having any excess H^+ to greatly lower the pH.

	CN^-	+	H^+	$\rightleftharpoons$	HCN
Moles initial	0.0300				0.0200
Moles added			0.0300		
Change	−0.0300		−0.0300		+0.0300
Moles at equilibrium	~0		~0		0.0500

c) We continue by finding out how many moles of HCl are added.

$$\text{moles HCl}_{added} = (0.0750 \text{ L})\left(\frac{1.00 \text{ mol HCl}}{1 \text{ L}}\right) = 0.0750 \text{ mol HCl added}$$

Let's see if excess H^+ remains after it reacts with the CN^- in the buffer solution.

	CN^-	+	H^+	$\rightleftharpoons$	HCN
Moles initial	0.0300				0.0200
Moles added			0.0750		
Change	−0.0300		−0.0300		+0.0300
Moles at equilibrium	~0		0.0450		0.0500

Excess H^+ remains after the reaction, so the buffering capacity has been exceeded.

At equilibrium, the solution contains 0.0450 mol H^+ and 0.0500 mol HCN. The HCN is so much weaker than the HCl (based on their K_a values) that its presence will not affect the final hydrogen

298

Chapter 18

ion concentration. Therefore, only the H^+ in HCl will determine the final pH. The concentration of H^+ is

$$[H^+] = \frac{0.0450 \text{ mol}}{(0.1000L + 0.0750 \text{ L})} = 0.257 \text{ } M$$

$$pH = -\log(0.257 \text{ } M)$$

$$pH = 0.590$$

In summary, here are the criteria for a suitable buffer:

- A buffer should not react with the system it is buffering.
- The pK_a of a buffer should be as close as possible to the pH you want to maintain.
- The buffering capacity of a buffer must be sufficient to accommodate the addition of a strong acid or base.

18.2   Acid–Base Titrations

Four main types of acid–base titrations are distinguished:

- Strong acid–strong base
- Strong acid–weak base
- Weak acid–strong base
- Weak acid–weak base (although this is typically not used)

Example 4   **Strong Acid–Strong Base Titration**

Calculate the pH after the following total volumes of 0.200 M HNO_3 have been added to 50.00 mL of 0.150 M NaOH.

a) 0.00 mL   c) 37.50 mL
b) 5.00 mL   d) 50.00 mL

First Thoughts

The reaction of a strong base with a strong acid is

$$NaOH(aq) + HNO_3(aq) \rightleftharpoons H_2O(l) + NaNO_3(aq)$$

The net ionic reaction is

$$OH^-(aq) + H^+(aq) \rightleftharpoons H_2O(l)$$

Use the net ionic equation to help you determine the limiting reactant when HNO_3 is added to the NaOH solution.

Solution

a) We can calculate the pH of the initial 0.150 M NaOH solution as we would in any other strong base equilibrium problem.

$$pOH = -\log[OH^-] = -\log(0.150)$$

$$pOH = 0.824$$

$$pH = 14.000 - 0.824$$

$$pH = 13.176$$

Copyright © Houghton Mifflin Company. All rights reserved.

b) To determine whether the H^+ or the OH^- runs out first, we must calculate the moles of reactants present.

$$\left(5.00 \text{ mL } H^+\right)\left(\frac{1 \text{ L}}{1000 \text{ mL}}\right)\left(\frac{0.200 \text{ mol } H^+}{1 \text{ L}}\right) = 0.00100 \text{ mol } H^+$$

$$\left(50.00 \text{ mL } OH^-\right)\left(\frac{1 \text{ L}}{1000 \text{ mL}}\right)\left(\frac{0.150 \text{ mol } OH^-}{1 \text{ L}}\right) = 0.00750 \text{ mol } OH^-$$

We now use our table (as in Example 3) to determine which species remain at equilibrium.

$$OH^- \quad + \quad H^+ \quad \rightleftharpoons \quad H_2O$$

	OH^-	H^+	H_2O
Moles initial	0.00750		
Moles added		0.00100	
Change	−0.00100	−0.00100	+0.00100
Moles at equilibrium	0.00650	~0	0.00100

The concentration of OH^- at equilibrium is

$$[OH^-] = \frac{0.00650 \text{ mol}}{\left(0.00500 \text{ L} + 0.05000 \text{ L}\right)} = 0.118 \ M$$

$$pOH = -\log(0.118) = 0.927$$

$$pH = 14.000 - 0.927$$

$$pH = 13.073$$

c) The moles of H^+ initially present are

$$\left(37.50 \text{ mL } H^+\right)\left(\frac{1 \text{ L}}{1000 \text{ mL}}\right)\left(\frac{0.200 \text{ mol } H^+}{1 \text{ L}}\right) = 0.00750 \text{ mol } H^+$$

$$OH^- \quad + \quad H^+ \quad \rightleftharpoons \quad H_2O$$

	OH^-	H^+	H_2O
Moles initial	0.00750		
Moles added		0.00750	
Change	−0.00750	−0.00750	+0.00750
Moles at equilibrium	~0	~0	0.00750

At this point, all of the NaOH has been neutralized by the HNO_3. This is called the **equivalence point** of the titration, the exact point at which the reactant has been neutralized by the titrant.

Because water is essentially the only species that remains at equilibrium, the autoprotolysis of water becomes significant.

$$H_2O(l) \rightleftharpoons OH^-(aq) + H^+(aq) \qquad K_w = 1.0 \times 10^{-14}$$

$$[OH^-] = [H^+] = 1.0 \times 10^{-7} \ M$$

$$pH = 7.00$$

(Note: The only case in which the equivalence point will have pH = 7.00 will be that of a strong acid–strong base titration!)

d) The moles of H^+ initially present are

$$\left(50.00 \text{ mL } H^+\right)\left(\frac{1 \text{ L}}{1000 \text{ mL}}\right)\left(\frac{0.200 \text{ mol } H^+}{1 \text{ L}}\right) = 0.0100 \text{ mol } H^+$$

$$OH^- \quad + \quad H^+ \quad \rightleftharpoons \quad H_2O$$

	OH^-	H^+	H_2O
Moles initial	0.00750		
Moles added		0.0100	
Change	–0.00750	–0.00750	+0.00750
Moles at equilibrium	~0	0.0025	0.00750

We have now exceeded the equivalence point and added excess H^+. The strong acid is now the dominant equilibrium.

$$[H^+] = \frac{0.0025 \text{ mol}}{\left(0.05000 \text{ L} + 0.05000 \text{ L}\right)} = 0.025 \; M$$

$$pH = - \log (0.025)$$
$$pH = 1.60$$

Further Insight

Here is a titration curve representing the titration of NaOH with HNO_3 (strong base with strong acid).

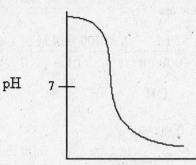

Example 5   **Strong Acid–Weak Base Titration**

Calculate the pH after the following total volumes of 0.100 M HCl have been added to 100.00 mL of 0.050 M NH_3.

a) 0.00 mL   c) 25.00 mL   e) 60.00 mL
b) 5.00 mL   d) 50.00 mL

First Thoughts

The reaction of a strong acid with a weak base is

$$HCl(aq) + NH_3(aq) \rightleftharpoons Cl^-(aq) + NH_4^+(aq)$$

The net ionic reaction is

$$H^+(aq) + NH_3(aq) \rightleftharpoons NH_4^+(aq)$$

Use the net ionic equation to help you determine the limiting reactant when the HCl is added to the NH_3 solution.

Solution

a) We can calculate the pH of the initial 0.050 M NH_3 solution as we would any other weak base equilibrium problem.

The mass-action expression is

$$K_b = \frac{[NH_4^+][OH^-]}{[NH_3]} \qquad K_b = 1.8 \times 10^{-5}$$

	NH_3	$\rightleftharpoons$	NH_4^+	+	OH^-
Initial	0.050 M		0 M		0 M
Change	$-x$		$+x$		$+x$
Equilibrium	$0.050 - x$		$+x$		$+x$
Assumption	0.050		$+x$		$+x$

$$1.8 \times 10^{-5} = \frac{x^2}{0.050}$$

$$x = [NH_4^+] = [OH^-] = 9.5 \times 10^{-4} \ M$$

$$pOH = -\log(9.5 \times 10^{-4}) = 3.02$$
$$pH = 14.00 - 3.02$$
$$pH = 10.98$$

b) To determine whether the H^+ or the NH_3 runs out first, we must calculate the moles of reactants present.

$$\left(5.00 \text{ mL } H^+\right)\left(\frac{1 \text{ L}}{1000 \text{ mL}}\right)\left(\frac{0.100 \text{ mol } H^+}{1 \text{ L}}\right) = 0.000500 \text{ mol } H^+$$

$$\left(100.00 \text{ mL } NH_3\right)\left(\frac{1 \text{ L}}{1000 \text{ mL}}\right)\left(\frac{0.0500 \text{ mol } NH_3}{1 \text{ L}}\right) = 0.00500 \text{ mol } NH_3$$

Use our table to determine which species remain after the H^+ and NH_3 react.

	NH_3	+	H^+	$\rightleftharpoons$	NH_4^+
Moles initial	0.00500				
Moles added			0.000500		
Change	-0.000500		-0.000500		$+0.000500$
Moles at equilibrium	0.00450		~ 0		0.000500

We have both the weak base and its conjugate acid left over, so we have produced a buffer and can solve for the pH using the Henderson-Hasselbalch equation:

$$pH = pK_a + \log\left(\frac{[Base]}{[Acid]}\right)$$

$$pH = -\log\left(\frac{1.0\times10^{-14}}{1.8\times10^{-5}}\right) + \log\left(\frac{0.00450}{0.000500}\right)$$

$$pH = 9.26 + 0.95$$

$$pH = 10.21$$

(Note: Moles can be used in place of the concentrations because both NH_3 and NH_4^+ are present in the same volume of solution.)

c) Use the same procedures as in part b. The moles of H^+ initially present is

$$\left(25.00 \text{ mL } H^+\right)\left(\frac{1 \text{ L}}{1000 \text{ mL}}\right)\left(\frac{0.100 \text{ mol } H^+}{1 \text{ L}}\right) = 0.00250 \text{ mol } H^+$$

Use the table to determine which species remain after the H^+ and NH_3 react.

	NH_3	+	H^+	$\rightleftharpoons$	NH_4^+
Moles initial	0.00500				
Moles added			0.00250		
Change	−0.00250		−0.00250		+0.00250
Moles at equilibrium	0.00250		~0		0.00250

$$pH = 9.26 + \log\left(\frac{0.00250}{0.00250}\right)$$

$$pH = 9.26$$

This point, at which one-half of the NH_3 has been neutralized and an equal amount of NH_4^+ has been formed, is called the **titration midpoint** (or halfway point). At the titration midpoint, **pH = pK_a.**

d) The moles of H^+ initially present is

$$\left(50.00 \text{ mL } H^+\right)\left(\frac{1 \text{ L}}{1000 \text{ mL}}\right)\left(\frac{0.100 \text{ mol } H^+}{1 \text{ L}}\right) = 0.00500 \text{ mol } H^+$$

	NH_3	+	H^+	$\rightleftharpoons$	NH_4^+
Moles initial	0.00500				
Moles added			0.00500		
Change	−0.00500		−0.00500		+0.00500
Moles at equilibrium	~0		~0		0.00500

This is the equivalence point, where exactly all of the NH_3 has been neutralized. NH_4^+ remains present and will continue to react with H_2O and reach a new state of equilibrium. The concentration of NH_4^+ is

$$[NH_4^+] = \frac{0.00500 \text{ mol}}{[0.05000 \text{ L} + 0.10000 \text{ L}]} = 0.0333 \ M$$

$$NH_4^+ \quad \rightleftharpoons \quad NH_3 \quad + \quad H^+$$

Initial	0.0333 M	0 M	0 M
Change	$-x$	$+x$	$+x$
Equilibrium	$0.0333 - x$	$+x$	$+x$
Assumption	0.0333	$+x$	$+x$

$$K_a = \frac{K_w}{K_b} = \frac{1.0 \times 10^{-14}}{1.8 \times 10^{-5}} = 5.6 \times 10^{-10} = \frac{x^2}{0.0333}$$

$$x = [NH_3] = [H^+] = 4.3 \times 10^{-6} \ M$$

$$pH = - \log(4.3 \times 10^{-6})$$

$$pH = 5.36$$

The pH is less than 7.00, which makes sense because we are titrating a weak base with a strong acid.

e) The moles of H^+ initially present is

$$\left(60.00 \text{ mL H}^+\right)\left(\frac{1 \text{ L}}{1000 \text{ mL}}\right)\left(\frac{0.100 \text{ mol H}^+}{1 \text{ L}}\right) = 0.00600 \text{ mol H}^+$$

$$NH_3 \quad + \quad H^+ \quad \rightleftharpoons \quad NH_4^+$$

Moles initial	0.00500		
Moles added		0.00600	
Change	-0.00500	-0.00500	$+0.00500$
Moles at equilibrium	~0	0.00100	0.00500

We have now exceeded the equivalence point and added excess H^+. The strong acid is now the dominant equilibrium.

$$[H^+] = \frac{0.00100 \text{ mol}}{\left(0.06000 \text{ L} + 0.05000 \text{ L}\right)} = 0.00909 \ M$$

$$pH = - \log (0.00909)$$

$$pH = 2.041$$

Further Insight

Here is a titration curve representing the titration of NH_3 with HCl (weak base with strong acid).

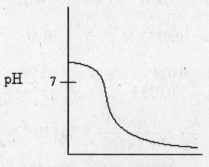

You use the same methods to solve for pH when you are titrating a weak acid with a strong base (such as $HC_2H_3O_2$ with NaOH). Here is a titration curve for this type of titration:

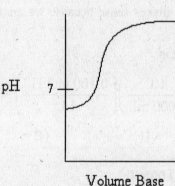

When performing titrations in the laboratory, it is very useful to use an **acid–base indicator** (also called a pH indicator). These indicators tell you the change in pH as you approach, reach, and pass the equivalence point. Here are some key points about indicators:

- Indicators are conjugate acid–base pairs of organic molecules that change color as they change between their acid and base forms.
- When performing a titration, choose an indicator with a pK_a as close as possible to the pH at the equivalence point.
- The color change of pH indicators are visible to ± 1 pH unit on either side of the indicator pK_a.
- The point at which you see the change in indicator color that signals the titration is finished is called the **titration end point.** You want the end point and equivalence point to be as close as possible, so only a few drops of indicator are added to the analyte solution.
- **Figure 18.14** in your text lists common pH indicators, their color changes, and pH ranges.

Example 6   **Picking an Indicator**

List three possible indicators in Figure 18.14 that could be used for the titrations in Examples 4 and 5.

First Thoughts

The indicators will change color when you approach and then reach the end point (or equivalence point). Choose an indicator with a pK_a as close as possible to the pH of the equivalence point in Examples 4 and 5.

Solution

Example 4

The pH at the equivalence point is 7.00. Three possible indicators we could use are

- Cresol red: analyte changes from colorless to red, orange, and yellow at pH 7
- Alizarin: analyte changes from colorless to red at pH 7
- Phenol red: analyte changes from colorless to red, orange, and yellow at pH 7

Example 5

The pH at the equivalence point is 5.36. Three possible indicators we could use are

- Eriochrome Black T: analyte changes from colorless to blue to red at pH 5.36
- Bromcresol purple: analyte changes from purple to yellow at pH 5.36
- Bromcresol green: analyte changes from colorless to blue-green at pH 5.36

18.3   Solubility Equilibria

The solubility of solids is an important factor to consider in equilibrium. Consider the dissociation of copper(I) bromide:

$$CuBr(s) \rightleftharpoons Cu^+(aq) + Br^-(aq)$$

The **solubility product** is the mass-action expression for the dissociation of a solid such as CuBr.

$$K_{sp} = [Cu^+][Br^-]$$

Table 18.3 in your text lists K_{sp} values for several common ionic solids. Using this information, we can determine the molar solubility of the solid (the solute being dissolved in solution).

Example 7   **Molar Solubility**

The K_{sp} value for calcium fluoride, CaF_2, is 4.0×10^{-11} at 25°C. Calculate the molar solubility (s) at this temperature.

First Thoughts

The dissociation reaction is

$$CaF_2(s) \rightleftharpoons Ca^{2+}(aq) + 2F^-(aq)$$

$$K_{sp} = [Ca^{2+}][F^-]^2$$

Solution

	CaF_2	$\rightleftharpoons$	Ca^{2+}	+	$2F^-$
Initial	s		0		0
Change	$-s$		$+s$		$+2s$
Equilibrium	0		s		$2s$

$$K_{sp} = [Ca^{2+}][F^-]^2$$
$$4.0 \times 10^{-11} = (s)(2s)^2 = 4s^3$$
$$s = \sqrt[3]{1.0 \times 10^{-11}}$$
$$s = 2.2 \times 10^{-4} \ M$$

Example 8   **Calculating K_{sp} of an Ionic Solid**

Calculate the K_{sp} value for Ag_3PO_4, which has a molar solubility of $1.6 \times 10^{-5} \ M$ at 25°C.

Solution

$$Ag_3PO_4 \quad \rightleftharpoons \quad 3Ag^+ \quad + \quad PO_4^{3-}$$

Initial	s	0	0
Change	$-s$	$+3s$	$+s$
Equilibrium	0	$3s$	s

$$K_{sp} = [Ag^+]^3[PO_4^{3-}]$$
$$K_{sp} = \left[3(1.6 \times 10^{-5} \ M) \right]^3 \left[1.6 \times 10^{-5} \ M \right]$$
$$K_{sp} = 1.8 \times 10^{-18}$$

When two ions are mixed in solution, a precipitate may or may not form. The best way to predict whether a precipitate will form is to compare the Q_{sp} of the solid dissociation to the K_{sp}. Here are the general rules:

- If $Q_{sp} > K_{sp}$, a precipitate forms until $Q_{sp} = K_{sp}$.
- If $Q_{sp} < K_{sp}$, no precipitate forms.

Example 9   **Will a Precipitate Form?**

Two solutions are mixed: one containing barium ions and one containing fluoride ions. The concentrations are $[Ba^{2+}] = 2.0 \times 10^{-3} \ M$ and $[F^-] = 1.1 \times 10^{-4} \ M$. Will a solid form at 25°C? K_{sp} for $BaF_2 = 2.4 \times 10^{-5}$.

First Thoughts

The dissociation reaction is

$$BaF_2(s) \rightleftharpoons Ba^{2+}(aq) + 2F^-(aq)$$

Solution

To determine whether a solid forms, we need to compare Q_{sp} to K_{sp}.

$$Q_{sp} = [Ba^{2+}][F^-]^2$$
$$Q_{sp} = (2.0 \times 10^{-3} \ M)(1.1 \times 10^{-4} \ M)^2$$
$$Q_{sp} = 2.4 \times 10^{-11}$$

Q_{sp} is less than K_{sp}, so solid BaF_2 will not form.

18.4   Complex-Ion Equilibria

In some solutions, a complex ion forms. A complex ion consists of at least one metal cation surrounded by Lewis bases called **ligands**. Examples of ligands include NH_3, Cl^-, F^-, CN^-, and H_2O. $Co(H_2O)_6^{2+}$ and $Ag(S_2O_3)_2^{3-}$ are examples of complex ions. When complex ions are formed from a metal ion and ligand, an equilibrium constant is associated with the reaction, called a **formation constant (K_f)** or **stability constant**. A good and complete example of complex-ion formation is shown in your text with EDTA.

Example 10   **Complex-Ion Formation**

Write the equilibria equations that show the stepwise formation of the complex ion $Ag(NH_3)_2^+$.

Solution

$$Ag^+ + NH_3 \rightleftharpoons Ag(NH_3)^+$$
$$Ag(NH_3)^+ + NH_3 \rightleftharpoons Ag(NH_3)_2^+$$

Exercises

Section 18.1

1. What is a buffer?

2. Predict whether a buffer solution containing 0.100 M NH_4Cl and 0.200 M NH_3 would be acidic, basic, or neutral. K_b for $NH_3 = 1.8 \times 10^{-5}$.

3. Verify your prediction in Exercise 2 by calculating the pH of this buffer solution.

4. Determine the pH of a buffer solution containing 0.110 M HOBr and 0.195 M NaOBr. K_a for HOBr = 2×10^{-9}.

5. Calculate the pH of a buffer solution containing 0.350 M HNO_2 and 0.150 M KNO_2. K_a for $HNO_2 = 4.0 \times 10^{-4}$.

6. A buffer solution containing 0.230 M $HC_2H_3O_2$ has a pH of 4.98. What was the original concentration of the conjugate base (acetate ion)? K_a for $HC_2H_3O_2 = 1.8 \times 10^{-5}$.

7. What is the ratio of $[F^-]/[HF]$ in a buffer solution with a pH of 3.80? K_a for HF = 7.2×10^{-4}.

8. What is the ratio of $[HOCl]/[OCl^-]$ in a buffer solution with a pH of 7.65? K_a for HOCl = 3.5×10^{-8}.

9. How many milliliters of 0.300 M NaOH must be added to 100.0 mL of a 0.250 M solution of HNO_2 to prepare a buffer that has a pH of 4.10? K_a for $HNO_2 = 4.0 \times 10^{-4}$.

10. How many milliliters of 0.130 M HCl must be added to 75.00 mL of a 0.275 M solution of NH_3 to prepare a buffer that has a pH of 8.70? K_b for $NH_3 = 1.8 \times 10^{-5}$.

11. How many grams of $NaC_2H_3O_2$ must be dissolved in a 0.350 M solution of $HC_2H_3O_2$ to make 200.0 mL of a buffer solution with a pH = 5.30? Assume that the volume of the solution remains constant when you add the $NaC_2H_3O_2$. K_a for $HC_2H_3O_2 = 1.8 \times 10^{-5}$.

12. Consider a 150.0 mL solution containing 0.300 M HF and 0.500 M KF. What is the buffer capacity toward a strong base like LiOH? In other words, how many moles of LiOH can be added to the solution before the buffer capacity is exceeded? K_a for HF = 7.2×10^{-4}.

13. Using the same buffer conditions as in Exercise 12, will you exceed the buffer capacity by adding 100.0 mL of 0.750 M LiOH to the solution? What will the pH be after this addition?

14. Which of the following statements is *false* concerning buffers?

   a) A buffer is the combination of a strong acid and its conjugate base or the combination of a strong base and its conjugate acid.
   b) A buffer is a chemical system that is able to resist changes in pH.
   c) For a monoprotic acid or base, the buffer capacity is greatest when $[HA] = [A^-]$.
   d) The buffer capacity of a system is a measure of the number of moles of strong acid or base that can be added while keeping the pH relatively constant.
   e) When selecting a buffer, the pK_a of the buffer should be as close as possible to the pH you want to maintain.

Section 18.2

15. What is the pH after 10.00 mL of 0.400 M KOH has been added to 80.00 mL of 0.250 M HCl?

16. What is the pH after 50.00 mL of 0.400 M KOH has been added to 80.00 mL of 0.250 M HCl?

17. What is the pH after 80.00 mL of 0.400 M KOH has been added to 80.00 mL of 0.250 M HCl?

18. What is the pH after 15.00 mL of 0.275 M NaOH has been added to 50.00 mL of 0.450 M $HC_2H_3O_2$? K_a for $HC_2H_3O_2 = 1.8 \times 10^{-5}$.

19. What is the pH after 40.91 mL of 0.275 M NaOH has been added to 50.00 mL of 0.450 M $HC_2H_3O_2$? K_a for $HC_2H_3O_2 = 1.8 \times 10^{-5}$.

20. What is the pH after 81.82 mL of 0.275 M NaOH has been added to 50.00 mL of 0.450 M $HC_2H_3O_2$? K_a for $HC_2H_3O_2 = 1.8 \times 10^{-5}$.

21. What is the pH after 100.00 mL of 0.275 M NaOH has been added to 50.00 mL of 0.450 M $HC_2H_3O_2$? K_a for $HC_2H_3O_2 = 1.8 \times 10^{-5}$.

22. What is the pH after 10.00 mL of 0.300 M HNO_3 has been added to 200.00 mL of 0.100 M CH_3NH_2? K_b for $CH_3NH_2 = 4.4 \times 10^{-4}$.

23. What is the pH after 33.33 mL of 0.300 M HNO_3 has been added to 200.00 mL of 0.100 M CH_3NH_2? K_b for $CH_3NH_2 = 4.4 \times 10^{-4}$.

24. What is the pH after 66.67 mL of 0.*300* M HNO_3 has been added to 200.00 mL of 0.100 M CH_3NH_2? K_b for $CH_3NH_2 = 4.4 \times 10^{-4}$.

25. What is the pH after 100.00 mL of 0.300 M HNO_3 has been added to 200.00 mL of 0.100 M CH_3NH_2? K_b for $CH_3NH_2 = 4.4 \times 10^{-4}$.

26. For the titration of 100.00 mL of 0.600 M $HClO_4$ with 0.800 M NaOH, what volume of NaOH (in milliliters) must be added to reach the equivalence point?

27. For the titration of 50.00 mL of 0.900 M NH_3 with 0.720 M HCl, what volume of HCl (in milliliters) must be added to reach the titration midpoint (halfway point)?

28. For the titration in Exercise 27, what volume of HCl (in milliliters) must be added to reach the equivalence point?

29. Using Figure 18.14 in your text, list two indicators that could be used for the titration in Exercise 26.

30. Using Figure 18.14 in your text, list two indicators that could be used for the titration in Exercise 27.

Consider the following titration curve to answer Exercises 31–33.

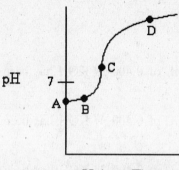

Volume Titrant

31. The curve represents which of the following titrations?
    a) a strong acid titrated with a strong base
    b) a strong base titrated with a strong acid
    c) a strong base titrated with a weak acid
    d) a weak base titrated with a strong acid
    e) a weak acid titrated with a strong base

32. Which point (A, B, C, D) is the equivalence point?

33. At which point (A, B, C, D) is pH = pK_a?

Consider the following titration curve to answer Exercises 34–36.

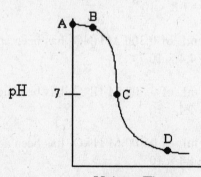

Volume Titrant

34. The curve represents which of the following titrations?
    a) a strong acid titrated with a strong base
    b) a strong base titrated with a strong acid
    c) a strong base titrated with a weak acid
    d) a weak base titrated with a strong acid
    e) a weak acid titrated with a strong base

35. At which point (A, B, C, D) do hydrogen ions (H^+) control the equilibrium?

36. Which compound is most likely to be the titrant?
    a) NaOH
    b) $HClO_4$
    c) NH_3
    d) $HC_2H_3O_2$
    e) H_2O

Section 18.3

37. Write the reaction for the dissolution of the following ionic compounds:
    a) AgBr
    b) PbI_2
    c) $Sr_3(PO_4)_2$

38. Write the mass-action expressions for each of the equilibrium reactions in Exercise 37.

39. The K_{sp} value for $PbBr_2$ is 4.6×10^{-6} at 25°C. Determine the molar solubility at this temperature.

40. Calculate the molar solubility of $Co(OH)_3$ at 25°C. K_{sp} for $Co(OH)_3 = 2.5 \times 10^{-43}$.

41. Determine the K_{sp} value for $NiCO_3$, which has a molar solubility of 3.7×10^{-4} M at 25°C.

42. Determine the K_{sp} value for Ag_2SO_4, which has a molar solubility of 0.0144 M at 25°C.

43. Two solutions are mixed: one containing silver ions and one containing bromide ions. The concentrations are $[Ag^+] = 3.5 \times 10^{-2}$ M and $[Br^-] = 2.9 \times 10^{-4}$ M. Will a solid form at 25°C? Justify your answer. K_{sp} for AgBr $= 5.0 \times 10^{-13}$.

44. Will a precipitate form when a 5.1×10^{-5} M Ba^{2+} solution and 4.0×10^{-4} M OH^- solution are mixed at 25°C? Justify your answer. K_{sp} for barium hydroxide $= 5.0 \times 10^{-3}$.

Section 18.4

45. What is a ligand? Give three examples of ligands.

46. What is a complex ion? Give an example of a complex ion.

47. Which useful compound can form complex ions with various metal ions such as calcium, zinc, aluminum, and nickel?

48. What is the name of the equilibrium constant used for complex-ion equilibria?

49. Write the equilibria equations that show the stepwise formation of the complex ion $Co(NH_3)_6^{2+}$.

50. Write the equilibria equations that show the stepwise formation of the complex ion $Ni(CN)_4^{2-}$.

Answers to Exercises

1. A buffer is a chemical system that is able to resist a change in pH. It is a combination of a weak acid and its conjugate base or the combination of a weak base and its conjugate acid.

2. The K_b for NH_3 is larger than the K_a for NH_4^+; the buffer solution will be basic.

3. 9.56

4. 8.9

5. 3.03

6. $0.40\,M\,C_2H_3O_2^-$

7. 4.54

8. 0.64

9. 69.5 mL NaOH

10. 124 mL HCl

11. 20.6 g

12. 0.045 mol of LiOH can be added before the buffer capacity is exceeded.

13. Yes, the buffer capacity is exceeded. pH = 13.08 after the addition of LiOH.

14. a

15. 0.725

16. 7.00

17. 12.875

18. 4.10

19. 4.74

20. 8.55

21. 12.52

22. 11.40

23. 10.64

24. 5.88

25. 1.477

26. 75.0 mL NaOH

27. 31.3 mL HCl

28. 62.5 mL HCl

29. phenolphthalein, cresol red

30. methyl red, bromcresol green

31. e

32. C

33. B

34. b

35. D

36. b

37. a)　$AgBr(s) \rightleftharpoons Ag^+(aq) + Br^-(aq)$

　　b)　$PbI_2(s) \rightleftharpoons Pb^{2+}(aq) + 2I^-(aq)$

　　c)　$Sr_3(PO_4)_2(s) \rightleftharpoons 3Sr^{2+}(aq) + 2PO_4^{3-}(aq)$

38. a)　$K_{sp} = [Ag^+][Br^-]$

　　b)　$K_{sp} = [Pb^{2+}][I^-]^2$

　　c)　$K_{sp} = [Sr^{2+}]^3[PO_4^{3-}]^2$

39. 0.010 M

40. 9.8×10^{-12} M

41. 1.4×10^{-7}

42. 1.19×10^{-5}

43. Yes, solid AgBr will form. Q_{sp} (1.0×10^{-5}) is greater than K_{sp} (5.0×10^{-13}).

44. No, solid Ba(OH)$_2$ will not form. Q_{sp} (8.2×10^{-12}) is less than K_{sp} (5.0×10^{-3}).

45. A ligand is a Lewis base such as H$_2$O, NH$_3$, or Cl$^-$.

46. A complex ion contains at least one metal ion surrounded by ligands, such as $Co(H_2O)_6^{2+}$.

47. EDTA

48. formation constant or stability constant (K_f)

49.
$$Co^{2+} + NH_3 \rightleftharpoons Co(NH_3)^{2+}$$
$$Co(NH_3)^{2+} + NH_3 \rightleftharpoons Co(NH_3)_2^{2+}$$
$$Co(NH_3)_2^{2+} + NH_3 \rightleftharpoons Co(NH_3)_3^{2+}$$
$$Co(NH_3)_3^{2+} + NH_3 \rightleftharpoons Co(NH_3)_4^{2+}$$
$$Co(NH_3)_4^{2+} + NH_3 \rightleftharpoons Co(NH_3)_5^{2+}$$
$$Co(NH_3)_5^{2+} + NH_3 \rightleftharpoons Co(NH_3)_6^{2+}$$

50.
$$Ni^{2+} + CN^- \rightleftharpoons Ni(CN)^+$$
$$Ni(CN)^+ + CN^- \rightleftharpoons Ni(CN)_2$$
$$Ni(CN)_2 + CN^- \rightleftharpoons Ni(CN)_3^-$$
$$Ni(CN)_3^- + CN^- \rightleftharpoons Ni(CN)_4^{2-}$$

Chapter 19

Electrochemistry

The Bottom Line

When you finish this chapter, you will be able to:

- Understand the difference between a voltaic cell and an electrolytic cell.
- Balance oxidation–reduction (redox) reactions.
- Distinguish between an oxidizing agent and a reducing agent.
- Calculate the voltage developed by an electrochemical cell under standard and nonstandard conditions.
- Calculate equilibrium constants from cell voltages.
- Determine how much electrical energy is necessary to reduce a substance in an electrolytic cell.

19.1   What Is Electrochemistry?

This section introduces the topic. The important concepts are outlined here:

- **Electrochemistry** is the study of the reduction and oxidation processes that occur at the interface between different phases of the system.
- A **voltaic cell** (also known as a **Galvanic cell**) is an electrochemical cell that produces electricity from a chemical reaction.
- An **electrolytic cell** is an electrochemical cell that requires external electrical energy to perform a chemical reaction.
- Electrochemical cells are composed of two parts where **half-reactions** occur. The **oxidation half-reaction** supplies electrons to the reaction. The **reduction half-reaction** utilizes these electrons. The overall reaction is the sum of the two half-reactions; it is called the **redox reaction.**
- **Electrodes** are either the **anode**, the site of the oxidation reaction, or the **cathode**, the site of the reduction reaction. They are commonly metal surfaces.

19.2   Oxidation States: Electronic Bookkeeping

Review **Section 4.7** of the **Study Guide.** It is important that you be able to identify a redox reaction and assign oxidation numbers. **Exercise 18** in Study Guide Chapter 4 demonstrates how to assign oxidation numbers. Using this information, you will be able to identify the oxidation reaction and the reduction reaction.

You should memorize these two important definitions:

- **Oxidizing agent:** a substance that is reduced and causes an oxidation.
- **Reducing agent:** a substance that is oxidized and causes a reduction.

Chapter 19

Exercise 1 — Redox Half-Reactions and Agents

For the following reaction, identify the oxidation half-reaction, the reduction half-reaction, the oxidizing agent, and the reducing agent.

$$Zn(s) + 2HCl(aq) \rightarrow ZnCl_2(aq) + H_2(g)$$

First Thoughts

Separate the aqueous species into ions. Use the rules in **Table 4.5** in your text to identify the oxidation number for all species.

$$\overset{0}{Zn} + \overset{+1}{2H^+} + \overset{-1}{2Cl^-} \rightarrow \overset{+2}{Zn^{2+}} + \overset{-1}{2Cl^-} + \overset{0}{H_2}$$

Oxidation reaction: $Zn \rightarrow Zn^{2+} + 2e^-$

Reduction reaction: $2e^- + 2H^+ \rightarrow H_2$

Zinc was oxidized, so it is the **reducing agent** (it caused the reduction of the hydrogen ion). Hydrogen ion is the **oxidizing agent** (it caused the oxidation of the zinc).

19.3   Redox Reactions

Table 19.3 in the textbook lists selected standard reduction potentials. These half-reactions had their **electromotive force (emf)** or **voltage** E° measured against the **standard hydrogen electrode (SHE)** under standard conditions: aqueous solutions, 1.0 M; gases, 1.0 atm; and temperature, - 25°C. By definition, the SHE reaction is

$$2H^+(aq) + 2e^- \rightarrow H_2(g) \quad E^\circ = 0.00 \text{ V}$$

Table 19.3 lists **standard reduction potentials (SRP),** in which all half-reactions are written as reductions. The value of E° measures how strongly the reduced species on the right-hand side of the reduction half-reaction pulls the electron toward itself.

The potential can be used to judge the spontaneity of the half-reaction by calculating ΔG°:

$$\Delta G^\circ = -nFE^\circ$$

where n is the number of moles of electrons transferred in the reaction and F is Faraday's constant (96,485 C). One coulomb (C) = 1 J/V. Recall from Chapter 14 that if $\Delta G^\circ < 0$, the reaction is spontaneous.

Example 2   Spontaneity and Potential

The reduction of the silver ion is

$$Ag^+(aq) + e^- \rightarrow Ag(s) \quad E^\circ = 0.80 \text{ V}$$

What is the free energy change for this process? Is the half-reaction spontaneous?

Solution

$$\Delta G^\circ = -nFE^\circ$$

$$= -\left(1 \text{ mol e}^-\right)\left(\frac{96485 \text{ C}}{\text{mol e}^-}\right)(0.80 \text{ J/C})$$

$$= -7.7 \times 10^4 \text{ J} = -77 \text{ kJ}$$

The free energy is negative, so the reaction is spontaneous.

Balancing Redox Reactions

Your textbook gives a thorough explanation of a seven-step process for balancing redox reactions. The next two examples will follow the textbook method.

Example 3   **Balancing a Redox Reaction in Acidic Solution**

Balance the following reaction:

$$Cr_2O_7^{2-}(aq) + Cl^-(aq) \rightarrow Cr^{3+}(aq) + Cl_2(g) \qquad \text{(acidic solution)}$$

Solution

Assign oxidation numbers and then divide the reaction into the two half-reactions.

$$\overset{+6}{Cr_2O_7^{2-}}(aq) + \overset{-1}{Cl^-}(aq) \rightarrow \overset{+3}{Cr^{3+}}(aq) + \overset{0}{Cl_2}(g)$$

Oxidation: $2Cl^- \rightarrow Cl_2$
Reduction: $Cr_2O_7^{2-} \rightarrow 2Cr^{3+}$

Balance O's with H_2O and H's with H^+:

$$2Cl^- \rightarrow Cl_2$$

$$14H^+ + Cr_2O_7^{2-} \rightarrow 2Cr^{3+} + 7H_2O$$

Balance charge by adding e^-:

$$2Cl^- \rightarrow Cl_2 + 2e^-$$

$$6e^- + 14H^+ + Cr_2O_7^{2-} \rightarrow 2Cr^{3+} + 7H_2O$$

Use a multiplier to equalize the number of electrons in each half-relation:

$$(2Cl^- \rightarrow Cl_2 + 2e^-) \times 3$$

$$(6e^- + 14H^+ + Cr_2O_7^{2-} \rightarrow 2Cr^{3+} + 7H_2O) \times 1$$

Add the two half-reactions and eliminate species that appear on both sides of the arrow:

$$6Cl^- \rightarrow 3Cl_2 + 6e^-$$

$$6e^- + 14H^+ + Cr_2O_7^{2-} \rightarrow 2Cr^{3+} + 7H_2O$$

$$\overline{14H^+(aq) + Cr_2O_7^{2-}(aq) + 6Cl^-(aq) \rightarrow 3Cl_2(g) + 2Cr^{3+}(aq) + 7H_2O(l)}$$

Example 4   **Balancing a Redox Reaction in Basic Solution**

Balance the following reaction:

$$CN^-(aq) + MnO_4^-(aq) \rightarrow CNO^-(aq) + MnO_2(s) \text{ (basic solution)}$$

First Thoughts

Follow the procedure in Example 3 to arrive at the answer in acidic solution; then add OH^- ions to neutralize the acid and make the solution basic.

Solution

Using the steps from Example 3, we arrive at the following half-reactions:

$$(CN^- + H_2O \rightarrow CNO^- + 2H^+ + 2e^-) \times 3$$

$$(3e^- + 4H^+ + MnO_4^- \rightarrow MnO_2 + 2H_2O) \times 2$$

$$2H^+ + 2MnO_4^- + 3CN^- \rightarrow 3CNO^- + 2MnO_2 + H_2O$$

Now, add the same amount of OH^- to each side as there is H^+. Each H^+ reacts with OH^- to make H_2O.

$$2H_2O + 2MnO_4^- + 3CN^- \rightarrow 3CNO^- + 2MnO_2 + H_2O + 2OH^-$$

Cancel out the extra waters and you will have the final balanced equation.

$$H_2O + 2MnO_4^- + 3CN^- \rightarrow 3CNO^- + 2MnO_2 + 2OH^-$$

Manipulating Half-Cell Reactions

The potentials for the half-reactions can be manipulated like thermodynamic quantities, with the exception that they are independent of the amount of substance present.

- $E_{rxn}^o = E_{red}^o + E_{ox}^o$; the half-cell potentials can be combined to calculate the potential of the cell.
- If you reverse the direction of a half-reaction, you reverse the sign of E^o. $E_{ox}^o = -E_{red}^o$.
- E^o does *not* depend on the coefficients of the chemical reaction.
- $E^o > 0$ for a spontaneous reaction.

Example 5 E_{rxn}^o from a Reaction

Determine the cell voltage under standard conditions for the following reaction:

$$2Li(s) + F_2(g) \rightarrow 2LiF(aq)$$

Solution

Separate the reaction into two half-reactions. Look up the SRPs in the table.

Oxidation: $Li(s) \rightarrow Li^+(aq) + e^-$ $E^o = +3.04$ V

Reduction: $F_2(g) + 2e^- \rightarrow 2F^-(aq)$ $E^o = +2.87$ V

$$E_{rxn}^o = 2.87 \text{ V} + 3.04 \text{ V} = 5.91 \text{ V}$$

19.4   Electrochemical Cells

In the chemical laboratory, voltaic cells consist of an anode, where the oxidation half-reaction occurs; a cathode, where the reduction half-reaction occurs; and a salt bridge. Cells may be composed of many different half-reactions. Example 6 illustrates one of them.

Example 6 E_{rxn}^o for a Spontaneous Reaction

Using Table 19.3 in your textbook to find the standard reduction potentials, construct a cell made from Fe^{2+}/Fe and Cl_2/Cl^- that would generate a spontaneous chemical reaction.

First Thoughts

The SRPs are available in Table 19.3. The reactions must be chosen so that the sum of the oxidation reaction plus the reduction reaction is positive. $E^o > 0$ for a spontaneous reaction.

Solution

From Table 19.3:

$$Fe^{2+}(aq) + 2e^- \rightarrow Fe(s) \qquad E^\circ = -0.44 \text{ V}$$
$$Cl_2(g) + 2e^- \rightarrow 2Cl^-(aq) \qquad E^\circ = +1.36 \text{ V}$$

If we make Fe be the oxidation reaction, then

$$Fe(s) \rightarrow Fe^{2+} + 2e^- \qquad E^\circ = +0.44 \text{ V}$$

Adding the reduction and oxidation half-reactions:

$$Fe(s) + Cl_2(g) \rightarrow Fe^{2+}(aq) + 2Cl^-(aq) \qquad E_{rxn}^\circ = 0.44 \text{ V} + 1.36 \text{ V} = 1.80 \text{ V; a spontaneous reaction}$$

Further Insight

If we mistakenly choose Cl^- to be the oxidation reaction, $E^\circ = -1.36$ V. Adding that value to E° for Fe would give −1.80 V, indicating a nonspontaneous reaction.

Cell Notation

A shorthand notation is used to describe an electrochemical cell. It consists of ABC: anode, bridge, and cathode. The anode is written on the left, the salt bridge in the middle, and the cathode on the right. The use of cell notation is illustrated in Example 7.

Example 7   Cell Notation

The Daniels cell is represented by the following reaction:

$$Zn(s) + Cu^{2+}(aq) \rightarrow Zn^{2+}(aq) + Cu(s) \qquad E^\circ = 1.10 \text{ V}$$

Use cell notation to represent this cell.

Solution

$$Zn(s)|\ Zn^{2+}(aq) \parallel Cu^{2+}(aq)|\ Cu(s)$$

The zinc is oxidized ($Zn \rightarrow Zn^{2+} + 2e^-$) and becomes the anode, the salt bridge is represented by $\parallel$, and the copper is reduced ($Cu^{2+} + 2e^- \rightarrow Cu$) and becomes the cathode. Notice that the two species for the electrode are separated by a vertical line.

The rest of this section discusses batteries (see **Table 19.5** in the textbook) and the chemistry of some common batteries. You should read this material.

19.5   Chemical Reactivity Series

The table of standard reduction potentials (**Table 19.3**) can also be used to rank reducing and oxidizing agents. This ranking produces a reactivity series (see **Table 19.6**). The strongest reducing agents (those that are most easily oxidized) have the most negative SRPs and are located at the top of the table. These metals are also the most reactive. The less reactive metals are located toward the bottom of the table.

Example 8   Reactivity Series

Arrange the following metals in order of increasing reactivity: Au, Al, Ag.

Solution

Au has an SRP of +1.50 V, Al has an SRP of −1.66 V, and Ag has an SRP of +0.80 V. Thus Au < Ag < Al.

19.6   Not-So-Standard Conditions: The Nernst Equation

The Nernst equation provides a method for calculating the cell potential under nonstandard conditions:

$$E = E^\circ - \frac{RT}{nF} \ln Q$$

where R = 8.3145 J/K mol, T is the Kelvin temperature, n is the number of moles of electrons transferred, F = 96,485 C/mol, and Q is the reaction quotient. If **T = 298 K**, a useful form of the Nernst equation is:

$$E = E^\circ - \frac{0.0257}{n} \ln Q$$

Using base-10 logarithms, again at T = 298 K, the equation becomes

$$E = E^\circ - \frac{0.0592}{n} \log Q$$

The effect of concentration changes in the reaction can be measured by these equations.

Example 9   **The Nernst Equation**

Determine E for the following cell:

$$Zn(s) \mid Zn^{2+}(0.0010 \ M) \parallel Cu^{2+}(0.100 \ M) \mid Cu(s) \quad E^\circ = 1.10 \ V$$

First Thoughts

First write the cell reaction to determine n; then calculate Q. Examining the two half-reactions, we see that $n = 2$.

Oxidation: $Zn \rightarrow Zn^{2+} + 2e^-$

Reduction: $Cu^{2+} + 2e^- \rightarrow Cu$

Solution

$Zn(s) + Cu^{2+}(aq) \rightarrow Cu(s) + Zn^{2+}(aq)$. The number of electrons transferred is two, so $n = 2$.

$$E = E^\circ - \frac{0.0592}{n} \log Q$$

$$= 1.10 \ V - \frac{0.0592}{2} \log \frac{0.0010}{0.100}$$

$$= 1.10 \ V + 0.059 \ V$$

$$= 1.16 \ V$$

The Nernst Equation and the Equilibrium Constant

Having the standard cell potential provides us with a convenient way to solve for the equilibrium constant. At equilibrium, $\Delta G = 0$ and thus $E = 0$. Here is the Nernst equation at equilibrium under standard conditions:

$$0 = E^\circ - \frac{0.0592}{n} \log K$$

$$\log K = \frac{nE^\circ}{0.0592} \quad \text{or} \quad K = 10^{\frac{nE^\circ}{0.0592}}$$

Example 10   **The Nernst Equation and Equilibrium**

Calculate the equilibrium constant at 298 K for the following reaction:

$$Zn(s) + Cu^{2+}(aq) \rightarrow Cu(s) + Zn^{2+}(aq) \quad E^\circ = 1.10 \text{ V}$$

Solution

$$K = 10^{\frac{nE^\circ}{0.0592}}$$

$$= 10^{\frac{(2)(1.10)}{0.0592}}$$

$$= 1.45 \times 10^{37}$$

19.7   Electrolysis

In electrolysis reactions, electrical energy is used to cause a chemical change. Substances, such as molten salts or water, can be separated into their respective components or metals and can be plated onto other metals. Faraday recognized that to reduce a metal, a certain amount of charge had to be passed through the electrolysis cell. For example, to reduce

$$Cr^{3+}(aq) + 3e^- \rightarrow Cr(s)$$

three moles of e^- are necessary. Three moles of e^- are equal to three Faradays of charge (1 F = 96,485 C). Example 11 illustrates this calculation.

Example 11   **Electrolysis**

A current of 3.50 A (3.50 C/s) is passed through a solution containing $Cr^{3+}(aq)$ for 60.0 min. How much chromium, in grams, will be deposited at the cathode? The reaction at the cathode is $Cr^{3+}(aq) + 3e^- \rightarrow Cr(s)$.

Solution

Your textbook gives a formula that performs the calculation directly. We will show the calculation in steps to illustrate the steps of the procedure.

1. Calculate the charge.

$$\text{charge (coulombs, C)} = \text{current (A)} \times \text{time (s)}$$

$$= (3.50 \text{ C/s})(60.0 \text{ min})(60 \text{ s/min})$$

$$= 1.26 \times 10^4 \text{ C}$$

2. Calculate the number of moles of electrons (Faradays).

$$\text{mol e}^- = \left(1.26 \times 10^4 \text{ C}\right)\left(\frac{1 \text{ mol e}^-}{96{,}485 \text{ C}}\right)$$

$$= 0.131 \text{ mol e}^-$$

3. Calculate the mass of chromium from the moles of chromium.

$$\text{mass Cr} = \left(0.131 \text{ mol e}^-\right)\left(\frac{1 \text{ mol Cr}}{3 \text{ mol e}^-}\right)\left(\frac{52.0 \text{ g}}{1 \text{ mol Cr}}\right)$$

$$= 2.27 \text{ g}$$

Exercises

Section 19.1

1. What is a voltaic cell?

2. How does an electrolytic cell differ from a galvanic cell?

3. True or false? A galvanic cell is different from a voltaic cell.

4. True or false? Redox reactions are composed of two half-reactions.

5. What reaction occurs at the anode of a voltaic cell?

6. What reaction occurs at the cathode of an electrolytic cell?

Section 19.2

7. Define oxidizing agent.

8. True or false? A reducing agent is a substance that is oxidized in a redox reaction.

9. Write the oxidation number of the underlined atom:

   a) K$\underline{Mn}$O$_4$ b) $\underline{Cr}$O$_4{}^{2-}$ c) $\underline{Al}_2$O$_3$

10. Write the oxidation number of the underlined atom:

    a) H$_3\underline{As}$O$_3$ b) $\underline{Sn}$F$_2$ c) $\underline{C}$O$_3{}^{2-}$

11. Write the oxidation number of the underlined atom:

    a) $\underline{W}$O$_4{}^{2-}$ b) $\underline{Mn}_2$O$_7{}^{2-}$ c) K$_2\underline{Cr}_2$O$_7$

12. Identify the oxidizing agent in the following reaction: $2Fe(s) + 6HCl(aq) \rightarrow 2FeCl_3(aq) + 3H_2(g)$.

13. Identify the reducing agent in the following reaction: $2Ag(s) + Ni^{2+}(aq) \rightarrow 2Ag^+(aq) + Ni(s)$.

Section 19.3

Balance the redox reactions in Exercises 14–18 in *acidic* solution.

14. $Cr_2O_7{}^{2-}(aq) + C_2O_4{}^{2-}(aq) \rightarrow Cr^{3+}(aq) + CO_2(g)$

15. $Ag(s) + NO_3{}^-(aq) \rightarrow NO_2(g) + Ag^+(aq)$

16. $MnO_4{}^-(aq) + HSO_3{}^-(aq) \rightarrow Mn^{2+}(aq) + SO_4{}^{2-}(aq)$

17. $Zn(s) + NO_3{}^-(aq) \rightarrow Zn^{2+}(aq) + N_2O(g)$

18. $Cu(s) + NO_3^-(aq) \rightarrow Cu^{2+}(aq) + NO(g)$

Balance the redox reactions in Exercises 19–23 in *basic* solution.

19. $Al(s) + OH^-(aq) \rightarrow Al(OH)_4^-(aq) + H_2(g)$

20. $CrO_4^{2-}(aq) + SO_3^{2-}(aq) \rightarrow Cr(OH)_3(s) + SO_4^{2-}(aq)$

21. $Zn(s) + Cu(OH)_2(s) \rightarrow Zn(OH)_4^{2-}(aq) + Cu(s)$

22. $HS^-(aq) + ClO_3^-(aq) \rightarrow S(s) + Cl^-(aq)$

23. $MnO_4^-(aq) + I^-(aq) \rightarrow MnO_2(s) + IO_3^-(aq)$

24. True or false? E° does not depend on the coefficients of the chemical reaction.

25. True or false? $E^\circ < 0$ for a spontaneous reaction.

26. Using the table of standard reduction potentials in the Appendix of the textbook, calculate E° for $Al(s) + Fe^{2+}(aq) \rightarrow Fe(s) + Al^{3+}(aq)$.

27. Is the following reaction spontaneous: $Cu(s) + Mg^{2+}(aq) \rightarrow Cu^{2+}(aq) + Mg(s)$? (You can find standard reduction potentials in the Appendix of the textbook.)

Section 19.4

28. What is cell notation? What does it consist of?

29. Use the half-reactions $Fe|Fe^{2+}$ and $Br_2|Br^-$ to construct a cell with a spontaneous reaction. What is E° for this cell?

30. Use the half-reactions $Na|Na^+$ and $Ag|Ag^+$ to construct a cell with a spontaneous reaction. What is E° for this cell?

31. Write the cell notation for the following half-reactions:

$$Ni(s) \rightarrow Ni^{2+}(aq) + 2e^- \text{ and } Pb^{2+}(aq) + 2e^- \rightarrow Pb(s)$$

32. Write the cell notation for the following half-reactions:

$$Al(s) \rightarrow Al^{3+}(aq) + 3e^- \text{ and } Zn^{2+}(aq) + 2e^- \rightarrow Zn(s)$$

33. Write the overall cell reaction for the following half-reactions: $Fe(s) \mid Fe^{3+}(aq) \parallel Cu^{2+}(aq) \mid Cu(s)$

34. Write the overall cell reaction for the following half-reactions: $Zn(s) \mid Zn^{2+}(aq) \parallel Ni^{2+}(aq) \mid Ni(s)$

Section 19.5

35. Using Table 19.3 in the text, identify the strongest reducing agent.

36. Using Table 19.3 in the text, identify the strongest oxidizing agent.

37. Which is the more reactive metal, Al or Fe?

38. Which is the more reactive metal, Cu or Ag?

39. Which is the more reactive metal, Li or Cd?

Section 19.6

For Exercises 40–45, use the table of standard reduction potentials in the Appendix of the text to find E^o for the half-cells.

40. Calculate E_{cell} at 298 K for $Zn(s) + Cu^{2+}(0.15\ M) \rightarrow Zn^{2+}(1.25\ M) + Cu(s)$.

41. Calculate E_{cell} at 298 K for $Mg(s) + Sn^{2+}(0.035\ M) \rightarrow Mg^{2+}(0.545\ M) + Sn(s)$.

42. Calculate E_{cell} at 298 K for $Mg(s) \mid Mg^{2+}(0.26\ M) \parallel Mg^{2+}(0.98\ M) \mid Mg(s)$.

43. What is the value of E_{cell} at equilibrium?

44. Calculate the equilibrium constant, K, for the following reaction at 298 K:
$2Fe(s) + 3Cu^{2+}(aq) \rightarrow 2Fe^{3+}(aq) + 3Cu(s)$.

45. Calculate the equilibrium constant, K, at 298 K for $Sn(s) \mid Sn^{2+}(aq) \parallel Cu^{2+}(aq) \mid Cu(s)$.

Section 19.7

46. How many Faradays are necessary to reduce one mole of Co^{3+} to Co?

47. An electrolytic cell contains molten $BaCl_2$. At which electrode will the barium be deposited?

48. Consider the following half-reaction: $Cr^{3+}(molten) + 3e^- \rightarrow Cr(s)$. If 1.25 F was supplied to the electrode, how many grams of chromium would be deposited?

49. Aluminum is prepared from molten cryolite, Al_2O_3. How many kilograms of Al would be produced by passing 175 A through a cell for 24.0 h? The reaction is $Al^{3+}(aq) + 3e^- \rightarrow Al(s)$.

50. An electrolytic cell contains $CuSO_4(aq)$. If the cell is operated at 0.66 A, how many minutes would the cell have to be operated to produce 0.50 g of copper?

Answers to Exercises

1. A voltaic cell is an electrochemical cell that produces electricity from a redox reaction.

2. An electrolytic cell performs a chemical reaction if electrical energy is supplied. A galvanic or voltaic cell produces electricity from a chemical reaction.

3. False. They are both the same.

4. True

5. The oxidation reaction occurs at the anode.

6. Reduction always occurs at the cathode in an electrochemical cell.

7. An oxidizing agent is a substance that causes an oxidation. The oxidizing agent itself is reduced.

8. True

9. a) +7; b) +6; c) +3

10. a) +3; b) +2; c) +4

11. a) +6; b) +6; d) +6

12. H^+ from HCl is the oxidizing agent. H^+ is reduced to H^0.

13. Ag is the reducing agent. Ag is oxidized to Ag^+.

14. $Cr_2O_7^{2-}(aq) + 3C_2O_4^{2-}(aq) + 14H^+ \rightarrow 2Cr^{3+}(aq) + 7H_2O(l) + 6CO_2(g)$

15. $Ag(s) + NO_3^-(aq) + 2H^+ \rightarrow Ag^+(aq) + NO_2(g) + H_2O(l)$

16. $2MnO_4^-(aq) + 5HSO_3^-(aq) + H^+(aq) \rightarrow 2Mn^{2+}(aq) + 5SO_4^{2-}(aq) + 3H_2O(l)$

17. $4Zn(s) + 2NO_3^-(aq) + 10H^+(aq) \rightarrow 4Zn^{2+}(aq) + N_2O(g) + 5H_2O(l)$

18. $3Cu(s) + 2NO_3^-(aq) + 8H^+(aq) \rightarrow 3Cu^{2+}(aq) + 2NO(g) + 4H_2O(l)$

19. $2Al(s) + 2OH^-(aq) + 6H_2O(l) \rightarrow 2Al(OH)_4^-(aq) + 3H_2(g)$

20. $2CrO_4^{2-}(aq) + 3SO_3^{2-}(aq) + 5H_2O \rightarrow 2Cr(OH)_3(s) + 3SO_4^{2-}(aq) + 4OH^-(aq)$

21. $Zn(s) + 2OH^-(aq) + Cu(OH)_2(s) \rightarrow Zn(OH)_4^{2-}(aq) + Cu(s)$

22. $3HS^-(aq) + ClO_3^-(aq) \rightarrow 3S(s) + Cl^-(aq) + 3OH^-(aq)$

23. $2MnO_4^-(aq) + I^-(aq) + H_2O(l) \rightarrow 2MnO_2(s) + IO_3^-(aq) + 2OH^-(aq)$

24. True

25. False. $E°> 0$ for a spontaneous reaction.

26. 1.22 V

27. Nonspontaneous; $E° = -2.72$ V

28. Cell notation is a shorthand used to describe an electrochemical cell. It consists of an anode, a bridge, and a cathode.

29. $Fe(s) \mid Fe^{2+}(aq) \parallel Br^-(aq) \mid Br_2(l)$; $E° = 1.51$ V

30. $Na(s) \mid Na^+(aq) \parallel Ag^+(aq) \mid Ag(s)$; $E° = 3.51$ V

31. $Ni(s) \mid Ni^{2+}(aq) \parallel Pb^{2+}(aq) \mid Pb(s)$

32. $Al(s) \mid Al^{3+}(aq) \parallel Zn^{2+}(aq) \mid Zn(s)$

33. $2Fe(s) + 3Cu^{2+}(aq) \rightarrow 2Fe^{3+}(aq) + 3Cu(s)$

34. $Zn(s) + Ni^{2+}(aq) \rightarrow Zn^{2+}(aq) + Ni(s)$

35. Li is the strongest reducing agent.

36. F_2 is the strongest oxidizing agent.

37. Al is more reactive.

38. Cu is more reactive.

39. Li is more reactive.

40. 1.07 V

41. 2.19 V

42. 0.017 V

43. $E = 0$

44. 2.5×10^{30}

45. 1.6×10^{16}

46. Three Faradays: $Co^{3+}(aq) + 3e^- \rightarrow Co(s)$

47. The reduction of $Ba^{2+}(aq) + 2e^- \rightarrow Ba(s)$ occurs at the cathode.

48. 21.7 g

49. 1.41 kg

50. 38 min

Chapter 20

Coordination Complexes

The Bottom Line

When you finish this chapter, you will be able to:

- Identify a coordination complex and the ligands within a complex.
- Determine the coordination number of complex ions and use that information to predict the geometry of a complex.
- Understand the various isomers that exist for coordination compounds.
- Write the formulas for coordination compounds.
- Name coordination compounds.
- Apply crystal field theory to various complexes with octahedral, tetrahedral, or square planar geometries.
- Understand two chemical reactions that occur with coordination complexes.

20.1   Bonding in Coordination Complexes

A **coordination complex** consists of a central atom or ion that is bonded to several other components. **Coordinate covalent bonding** occurs when both bonding electrons in a metal complex originate from one atom. Ligands form a coordinate covalent bond with a metal or ion within the metal complex. An example of a complex ion is $[Co(NH_3)_6]^{2+}$. Here is the general equation that shows the formation of a coordination complex by donation of a lone pair of electrons from a ligand, L, to a metal center, M:

$$M + :L \rightarrow M\text{-}L$$

20.2   Ligands

Lewis bases, known as **ligands,** coordinate to the metal in a coordination complex. Here are some key points about ligands:

- Ligands donate a lone pair of electrons to the metal or metal ion to form the coordinate covalent bond. **Table 20.1** lists some common ligands.
- A ligand can use more than one nonbonded pair of electrons to bond to a metal center. It may use each of those pairs to form independent coordinate covalent bonds to the metal.
- A **bidentate ligand** forms two bonds to the metal. A **tridentate ligand** forms three bonds to the metal.
- The polydentate ligands are known as **chelates** because of how they bond to the metal center so tightly.

Example 1   **Identifying Ligands**

Which of the following species could act as ligands?

a) Cl^- c) SiH_4
b) Mg^{2+} d) H_2O

First Thoughts

Draw the Lewis structures. Look for the species that have at least one nonbonded pair of electrons (lone pairs) on one or more atoms.

Solution

Species a and d could act as ligands.

20.3   Coordination Number

The number of donor atoms to which a given metal ion bonds is called the **coordination number.** Coordination numbers can range from 2 to 8, although coordination numbers of 4 and 6 are the most common.

Example 2   **Coordination Number**

What is the coordination number for each of the following complexes?

a) $[Co(H_2O)_6]^{2+}$ b) $[Ag(NH_3)_2]^+$ c) $[Ni(CN)_4]^{2-}$

Solution

a) 6 (six H_2O ligands coordinated to cobalt metal)
b) 2 (two NH_3 ligands coordinated to silver metal)
c) 4 (four CN ligands coordinated to nickel metal)

20.4   Structure

It is important to visualize these metal complexes three-dimensionally because this perspective can affect the complex's reactivity with other compounds. Here are the general rules for predicting the geometry of metal complexes:

- For complexes with a coordination number of 2, the geometry is **linear** with bond angles of 180°.
- For complexes with a coordination number of 4, two geometries are observed: tetrahedral and square planar. **Tetrahedral** geometry follows the regular VSEPR rules when four electron pairs surround the central atom. The bond angles are close to 109°. **Square planar** geometry results when the metal ion has an outer nd^8 electron configuration such as Ni^{2+}.
- For complexes with a coordination number of 6, the geometry is **octahedral** with bond angles of 90°.

Example 3   **Complex Geometry**

Predict the geometry of the following complexes:

a) $[Fe(CN)_6]^{4-}$ b) $[NiCl_4]^{2-}$ c) $[Fe(C_2O_4)_3]^{3-}$

First Thoughts

To determine the geometry of the complexes, first figure out the coordination number.

a) $[Fe(CN)_6]^{4-}$ has a coordination number of 6.
b) $[NiCl_4]^{2-}$ has a coordination number of 4.
c) $[Fe(C_2O_4)_3]^{3-}$ has a coordination number of 6 ($C_2O_4^{2-}$ is a bidentate ligand).

Solution

a) octahedral
b) square planar (Ni^{2+} has eight *d*-orbital electrons)
c) octahedral

20.5   Isomers

Like organic molecules, coordination compounds can form isomers. Isomers have the same formula but different properties. **Figure 20.11** in your text illustrates how isomers of coordinated compounds are divided and subdivided. In summary:

- **Linkage isomers:** The ligand binds through different electron donor atoms.
- **Ionization isomer:** The ligand and a counter ion exchange roles (are interchanged with each other).
- **Coordination sphere isomers:** Different ligands are in coordination spheres of cations and anions.
- **Geometric isomers:** All of the atoms are attached with the same connectivity but the geometric orientation differs.

Table 20.2 in your text illustrates each type of isomer.

20.6   Formulas and Names

Coordination compounds can be named using the IUPAC rules. **Table 20.3** in your text lists the rules for **writing** the formulas of coordination compounds. **Table 20.4** lists the rules for **naming** coordination compounds. Study both tables carefully.

Example 4   **Writing the Formulas of Coordination Compounds**

Write the formula for each of the following coordination compounds.

a) pentaamminebromocobalt(III) chloride
b) sodium hexacyanoferrate(III)

First Thoughts

Use Tables 20.1 and 20.3 in your text to help write the formulas in the proper order and to determine the ligands.

Solution

a) The metal is written first, followed by the ligands.

> The metal is cobalt with a +3 oxidation state (Co^{3+}).
>
> ammine = NH_3; penta = 5 (so five ammonia ligands are present)
>
> bromo = Br (which has a −1 oxidation state)
>
> Therefore, the complex ion is
>
> $$Co^{3+} \qquad (NH_3)_5 \qquad Br^- \quad \Rightarrow \quad [Co(NH_3)_5Br]^{2+}$$
>
> "Chloride" stands for chlorine with a −1 oxidation state (Cl^-). To balance the 2+ charge on the complex ion, two chlorine ions are needed.
>
> The formula for the coordination compound is
>
> $$[Co(NH_3)_5Br]Cl_2$$

b) Sodium has a +1 oxidation state (Na$^+$).

The metal in the complex is iron with a +3 oxidation state (Fe^{3+}).

cyano = CN$^-$; hexa = 6 (so six cyanide ligands are present)

The complex ion is

$$Fe^{3+} \qquad (CN^-)_6 \qquad \Rightarrow \qquad [Fe(CN)_6]^{3-}$$

To balance the 3– charge on the complex ion, three sodium ions are needed.

The formula for the coordination compound is

$$Na_3[Fe(CN)_6]$$

Example 5   Naming Coordination Compounds

Name the following coordination compounds.

a) [Ni(NH$_3$)$_3$I]Cl
b) Li$_3$[CoF$_6$]
c) [Fe(en)$_2$(OH)$_2$]$_2$CO$_3$

First Thoughts

Use Tables 20.1 and 20.4 in your text to help you name the ligands and compounds. In addition, to determine the oxidation state of the metal ion, figure out the charges on all ligands and counter ions.

Solution

a) The oxidation state of Cl is –1, so the charge on the complex ion must be 1+ to balance out the charges. Looking within the complex ion, NH$_3$ is neutral and the oxidation state of I is –1; therefore the oxidation state of Ni must be +2 (to make the overall charge of the complex ion 1+).

Now that we have determined the oxidation states, let's name the compound.

Name the ligands first in alphabetical order and include prefixes when necessary.

triammineiodo

Next add the name of the metal with its charge.

triammineiodonickel(II)

Then add the chloride anion.

triammineiodonickel(II) chloride

b) There are three lithium ions with an oxidation state of +1, so the overall charge is 3+. Therefore, the charge on the complex ion must be 3–. Looking within the complex ion, the oxidation state of F is –1; because six of these ions are present, the overall charge is 6–. As a result, the oxidation state of cobalt must be +3 (to make the overall charge of the complex ion 3–).

+3 – 6 = –3 (overall charge of complex ion)

The cation is not a complex ion so it is named first.

lithium

We now focus on the complex ion, which has one type of ligand.

lithium hexafluoro

Finally we add in the metal with its charge. Because the complex is negatively charged, we add -*ate* to the end of cobalt's name.

lithium hexafluorocobaltate(III)

c) The oxidation state of CO_3 is −2 (carbonate ion), so the overall charge of the complex ion must be 2+. Looking within the complex ion, ethylenediamine (en) is neutral and the oxidation state of OH is −1. Because two OH⁻ ions are present, the overall charge becomes 2−. As a result, the oxidation state of Fe must be +3, which gives an overall charge of 1+. Because there are two of these complex ions, the charge becomes 2+ (and thus balances CO_3^{2-}).

Name the ligands first in alphabetical order and include prefixes when necessary.

bisethylenediaminedihydroxo

Next add the name of the metal with its charge.

bisethylenediaminedihydroxoiron(III)

Then add the carbonate anion.

bisethylenediaminedihydroxoiron(III) carbonate

20.7   Color and Coordination Compounds

Compounds containing transition metals tend to have colors. In most cases, this color is attributable to partially filled *d* orbitals, which are arranged such that they have a slight difference in energy. The introduction of a photon of energy can cause an electron in the lower-energy orbitals to jump to an empty (or partially filled) higher-energy *d* orbital. When the photon is absorbed, a certain wavelength of light is absorbed, and the color of the substance is determined by the wavelengths of light that remain.

Crystal field theory helps us better understand the nature of the *d*-orbital splitting when the electrons are excited. Distortions in the *d* orbitals of a transition metal occur when negatively charged anions get close to it.

In an octahedral complex, the *d*-orbital energies of the central metal are split into a lower-energy set and a higher-energy set, as a consequence of how the ligands approach the individual *d* orbitals with different orientations. See **Figure 20.23** in your text for a visualization of this concept. The following diagram depicts the crystal field splitting of the metal *d* orbitals with an octahedral geometry:

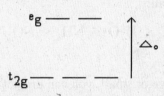

The energy difference between the t_{2g} orbitals and e_g orbitals is called the **crystal field splitting energy,** symbolized as Δ_o. The first three *d*-orbital electrons will fill the t_{2g} orbitals first and spread out between the three orientations (Hund's rule). After that, additional electrons could pair up and fill the remaining spots in the t_{2g} orbitals first before moving up to the e_g orbitals. This **low-spin** configuration has a large splitting energy (Δ_o).

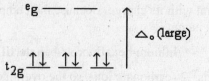

Alternatively, the additional electrons could move up to the e_g orbitals first before pairing up with the original electrons in the t_{2g} orbitals. This **high-spin** configuration has a smaller splitting energy.

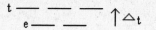

In a tetrahedral complex, the diagram of the crystal field splitting looks as follows:

The splitting energy of a tetrahedral complex is smaller than that of an octahedral complex (about 4/9 the size). High-spin configurations are almost always the only possibility for tetrahedral complexes.

In a square planar complex, the diagram of the crystal field splitting has the following appearance:

$$\uparrow \Delta_o$$

Note that the energy split between the last two d orbitals is the same as that for an octahedral complex. Low-spin configurations are almost always observed for square planar complexes.

The size of the energy splitting (Δ) depends on the geometry of the complex (as discussed earlier) and the nature of the ligands around the metal center. We use the **spectrochemical series** to help predict whether Δ will be large or small:

$$Cl^- < F^- < OH^- \leq H_2O < NH_3 < NO_2^- < CN^- < CO$$

(small Δ)   (large Δ)

Once you know whether Δ is large or small, you can predict whether the wavelength (λ) of light produced will be long or short using the equation from Chapter 6:

$$\Delta E = \frac{hc}{\lambda}$$

Example 6   Crystal Field Diagram I

Draw a crystal field diagram including the electrons for $[Fe(CN)_6]^{4-}$.

First Thoughts

The complex ion has a coordination number of 6, so we would expect an octahedral crystal field split. Looking at the spectrochemical series, we find that CN^- will produce a large Δ, resulting in a low-spin configuration. The oxidation state of Fe in this complex is +2. Fe^{2+} has a d^6 configuration.

Solution

The crystal field diagram follows:

$$e_g \ \underline{\ \ } \ \underline{\ \ }$$

Energy ↑ Δ_o (large) ↑

$$t_{2g} \ \underline{\uparrow\downarrow} \ \underline{\uparrow\downarrow} \ \underline{\uparrow\downarrow}$$

Further Insight

There are no unpaired electrons in this configuration, so the complex is diamagnetic.

Example 7   **Crystal-Field Diagram II**

Predict the number of unpaired electrons in $[NiCl_4]^{2-}$. Is this complex paramagnetic or diamagnetic?

First Thoughts

You must draw a crystal field diagram to determine the number of unpaired electrons. The complex ion has a coordination number of 4, so we would expect a square planar crystal field split. The oxidation state of nickel in this complex is +2, verifying that Ni^{2+} has a d^8 configuration. Low-spin configurations are expected for square planar complexes.

Solution

$$\underline{\ \ }$$

Energy ↑ $\underline{\uparrow\downarrow}$

$$\underline{\uparrow\downarrow}$$

$$\underline{\uparrow\downarrow} \ \underline{\uparrow\downarrow}$$

$[NiCl_4]^{2-}$ has no unpaired electrons and is diamagnetic.

Further Insight

Not all complexes with a coordination number of 4 are low spin. Remember, tetrahedral complexes like $[CoCl_4]^{2-}$ has a high-spin configuration. How many unpaired electrons are in this complex? You should get three unpaired electrons, which means this complex is paramagnetic.

20.8   Chemical Reactions

Transition metal coordination complexes also participate in chemical reactions. Two interesting types of reactions are described here:

- **Ligand exchange reaction.** This reaction involves exchanging the ligands in a complex ion for other ligands. Here is an example from the text:

$$[Ni(H_2O)_6]^{2+} + 6NH_3 \rightarrow [Ni(NH_3)_6]^{2+} + 6H_2O$$

Some coordination complexes exchange their ligands very fast; they are **labile**. Coordination complexes that exchange their ligands more slowly are known as **inert**.

- **Electron transfer reaction.** Because transition metals can have more than one oxidation state, they can undergo reactions that involve transferring electrons and thus changing their oxidation state within the complex.

Exercises

Section 20.1

1. What is a coordination complex? Give an example.

2. What type of bonding occurs between a ligand and a metal within the metal complex?

Section 20.2

3. What is a ligand?

4. Which of the following statements is *true* concerning ligands?
    a) Two examples of common ligands are Fe^{3+} and Ni^{2+}.
    b) Ligands always ionically bond with a metal in a metal complex.
    c) Ligands donate a lone pair of electrons to a metal in a metal complex.
    d) Ligands can use only one nonbonded pair of electrons to bond to a metal center in a complex.
    e) A bidentate ligand forms three bonds to the metal in a complex.

5. Which of the following species could act as a ligand?
    a) OH^-
    b) H_2
    c) Na^+

6. Which of the following species could act as a ligand?
    a) cuprate
    b) oxalate
    c) stannate

Section 20.3

7. What is the coordination number in a complex?

8. What are the two most common coordination numbers?

9. What is the coordination number for the complex $[Pt(NH_3)_4]^{4+}$?

10. What is the coordination number for the complex $[Co(NH_3)_6]^{3+}$?

11. What is the coordination number for the complex $[PtCl_4]^{2-}$?

12. What is the coordination number for the complex $[Ag(H_2O)_2]^+$?

13. What is the coordination number for the complex $[Cr(CN)_4]^{2-}$?

14. What is the coordination number for the complex $[Fe(en)_3]^{3+}$?

15. A complex has a coordination number of 6 and consists of iron with a +3 oxidation state and thiocyanate ligands. What is the formula of this complex?

Section 20.4

16. Match each of the following coordination numbers with its corresponding geometry (there may be more than one answer).
    1. 6   a) linear
    2. 4   b) tetrahedral
    3. 2   c) square planar
                           d) octahedral

17. Predict the geometry for the complex in Exercise 9.

18. Predict the geometry for the complex in Exercise 10.

19. Predict the geometry for the complex in Exercise 11.

20. Predict the geometry for the complex in Exercise 12.

21. Predict the geometry for the complex in Exercise 13.

22. Predict the geometry for the complex in Exercise 14.

23. Predict the geometry for the complex in Exercise 15.

Section 20.5

24. Which type of isomer exchanges the ligand and counter ion?

25. Which type of isomer has different ligands in coordination spheres of cations and anions?

26. Which type of isomer has the ligand binding through different electron donor atoms?

27. Which type of isomer has all of the atoms attached with the same connectivity but differing geometric orientations?

28. Draw geometric isomers of the complex $[Pt(NH_3)_4Br_2]^{2+}$.

29. Draw geometric isomers of the complex $[Co(NH_3)_3I_3]$.

30. Is SCN^- capable of linkage isomerism? Explain your answer.

Section 20.6

31. Write the formula for sodium tetrabromocobaltate(II).

32. Write the formula for triamminecyanoplatinum(II) bromide.

33. Write the formula for potassium dihydroxodioxalatoferrate(III).

34. What is the name of the coordination compound $Na_3[CoBr_6]$?

35. What is the name of the coordination compound $[Fe(NH_3)_5NO_2]Cl_2$?

36. What is the name of the coordination compound $[Cu(H_2O)_6]SO_4$?

37. What is the name of the coordination compound $[Co(en)_2(SCN)_2]NO_3$?

38. What is the name of the coordination compound $Li_3[Fe(C_2O_4)_3]$?

Section 20.7

39. Which of the following statements is *false* concerning crystal field theory?
    a) Distortions in the *d* orbitals of a transition metal occur when negatively charged anions get close to it.
    b) The splitting energy of a tetrahedral complex is larger than the splitting energy in an octahedral complex.
    c) Low-spin electron configurations are almost always observed for square planar complexes.
    d) The *d*-orbital energies of the central metal in a complex split into lower-energy sets and higher-energy sets for all geometries.
    e) High-spin electron configurations are almost always the only possibility for tetrahedral complexes.

40. What is the difference between a low-spin configuration and a high-spin configuration?

41. Draw a crystal field diagram including the electrons for $[Pt(NH_3)_4]^{4+}$. How many unpaired electrons are present?

42. Draw a crystal field diagram including the electrons for $[PtCl_4]^{2-}$. How many unpaired electrons are present?

43. Draw a crystal field diagram including the electrons for $[Cr(NO_2)_6]^{4-}$. How many unpaired electrons are present? Is the complex paramagnetic or diamagnetic?

44. Draw a crystal field diagram including the electrons for $[CoBr_6]^{4-}$. How many unpaired electrons are present? Is the complex paramagnetic or diamagnetic?

Use the following crystal field diagram to answer Exercises 45–48.

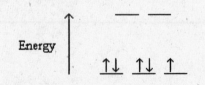

45. Is the complex paramagnetic or diamagnetic?

46. Does the diagram depict a low-spin or high-spin case? Justify your answer.

47. What is the geometry of the complex ion?
    a) tetrahedral
    b) square planar
    c) octahedral

48. Which of the following is *most likely* the complex ion?
   a) $[Fe(en)_3]^{3+}$
   b) $[Fe(CN)_4]^{2-}$
   c) $[Mn(NO_2)_4]^{2-}$
   d) $[MnI_6]^{4-}$
   e) $[Ag(H_2O)_2]^+$

Section 20.8

49. What are two common chemical reactions involving transition metal coordination complexes?

50. What is the difference between a labile coordination complex and an inert coordination complex?

Answers to Exercises

1. A coordination complex consists of a central atom or ion that is bonded to several other components. An example of a complex ion is $[Ni(H_2O)_6]^{2+}$.

2. coordinate covalent bonding

3. A ligand is a Lewis base. Ligands can coordinate to the metal ion in a coordination complex.

4. c

5. a

6. b

7. The coordination number is the number of donor atoms to which a given metal ion bonds in a complex.

8. 4 and 6

9. 4

10. 6

11. 4

12. 2

13. 4

14. 6

15. $[Fe(SCN)_6]^{3-}$

16. 1. d; 2. b or c; 3. a

17. tetrahedral

18. octahedral

19. square planar

20. linear

21. tetrahedral

22. octahedral

23. octahedral

24. ionization isomer

25. coordination sphere isomer

26. linkage isomer

27. geometric isomer

28.

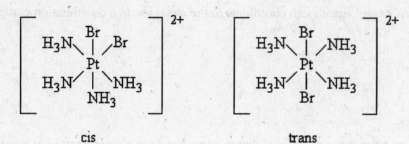

cis   trans

29.

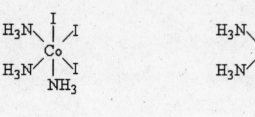

cis   trans

30. Yes, SCN⁻ can form linkage isomerism because it can bond to the metal ion in two different ways (through the S or N atoms).

31. $Na_2[CoBr_4]$

32. $[Pt(NH_3)_3CN]Br$

33. $K_3[Fe(OH)_2(C_2O_4)_2]$

34. sodium hexabromocobaltate(III)

35. pentaamminenitroiron(III) chloride

36. hexaaquacopper(II) sulfate

37. bisethylenediaminedithiocyanocobalt(III) nitrate

38. lithium trioxalatoferrate(III)

39. b

40. A low-spin configuration yields the minimum number of unpaired electrons in a crystal field diagram. A high-spin configuration gives the maximum number of unpaired electrons in a crystal field diagram.

41. There are four unpaired electrons.

42. There are no unpaired electrons.

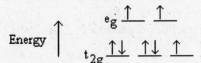

43. There are two unpaired electrons. The complex is paramagnetic.

44. There are three unpaired electrons. The complex is paramagnetic.

45. paramagnetic

46. The diagram is a low-spin case because the electrons are paired up instead of being spread out between all of the *d* orbitals (yielding the minimum number of unpaired electrons).

47. c

48. a

49. ligand exchange reaction, electron transfer reaction

50. Labile complexes exchange their ligands very fast in a ligand exchange reaction; inert complexes exchange their ligands more slowly.

Chapter 21

Nuclear Chemistry

The Bottom Line

When you finish this chapter, you will be able to:

- Distinguish between the different types of radioactive decay.
- Balance nuclear reactions.
- Calculate decay constants and half-lives for nuclear reactions.
- Calculate mass defect and binding energy for nuclear reactions.
- Understand the difference between nuclear fission and fusion.
- Appreciate the medical uses of radioisotopes.

21.1 Isotopes and More Isotopes

Recall from Chapter 2 that we can represent an isotope of an element by **nuclide notation:**

$$^{A}_{Z}X$$

where A is the mass number = number of protons + number of neutrons, and Z is the atomic number = number of protons. For example, one isotope of carbon is

$$^{13}_{6}C$$

This isotope has 6 protons and 7 neutrons in the nucleus. We can further simplify this notation by writing just ^{13}C, as the symbol C for carbon indicates that it has an atomic number of 6.

Example 1   **Decoding Isotopes**

How many protons and neutrons does each of the following nuclides contain: ^{31}P, ^{198}Au, ^{238}U?

Solution

Looking in a periodic table, we find: ^{31}P has $Z = 15$ protons and $31 - 15 = 16$ neutrons. ^{198}Au has $Z = 79$ protons and $198 - 79 = 119$ neutrons. ^{238}U has $Z = 92$ protons and $238 - 92 = 146$ neutrons.

21.2 Types of Radioactive Decay

Radioactive decay is a process by which a nucleus disintegrates to form another nucleus and gives off radiation as particles and energy. An important fact about nuclear reactions is that the sum of the superscripts on both sides of the equation is equal. The same is true for the subscripts. For example,

$$^{238}_{92}U \rightarrow \ ^{234}_{90}Th + \ ^{4}_{2}He$$

All elements having $Z > 83$ are radioactive and will decay in some way, forming other nuclides. The textbook discusses five types of radioactive decay, which are summarized in **Table 21.2.**

- **Alpha-particle emission:** results in the release of an $^{4}_{2}He$

$$^{212}_{84}Po \rightarrow \ ^{208}_{82}Pb + \ ^{4}_{2}He$$

- **Beta-particle emission:** results in the release of an electron, $_{-1}^{0}\beta$

$$_{6}^{14}C \rightarrow \ _{7}^{14}N + \ _{-1}^{0}\beta$$

- **Gamma ray emission:** results in the release of high-energy radiation $\left(_{0}^{0}\gamma\right)$ that accompanies nuclear reactions

$$_{84}^{212}Po \rightarrow \ _{82}^{208}Pb + \ _{2}^{4}He + \ _{0}^{0}\gamma$$

- **Positron emission:** results in the release of a positive electron, $_{1}^{0}\beta$

$$_{43}^{95}Tc \rightarrow \ _{42}^{95}Mo + \ _{1}^{0}\beta$$

- **Electron capture (EC):** occurs when an inner-orbital electron and a proton from the nucleus form a neutron

$$_{19}^{49}K + \ _{-1}^{0}\beta \rightarrow \ _{18}^{49}Ar$$

Example 2 Radioactive Decay

Write nuclear equations for each of the following events:

a) $_{20}^{47}Ca$ emits a beta particle.

b) $_{13}^{25}Al$ emits a positron.

c) $_{95}^{241}Am$ emits an alpha particle.

d) $_{37}^{81}Rb$ undergoes electron capture.

Solution

a) $_{20}^{47}Ca \rightarrow \ _{-1}^{0}\beta + \ _{19}^{47}K$

b) $_{13}^{25}Al \rightarrow \ _{-1}^{0}\beta + \ _{14}^{25}Si$ (balanced)

c) $_{95}^{241}Am \rightarrow \ _{2}^{4}He + \ _{93}^{237}Np$

d) $_{37}^{81}Rb + \ _{-1}^{0}\beta \rightarrow \ _{36}^{81}Kr$

A radioactive **decay series** is a sequence of nuclear reactions in which one radioactive nucleus decays to a second, which decays to a third, and so on, until a stable lead nucleus is reached. Your textbook illustrates this process with uranium nuclides, one of the naturally occurring radioactive decay series (see **Figure 21.5**).

21.3 Interaction of Radiation with Matter

Alpha, beta, and gamma radiation differ in their ability to penetrate matter and produce effects on the molecular level (see **Table 21.3 and Figure 21.5** in the textbook). All three are known as ionizing radiation—that is, radiation that is capable of producing ions by knocking out electrons from atoms. Ionizing radiation can cause serious health effects, leading to cancer and death, depending on the amount and length of exposure. The mechanism of the biological damage involves the formation of free radicals that damage DNA (see **Figure 21.6** in the textbook). The **rem** is the unit of measure for estimating biological damage. **Table 21.4** lists other units used for measuring radiation. Health effects of radiation exposure are listed in **Table 21.5**.

21.4   The Kinetics of Radioactive Decay

Radioactive decay follows first-order kinetics (see Chapter 15). Consequently, the half-life of a radioactive isotope is given by

$$t_{1/2} = \frac{0.693}{k}$$

The equation relates the amount of substance remaining, $[A_t]$; at any time, t, to the rate constant, k, and the initial amount of substance, $[A_i]$:

$$\ln \frac{[A_t]}{[A_i]} = -kt$$

Example 3   Half-Life Calculations I

Technetium is commonly used as part of stress tests. It has a half-life of 6.0 hours. How long would it take for 90% of a Tc dose to decay?

Solution

If 90% decays, then 10% remains at time t. The half-life allows us to calculate the rate constant.

$$\ln \frac{[A_t]}{[A_i]} = -kt \text{ and } k = \frac{0.693}{t_{1/2}}$$

$$\text{Substituting, } \ln \frac{[A_t]}{[A_i]} = -\left(\frac{0.693}{t_{1/2}}\right)t$$

$$\ln \frac{0.10}{1.00} = -\left(\frac{0.693}{6.0 \text{ h}}\right)t$$

$$-2.3 = -0.12\, t$$

$$t = 20.\, \text{h}$$

Example 4   Half-Life Calculations II

Phosphorus-32 is radioactive and has a half-life of 14.3 days. Suppose a sample was used 6.5 days after it was prepared. What fraction of the originally prepared isotope remains at that point?

Solution

If A_i is the original number of ^{32}P nuclei, and A_t is the number after 6.5 d, then the fraction remaining is A_t/A_i. As in Example 3, we can substitute the half-life for k and use the following equation:

$$\ln\frac{[A_t]}{[A_i]} = -\left(\frac{0.693}{t_{1/2}}\right)t$$

$$\ln\frac{[A_t]}{[A_i]} = -\left(\frac{0.693}{14.3\,d}\right)(6.5\,d)$$

$$= -0.315$$

$$\frac{[A_t]}{[A_i]} = 0.73 \text{ or } 73\%$$

21.5   Mass and Binding Energy

The **mass defect** is the mass difference between the individual protons and neutrons and the composite nucleus. The **binding energy** is the energy required to dismantle the nucleus into its individual protons and neutrons. The binding energy is always expressed as a positive number.

If we calculate the binding energy per nucleon (proton or neutron) and then plot it versus the number of nucleons in the nucleus, the resulting curve shows the relative stability of the nucleus (see **Figure 21.9** in the textbook). The maximum stability is reached at the ^{56}Fe nucleus.

Example 5   Mass Defect and Binding Energy

Calculate the mass defect and the binding energy for $^{19}_9F$ (nuclear mass = 18.9984 g/mol). The mass of the proton is 1.00727 g/mol, and the mass of the neutron is 1.008665 g/mol.

Solution

The mass defect is calculated by summing up the masses of the protons and neutrons in the nucleus and then subtracting the mass of the isotope (which is always smaller).

Protons: (9 mol)(1.00727 g/mol) = 9.06543 g

Neutrons: (10 mol)(1.008665 g/mol) = 10.08665 g

Total mass = 9.06543 g + 10.08665 g = 19.15208 g

Mass defect (Δm) = 19.15208 − 18.9984 = 0.1537 g

The binding energy is then

$$\Delta E = |\Delta m|c^2 = (0.1537 \times 10^{-3}\,kg)(2.9979 \times 10^8\,m/s)^2 = 1.381 \times 10^{13}\,J$$

Note: The mass must be converted to kilograms so that the answer will be in joules (= kg m^2/s^2).

Example 6   The Energy of Nuclear Reactions

Calculate the energy change for the nuclear reaction

$$^{238}_{92}U \rightarrow \, ^{234}_{90}Th + \, ^4_2He$$

given the following masses: ^{238}U, 238.0003 g/mol; ^{234}Th, 233.9942 g/mol; ^4He, 4.0015 g/mol.

Solution

First, calculate Δm for the reaction. Then, use the Einstein equation to calculate the energy change.

$$\Delta m = [233.9942 \text{ g/mol} + 4.0015 \text{ g/mol}] - 238.0003 \text{ g/mol} = -0.0046 \text{ g/mol} = -4.6 \times 10^{-6} \text{ kg/mol}$$

$$\Delta E = |\Delta m| c^2 = (4.6 \times 10^{-6} \text{ kg})(2.9979 \times 10^8 \text{ m/s})^2 = 4.1 \times 10^{11} \text{ J/mol}$$

21.6 Nuclear Stability and Human-Made Radioactive Nuclides

The determining factor as to whether a nucleus is stable is the neutron-to-proton ratio (n/p). **Figure 21.10** shows a band of stable nuclei. Stable atoms of elements with a low atomic number have an n/p ratio of about 1. As the atomic number increases, the n/p ratio increases, which seems to support the idea that more neutrons are needed to dilute the repulsive forces of the many protons.

The following observations are useful in deciding whether a nuclei is stable:

- Nuclei that contain **magic numbers** 2, 8, 20, 50, 82, or 126 are more stable.

- Nuclei with even numbers of both protons and neutrons are more stable than those with odd numbers.

- All isotopes of elements with $Z > 83$ are radioactive (as are all isotopes of $_{43}$Tc and $_{61}$Pm).

Example 7   **Nuclear Stability**

Predict whether each of the following isotopes is stable or radioactive.

a) ^{40}Ca

b) ^{92}Tc

c) ^{242}Cm

Solution

a) Ca ($Z = 20$) has 20 protons and 20 neutrons. It has double magic numbers, so it is stable.

b) Tc ($Z = 43$) has 43 protons and 49 neutrons, so it is radioactive.

c) Cm ($Z = 96$) is radioactive; $Z > 83$.

Elements beyond uranium in the periodic table have been produced using bombardment reactions. Consider the bombardment of ^{14}N by an alpha particle:

$$^{14}_{7}\text{N} + ^{4}_{2}\text{He} \rightarrow ^{17}_{8}\text{O} + ^{1}_{1}\text{p}$$

These reactions involve bombarding a nucleus with a smaller particle or smaller nucleus to make a larger nucleus. The energy needed to accelerate these particles is generated by linear accelerators or cyclotrons. To date, elements with atomic numbers up to $Z = 118$ have been created in the laboratory.

21.7 Splitting the Atom: Nuclear Fission

Nuclear fission is the process in which a heavy nucleus ($A > 200$) divides into smaller nuclei of intermediate mass and generates one or more neutrons. For example,

$$^{235}_{92}\text{U} + ^{1}_{0}\text{n} \rightarrow ^{90}_{35}\text{Br} + ^{143}_{57}\text{La} + 3\ ^{1}_{0}\text{n}$$

Nuclear fusion involves the combination of small nuclei into larger ones. One of the fusion reactions that occurs on the Sun is

$$\,_1^1H + \,_1^2H \rightarrow \,_2^3He$$

Both nuclear fission and nuclear fusion release tremendous amounts of energy. Fission reactions of ^{235}U and ^{239}Pu are used to power nuclear reactors, thereby generating electricity.

Here are some important facts about nuclear fission:

- Fission reactions release large amounts of energy and generate product nuclei that are smaller than the starting nuclei.

- Fission typically needs neutrons with sufficient energy to start the reaction.

- Once the fission reaction begins, it releases more neutrons that may initiate other fission reactions, creating a **chain reaction.** A chain reaction will occur only if a **critical mass** of fissionable material is present.

- Nuclei can split in more than one way, leading to the formation of many fission products that are usually radioactive. This process generates radioactive nuclear wastes that must be disposed of.

21.8   Medical Uses of Radioactive Isotopes

Radioactive isotopes are now used routinely in medicine for diagnosis and imaging.

- **Radiopharmaceuticals** are radioactive tracers that are used in imaging organs. ^{131}I and ^{99}Tc are examples of these isotopes.

- **Positron emission tomography (PET)** scanning produces images of chemical processes in action. Scans are made of a patient who has been transfused with radioactively labeled chemicals that are taken up by the organ of interest.

Exercises

Section 21.1

1. What is nuclide notation?

2. True or false? A is the symbol for the mass number, which is equal to the atomic mass of the element.

For Exercises 3–5, indicate the number of protons and neutrons in each nuclide.

3. $^{55}_{25}\text{Mn}$

4. ^{201}Hg

5. cobalt-60

6. Give the symbol for the electron in a nuclear reaction.

7. What is the symbol for an alpha particle?

Section 21.2

8. How do nuclear reactions differ from chemical reactions?

9. What are the steps in balancing nuclear reactions?

10. What is the difference between an electron and a positron?

For Exercises 11–15, identify X in each nuclear reaction.

11. $^{26}_{12}\text{Mg} + ^{1}_{1}\text{H} \rightarrow \text{X} + ^{23}_{11}\text{Na}$

12. $^{40}_{19}\text{K} \rightarrow ^{40}_{18}\text{Ar} + \text{X}$

13. $^{135}_{53}\text{I} \rightarrow ^{135}_{54}\text{Xe} + \text{X}$

14. $^{59}_{27}\text{Co} + ^{0}_{1}\text{n} \rightarrow \text{X} + ^{4}_{2}\text{He}$

15. $\text{X} \rightarrow ^{187}_{76}\text{Os} + ^{0}_{-1}\beta$

16. What type of radiation accompanies all radioactive decay reactions?

17. The process of an inner-orbital electron and a proton forming a neutron is known as
_____.

18. Is there a difference between $^{0}_{-1}\text{e}$ and $^{0}_{-1}\beta$?

19. Write a reaction equation for alpha-particle emission from a plutonium-242 nucleus.

20. Write a reaction equation for potassium-38 undergoing positron emission.

21. Write a reaction equation for bismuth-214 undergoing beta decay.

22. Write a reaction equation for gold-195 undergoing electron capture.

Section 21.3

23. Which radiation resulting from radioactive decay is least penetrating? Which is most penetrating?

24. What is ionizing radiation? How does ionizing radiation cause tissue damage?

25. True or false? Microwaves are a type of ionizing radiation.

26. What two factors determine the extent of tissue damage from exposure to radiation?

27. What is a rem?

28. What is the ultimate consequence of exposure to large amounts of radiation?

Section 21.4

29. Tl-206 decays to Pb-206 with a half-life of 4.20 min. What is the rate constant for this decay?

30. The rate constant for the decay of strontium-90 is 2.47×10^{-2} yr. What is the half-life, $t_{1/2}$, for Sr-90?

31. The half-life, $t_{1/2}$, for tritium, ^3H, is 12.3 yr. If 1.00 g were prepared today, how long would it take for the tritium to decay to 0.125 g?

32. Given the information for tritium in Exercise 31, how long would it take for 95% of the tritium to decay?

33. A 250-mg sample of Co-60 (half-life = 5.26 yr) was prepared 12 years ago. How much of the sample remains?

34. An old bottle of scotch was analyzed and found to have a tritium content that was 21% of the tritium content of water in the area where it was made. The half-life of tritium is 12.3 yr. Estimate the age of the scotch.

35. How much time is required for a 6.50-mg sample of chromium-51 to decay to 2.25 mg, if it has a half-life of 27.8 days?

Section 21.5

For Exercises 36–40, use the following information: $m_{proton} = 1.00727$ g/mol, $m_{neutron} = 1.008665$ g/mol, and $c = 2.9979 \times 10^8$ m/s.

36. Calculate the binding energy for ^{12}C (nuclear mass = 11.996708 g/mol).

37. What is the binding energy of Ba-137 (nuclear mass = 136.905812 g/mol)?

38. The binding energy per nucleon is simply the binding energy for the nuclide divided by the total number of protons and neutrons (nucleons) in the nucleus. What is the binding energy per nucleon for ^{201}Hg (nuclear mass = 201.970277 g/mol)?

39. Is ^{201}Hg more or less stable than ^{56}Fe? (*Hint:* Use Figure 21.10 in the text.)

40. Given the following masses (given in g/mol): H, 2.01410; ^{3}H, 3.01605; ^{4}He, 4.00260; calculate the energy change for $^2_1H + ^3_1H \rightarrow ^4_2He + ^1_0n$.

Section 21.6

41. What are "magic numbers," and what do they indicate?

42. Nuclides with double magic numbers are (as stable, more stable, less stable) than nuclides with one set of magic numbers.

43. Why do heavier nuclides have an n/p ratio greater than 1?

44. Which of the following is more stable: ^{16}O or ^{17}O?

45. Which of the following is radioactive: ^{209}Bi or ^{210}At?

Section 21.7

46. What is the difference between nuclear fission and nuclear fusion?

47. Define chain reaction and critical mass.

48. Which isotopes are used in commercial reactors to generate electricity?

49. What is one of the major concerns about the by-products of fission reactions used to generate electricity?

Section 21.8

50. How are radioactive tracers used to image brain functions?

Answers to Exercises

1. An example of nuclide notation is $^{14}_{6}C$, where the superscript indicates the mass number and the subscript indicates the atomic number.

2. False. A is the symbol for the mass number, which equals the sum of the number of protons and neutrons, not the atomic mass.

3. 25 protons, 30 neutrons

4. 80 protons, 121 neutrons

5. 27 protons, 33 neutrons

6. $^{0}_{-1}\beta$, also called a beta particle

7. $^{4}_{2}He$

8. Nuclear reactions involve converting elements from one to another; chemical reactions rearrange atoms by breaking and making chemical bonds.

9. In nuclear reactions, the sum of the mass numbers and the sum of the atomic numbers must be the same on both the reactant and product sides.

10. An electron has a -1 charge; a positron has $+1$ charge.

11. $^{4}_{2}He$

12. $^{0}_{1}\beta$

13. $^{0}_{-1}\beta$

14. $^{55}_{26}Fe$

15. $^{187}_{75}Re$

16. gamma radiation

17. electron capture

18. There is no difference. An electron is the same as a beta particle.

19. $^{242}_{94}Pu \rightarrow ^{4}_{2}He + ^{238}_{92}U$

20. $^{38}_{19}K \rightarrow ^{0}_{1}\beta + ^{38}_{18}Ar$

21. $^{214}_{83}Bi \rightarrow \, ^{0}_{-1}\beta + \, ^{214}_{84}Po$ (corrected)

22. $^{195}_{79}Au + \, ^{0}_{-1}\beta \rightarrow \, ^{195}_{78}Pt$

23. Alpha radiation is least penetrating. Gamma radiation is most penetrating.

24. Alpha, beta, and gamma radiation is ionizing radiation, which is capable of forming ions by knocking electrons out of atoms. It can cause biological damage by forming free radicals that damage DNA.

25. False. Microwaves are electromagnetic radiation that causes ionization of atoms.

26. The extent of damage depends on the amount of exposure to radiation and the length of time that the exposure occurs.

27. A rem is a unit of radiation dose that is related to biological damage.

28. ultimately death

29. 0.165 min^{-1}

30. 28.1 yr

31. 36.9 yr

32. 53 yr

33. 51.4 mg

34. 27.7 yr old

35. 42.6 days

36. 8.889×10^{12} J

37. 1.081×10^{14} J

38. 2.95×10^{11} J/nucleon

39. ^{201}Hg is less stable. ^{56}Fe is at the maximum of the curve (it has the highest stability).

40. 1.699×10^{12} J/mol

41. The magic numbers are 2, 8, 20, 50, 82, and 126. If nuclides have atomic numbers or mass numbers that are a magic number, then they are stable isotopes.

42. more stable

43. Neutrons in heavier nuclides dilute the repulsive forces of the many protons present in the nucleus.

44. ^{16}O; it has an even number of protons and neutrons and a magic number of protons.

45. ^{210}At is radioactive. Isotopes for which $Z > 83$ are radioactive.

46. Nuclear fission is a process where a heavy nucleus divides into small nuclei. Nuclear fusion is the process of combining small nuclei into larger ones.

47. A chain reaction is a self-sustaining fission reaction. Critical mass refers to the amount of fissionable material that must be present to maintain a chain reaction.

48. ^{235}U and ^{239}Pu

49. The products of fission reactions are all highly radioactive. Storage and disposal of these products are primary concerns of the nuclear reactor industry.

50. Glucose labeled with radioactive carbon is administered to a patient in a PET scanner. As the glucose is metabolized, it reaches the brain, allowing brain function to be recorded by the scanner.

Chapter 22

The Chemistry of Life

The Bottom Line

When you finish this chapter, you will be able to:

- Describe the characteristics of DNA, and determine the sequence of a new strand of DNA from a template strand.
- Compare and contrast genes and proteins, and describe how they function in living things.
- Describe the roles that enzymes play in living things.
- Classify the reactions that enzymes catalyze into six categories.
- Differentiate between carbohydrates and lipids, and describe their roles in living things.
- Appreciate the complexity of life and chemistry's role in creating a better understanding of life.

22.1   DNA: The Basic Structure

DNA is the fundamental basis of all life. DNA (or **deoxyribonucleic acid**) is the "blueprint" for all organisms. Here are the key points about DNA that you should know:

- DNA comprises a series of polymers composed of nucleotides.
- Each nucleotide consists of a phosphate group bonded to a deoxyribose sugar group, which is in turn bonded to one of four nitrogenous organic bases.
- The four nitrogenous bases are adenine, guanine, thymine, and cytosine (shown in **Figure 22.2** in your text). These bases hydrogen-bond with one another between two strands of DNA chains. Adenine always base-pairs with thymine; cytosine always base-pairs with guanine.
- To improve hydrogen bonding, the two DNA strands wind around each other to form a double helix.
- In DNA replication, the existing DNA strands unravel from the helix and determine the sequence of the new DNA strands. Two identical strands of DNA are created: one for the original cell and one for the newly created cell.

Example 1   **DNA Sequence**

What will be the sequence of a new double-helical DNA strand if the strand shown below serves as the template during DNA replication?

TTAGCACGGATACGAA

First Thoughts

The letters in the strand template represent the following bases:

    T: thymine
    A: adenine
    G: guanine
    C: cytosine

Adenine always base-pairs with thymine, and cytosine always base-pairs with guanine.

Solution

Template: TTAGCACGGATACGAA

New sequence: AATCGTGCCTATGCTT

22.2   Proteins

DNA is found in the nucleus of a cell, where it specifies which proteins are made by that cell. **Proteins** are polymers that are created when amino acids undergo condensation reactions. **Figure 22.10** in your text shows some of the important amino acids found in living cells. When these amino acids add together, they form a **polypeptide chain.** The sequence in which the amino acids form the polypeptide chains determines the function of the protein. The base sequence of DNA can specify the amino acid sequence of a protein.

22.3   How Genes Code for Proteins

Here are the main points of this section summarized from the text:

- A **gene** is a section of DNA that "encodes" a specific protein. The base sequence of a gene determines the amino acid sequence of a protein in a process known as **gene expression.**

- Gene expression occurs in two phases: transcription and translation. In **transcription,** an RNA copy of the gene is made (this process is illustrated in **Figure 22.15**). In **translation,** the RNA copy directs the production of the protein (this process is shown in **Figure 22.16**). These two phases are discussed in detail in your text. Make sure you understand them.

- RNA (ribonucleic acid) is just like DNA except it contains ribose instead of deoxyribose and the nitrogenous base uracil instead of thymine. RNA can form base pairs with a complementary strand of DNA.

- Proteins are made by translating the genetic code from messenger RNA (m-RNA). Transfer RNA (t-RNA) supplies the specific amino acids needed to the growing polypeptide chain. The formation of the polypeptide chain occurs at the ribosomes within the cytoplasm of a cell.

- Proteins fold in several different ways:

  Secondary structures are the regions of a polypeptide chain that fold into an α-helix or a β-pleated sheet.

  Tertiary structures are the result of folding the secondary structures within a polypeptide chain into a three-dimensional shape. They can be globular or linear in arrangement.

  Quaternary structures result when two or more polypeptide chains are folded into a specific shape. Not all proteins have quaternary structure.

- Folded proteins can be "unfolded" by **denaturing** them with heat, changing the pH around them, or altering the concentration of salt ions around them.

22.4   Enzymes

Special types of proteins in the body called **enzymes** speed up (that is, catalyze) chemical reactions, causing them to occur much faster. Here are the key points to remember about enzymes:

- Enzymes contain specific sites to which only certain chemicals can bind. One of the binding sites on an enzyme is the **active site;** it is where the **substrate** binds. The enzyme then catalyzes the reaction of the substrate and produces the product, which is released from the enzyme. This process is summarized in **Figure 22.23** in your text.

- The enzymes in the human body must be regulated and controlled. This regulation involves chemical modification of the enzymes. The major modifications are listed in your text. Be sure you understand what each modification entails.

The reactions that enzymes catalyze can be classified into six categories:

1. **Oxidoreductases:** enzymes that catalyze redox reactions
2. **Transferases:** enzymes that catalyze the transfer of groups from one molecule to another
3. **Hydrolases:** enzymes that catalyze hydrolysis reactions
4. **Lyases:** enzymes that catalyze elimination reactions
5. **Isomerases:** enzymes that catalyze the interconversion between isomers
6. **Ligases:** enzymes that catalyze the formation of new bonds linking substrates together

22.5   The Diversity of Protein Functions

Your text summarizes the relationship between genes and proteins perfectly: "Genes hold the instructions while proteins do the work." Proteins perform many tasks. For instance, they can function as

- Enzymes
- Transporters
- "Movers and shakers" (in the form of contractile proteins)
- Scaffolding and structure (in the form of structural proteins)
- Messengers (in the form of hormones such as insulin)
- Receptors
- Gates and pumps
- Controllers (in the form of regulatory proteins)
- Defenders (in the form of antibodies)

22.6   Carbohydrates

Carbohydrates are hydrates of carbon with the general formula $C_x(H_2O)_y$. Here are the key points about carbohydrates:

- The simplest carbohydrates are **sugars.** Triose sugars contain three carbon atoms per molecule, tetrose sugars contain four carbon atoms per molecule, pentose sugars contain five carbon atoms per molecule, and so forth.
- Carbohydrates are energy storage compounds.
- Carbohydrates can also be classified by functional group:

    Aldehyde = aldose (such as glucose)
    Ketone = ketose (such as fructose)

- An aldohexose is a carbohydrate containing six carbons with an aldehyde functional group. Glucose is an aldohexose.
- Sucrose is formed from a condensation reaction that links glucose and fructose together (see **Figure 22.24**).
- Sugars can also be classified as saccharides. Glucose and fructose are **monosaccharides**—they are single sugars. Sucrose is a **disaccharide**—it is made from two monosaccharides. Several monosaccharides linked into a giant polymer are called a **polysaccharide.**

- Some common polysaccharides are listed below:

  Starch: glucose polymer found in plants
  Glycogen: glucose polymer with a different arrangement than starch; found in animals and humans
  Cellulose: glucose polymer that makes plant cell walls

22.7   Lipids

A **lipid** is a compound that is soluble in nonpolar solvents. Lipids include fats, oils, some hormones, and some vitamins. Here are some key points about lipids:

- Lipids are important energy storage molecules, act as intercellular signaling molecules, and perform many structural roles in organisms.
- **Fatty acids** are some of the simplest lipids; they contain one or more C=C double bonds. **Table 22.3** lists some of the common fatty acids.
- A **triglyceride** is a combination of three fatty acids. Fats are triglycerides that are solid at room temperature; **oils** are triglycerides that are liquid at room temperature.
- One of the major classes of derived lipids is **steroids.** Examples of steroids include cholesterol and estrogen.

22.8   The Maelstrom of Metabolism

Figure 22.30 in your text summarizes the major sequences of the chemical reactions of life. Each particular sequence is called a **biochemical pathway.** The entire network of chemical reactions is called **metabolism.** Breaking down the chemicals in food into simpler forms is called **catabolism.** The reverse of this process is **anabolism.**

22.9   Biochemistry and Chirality

Many of the molecules in our body are chiral (see Chapter 12 to review the concept of chirality). The non-superimposable forms of molecules are called **enantiomers.** Chirality is important because it controls the way in which molecules interact. An example of the effect of this property is given in your text for the drug thalidomide.

22.10   A Look to the Future

A major reason for learning chemistry and understanding how it affects the body is to try to prevent and treat diseases. Major causes of illness and disease include infection, abnormal growth of tissues, abnormal production of important biochemicals, genetic diseases, and aging. Some current advances that target these causes include antibiotics, anticancer agents, hormones, gene therapy, and modulating neutrotransmitters.

Exercises

Section 22.1

1. What does DNA stand for?

2. What are the four nitrogenous bases in DNA? How do they base-pair?

3. Which of the following statements about DNA is *false?*
   a) DNA is the "blueprint" for all organisms.
   b) DNA is a series of polymers composed of nucleotides.
   c) To improve hydrogen bonding, two DNA strands interact with each other to form a β-pleated sheet.
   d) Each nucleotide is composed of a phosphate group bonded to a deoxyribose sugar group, bonded to one of four nitrogenous organic bases.
   e) Existing DNA strands determine the sequence of new DNA strands during DNA replication.

4. What will be the sequence of a new double-helical DNA strand if the strand below serves as the template during DNA replication?

   GACTTGCCACACGTAG

5. Consider the following DNA strand template:

   CAGTCAGTTGACCCGA

   What is the sequence of a new DNA strand based on this template?

6. A new DNA strand has the following sequence:

   AATTCCCGTACTAGGC

   What are the missing bases from the original DNA template?

   TT??GGG????ATCC?

Section 22.2

7. Where is DNA located in a cell?

8. _____ are polymers that form when amino acids undergo condensation reactions.

9. True or false? The sequence in which the amino acids form polypeptide chains determines the function of the protein.

10. True or false? The base sequence of DNA can specify the amino acid sequence of a protein.

Section 22.3

11. What is a gene?

12. What are the two phases of gene expression?

13. How is RNA different than DNA?

14. True or false? Proteins are made by translating the genetic code from transfer RNA. Messenger RNA supplies the specific amino acids to the growing polypeptide chain.

15. Which of the following is *not* a way that proteins can fold?
   a) tetrahedral
   b) α-helix
   c) β-pleated sheet
   d) globular
   e) linear

16. What does it mean to "denature" a protein?

Section 22.4

17. What types of proteins in the body speed up chemical reactions?

18. One of the binding sites on an enzyme is the _____ _____, where the substrate binds.

19. What are the six main enzyme-reaction categories?

20. What type of enzymes catalyzes redox reactions?

21. What type of enzymes catalyzes elimination reactions?

22. What type of enzymes catalyzes the formation of new bonds that link substrates together?

Section 22.5

23. List four functions of proteins.

24. Why are contractile proteins important?

25. Why are structural proteins important? Give two examples that are made of structural proteins.

Section 22.6

26. What are carbohydrates?

27. What is the name of a carbohydrate that contains six carbon atoms with a ketone functional group?

28. What is the name of a carbohydrate that contains five carbon atoms with an aldehyde functional group?

29. _____ is formed from a condensation reaction that links glucose and fructose together.

30. What are three common polysaccharides?

Section 22.7

31. A _____ is a compound that is soluble in nonpolar solvents.

32. True or false? Lipids are important energy storage molecules and intercellular signaling molecules.

33. Match the following examples with the appropriate classification.
    1) glucose   a) carbohydrate
    2) fats   b) protein
    3) transferase   c) lipid

34. Match the following examples with the appropriate classification.
    1) oils   a) carbohydrate
    2) sucrose   b) protein
    3) hair   c) lipid

35. Match the following examples with the appropriate classification.
    1) glycine   a) monosaccharide
    2) fructose   b) amino acid
    3) cholesterol   c) steroid

Section 22.8

36. Figure 22.30 in your text summarizes the major sequences of the chemical reactions of life. Each particular sequence is called a _____ _____.

37. What is the difference between catabolism and anabolism?

Section 22.9

38. The non-superimposable forms of molecules in the human body are called _____.

39. Which drug was responsible for birth abnormalities caused by an enantiomer within its racemic mixture?

Section 22.10

40. What are five major causes of illness and disease among humans?

41. _____ is when the human body becomes home to harmful microorganisms.

42. What is one cause of the degeneration of maintenance and repair functions within the body?

43. Penicillin is an example of what type of drug that helps fight illness?

44. What type of hormone must be injected into diabetics to help control their blood sugar?

45. What is gene therapy?

Answers to Exercises

1. deoxyribonucleic acid

2. adenine, guanine, thymine, and cytosine; adenine base-pairs with thymine; cytosine base-pairs with guanine

3. c

4. CTGAACGGTGTGCATC

5. GTCAGTCAACTGGGCT

6. A, A, C, A, T, G, G

7. nucleus

8. Proteins

9. True

10. True

11. A gene is a section of DNA that "encodes" a specific protein.

12. transcription and translation

13. RNA stands for ribonucleic acid. It contains ribose instead of deoxyribose. RNA also contains uracil instead of thymine.

14. False

15. a

16. Denaturing a protein means that the protein has been forced to unfold.

17. enzymes

18. active site

19. oxidoreductases, transferases, hydrolases, lyases, isomerases, ligases

20. oxidoreductases

21. lyases

22. ligases

23. enzymes, transporters, messengers, receptors

24. Contractile proteins make up the muscles in the body. They respond to stimuli from nerves by undergoing conformational changes that cause them to be ratcheted past one another and cause the whole structure of a muscle to contract.

25. Structural proteins maintain the shape of structure of cells, tissues, organs, and the body as a whole. Examples are hair and skin.

26. Carbohydrates are hydrates of carbon, usually known as sugars. They are energy storage compounds.

27. ketohexose

28. aldopentose

29. Sucrose

30. starch, glycogen, cellulose

31. lipid

32. True

33. 1. a; 2. c; 3. b

34. 1. c; 2. a; 3. b

35. 1. b; 2. a; 3. c

36. biochemical pathway

37. Catabolism is the process that breaks down the chemicals in food into simpler forms. Anabolism is the reverse of this process.

38. enantiomers

39. thalidomide

40. infection, abnormal growth of tissues, abnormal production of important biochemicals, genetic diseases, aging

41. Infection

42. aging

43. antibiotics

44. insulin

45. Gene therapy includes methods used to return genes to their normal state (that is, methods to repair genes that have been damaged).